Annotated Instructor's Edition
Laboratory Investigations
for Biology
Second Edition

Jean Dickey
Clemson University

San Francisco • Boston • New York
Cape Town • Hong Kong • London • Madrid • Mexico City
Montreal • Munich • Paris • Singapore • Sydney • Tokyo • Toronto

Executive Editor : Beth Wilbur
Senior Project Editor: Ginnie Simione Jutson
Marketing Manager: Josh Frost
Production Editor: Steven Anderson
Production Service: Matrix Productions
Copyeditor: Anna Trabucco
Proofreader: Mabel Bilodeau
Text Designer: Gary Head
Cover Designer: Yvo Riezebos
Cover Photograph: Getty/Art Wolfe
Photo Image Manager: Travis Amos
Illustrations: Judy and John Waller, Karl Miyajima, Mark Konrad
Manufacturing Buyer: Vivian McDougal
Compositor and Prepress: Atlis Graphics & Design
Cover Printer: Phoenix Color
Printer: Banta Company

ISBN 0-8053-6789-6 (student edition)
ISBN 0-8053-6792-6 (instructor's edition)

Benjamin Cummings

5 6 7 —BAM—06 05

www.aw.com/bc

Contents

See the Preface for an explanation of lab types: Investigative, Traditional, or Observational.

iii

See the Preface for an explanation of lab types: Investigative, Traditional, or Observational.

See the Preface for an explanation of lab types: Investigative, Traditional, or Observational.

See the Preface for an explanation of lab types: Investigative, Traditional, or Observational.

See the Preface for an explanation of lab types: Investigative, Traditional, or Observational.

See the Preface for an explanation of lab types: Investigative, Traditional, or Observational.

See the Preface for an explanation of lab types: Investigative, Traditional, or Observational.

Reviewers of the Second Edition

Richard R. Jurin, *University of Northern Colorado*
Craig Laufer, *Hood College*
Elisabeth Martin, *Black Hawk College*
Wilma Patterson, *South Mountain College*
J. Michael Reynolds, *Los Angeles Mission College*

Reviewers of the First Edition

Harold Adams, *Dabbet S. Lancaster Community College*
W. Sylvester Allred, *Northern Arizona University*
Jane Aloi, *Saddleback College*
George Boone, *Susquehanna University*
Jerry Button, *Portland Community College*
Karen Campbell, *Albright College*
Roger Christianson, *Southern Oregon State College*
Robert Creek, *Eastern Kentucky University*
Jean DeSaix, *University of North Carolina*
Terrance Farrell, *Stetson University*
Dino Fiabane, *Community College of Philadelphia*
Kathleen Fisher, *San Diego State University*
Robert Howe, *Suffolk University*
Alan Jaworski, *University of Georgia*
Ted Johnson, *St. Olaf College*
Lee Ann Kirkpatrick, *Glendale Community College*
Mary Rose Lamb, *University of Puget Sound*
Richard Liebaert, *Linn-Benton Community College*
Ivo Lindauer, *University of Northern Colorado*
Raymond Lynn, *Utah State University*
Presley Martin, *Drexel University*
William McComas, *University of Iowa*
Steven McCullough, *Kennesaw State College*
Richard Mortensen, *Albion College*
Paula Piehl, *Potomac State College of West Virginia*
Brian Reeder, *Morehead State University*
Annette Schaefer, *New York City Technical College, City University of New York*
Thomas Shellberg, *Henry Ford Community College*
Guy Steucek, *Millersville University of Pennsylvania*

All laboratory learning emphasizes activities: Hands-on experiences illustrate concepts or organisms presented in lecture, extend coverage of a topic beyond the textbook, and engage students in the process of scientific inquiry. But "hands-on" should be accompanied by "minds-on." Whether labs are observational, traditional exercises with a known outcome, or student-planned investigations, we can use them to teach students to think as well as do. This laboratory manual emphasizes science process skills as well as content. A unique feature of this manual is that student teams design and perform their own investigations for 8 of the 20 lab topics.

It's going well, but we'll need some plutonium to complete the experiment.

Is this your vision of investigative labs? Does the thought of having introductory students design and perform their own experiments conjure up some unpleasant scenarios? True, introductory students frequently have few laboratory skills and may lack ideas for possible experiments. Prep requirements can be unpredictable too. It may never be quite as easy to run open-ended investigative labs as it is to run traditional or observational labs.

But this Lab Investigations manual provides you with a realistic, practical approach to investigative labs that can be implemented even in large-enrollment courses. This manual and its supplements provide you with the instructional support necessary to meet all the teaching challenges investigative labs may pose.

Why should I use an investigative approach?

As lab instructors we've all had students who have no understanding of what they're doing; they just want to be told what to do next. They plow through one procedure after another, filling in data tables and answer blanks, always seeking the "right" answer, the fact to be memorized later. And though we tell them that there is no "right" answer to a scientific experiment, we ultimately reveal what they "should have gotten" if their data don't match the expected outcome. Thus students can mistakenly learn that science (or the instructor) always has the answer. Science equals facts.

Instead, laboratory learning can teach students to get from one fact to the next by knowing how to read a map rather than by mindlessly following directions. Certainly there are correct answers to many questions that are posed by laboratory material, and there are expected outcomes for most traditional experiments. But whether traditional or investigative, all labs can be designed to offer students the opportunity to think instead of just do, to arrive at answers by integrating background information with observations and results, rather than have answers dropped in their laps.

Investigative labs simply take this idea one step further. In these labs, students can actively participate in *every* phase of an experiment. Because students design their own experiments, even the instructor may not *know* what the results will be. When students engage in the process of how scientific knowledge is acquired, science gains a new meaning in the present, in their lives. As one of our students said after a semester of investigative labs, "Now I understand why it takes so long for the FDA to approve a new drug."

Does it take special skills to teach investigative labs successfully?

It doesn't take special knowledge to teach investigative labs—many of our lab instructors are first-year graduate students—but it does take an openness to students' ideas and a willingness to be more of a mentor than an authority. You have to think like a scientist and a teacher at the same time. Prepare by understanding the methods thoroughly. During the lab session, ask questions and give advice (but not directions) as needed. Let students know that you're there to help them think through problems, but that you don't know all the answers. The Instructor Notes for the investigative labs offer many tips and additional suggestions on facilitating the lab.

Can my students design experiments?

Many students are uncomfortable with the idea of designing their own experiments; they simply have never been asked to do such a thing before. Once they learn, however, they become increasingly confident. All of the investigative labs are organized to support and guide students towards success.

- Lab Topic 1 lays the foundation for student-designed investigations by teaching the components of an experiment.

- Each investigative lab topic begins with a brief demonstration experiment that teaches students how to use one or two methods. (See Lab Topic 6 for an example.)

- The introduction and demonstration experiment for each lab suggest possible independent variables.

- The Instructor Notes at the end of each investigative lab give hints for guiding students through the process of selecting an independent variable and suggest possible investigations.

- All investigative labs include a standard proposal form for students to describe their experimental design (for example, see Exercise 2.4).

- I recommend having students work in teams of three or four for the investigative labs, as working in groups encourages collaborative thought. Teams are often able to solve their own problems, so they rely less on the instructor, and collaboration enables students to perform a more thorough experiment in the small amount of time available.

How can I evaluate students on investigative labs?

Since investigative labs have a different emphasis from traditional labs, grading is also different. Some suggestions for evaluation:

- We collect one proposal from each team and give it a quick-look grade of 10, 8, 6, or 0 points, so grading is not time-consuming.

- After students have performed their experiments, there are two options:
  (a) to have students complete an additional exercise that has them present and interpret their results. (See Exercise 2.5 for an example.) This section is designed in a question-answer format that requires less writing than a lab report and is easier to grade.
  (b) to have students write formal lab reports on their investigations. A Guide to Writing a Scientific Report is included as an appendix to this manual. This guide consists of a sample lab report and a checklist of elements. The checklist can be used to help standardize grading in multiple-instructor courses. You can assign a point value or range of points to each item. With this system, students know in advance how their papers will be evaluated, and instructors are given uniform grading criteria to follow.

Whichever format is used, it is intended that students formalize and interpret their data outside of lab time.

Is there enough time in a 2-hour period to do investigations?

Although the labs are designed for a 3-hour lab period, the Instructor Notes for all the labs suggest how to fit the lab topic into a 2-hour lab period. The 2-hour labs inevitably require some reduction of content and choice, but investigative labs can be done.

Optional schedule. If you want to increase the emphasis on process and communication skills in your lab program, have students do fewer investigations and spend more time on the experiments they do. For example, after students design their experiments, have each team give an oral presentation to the class. With minimal instructor participation, class members critique each other's proposals. The cooperative goal is to have good experiments performed by all teams. Thus students participate in designing several experiments rather than just their own team's. Since this requires extra time, the experiments are not performed until the following week. This schedule works well for a 2-hour period. It also gives the prep staff extra time to anticipate students' needs. After students have obtained results, they can give oral presentations to their classmates, followed by discussion.

Won't this drive the prep staff crazy?

The prep for the investigative lab topics is reasonably predictable. There is a limited number of methods available to the students, and prep directions for each one are given in the Prep Guide, which is free to adopters of this manual. The Prep Guide and the Instructor Notes in this manual have suggestions for additional materials that should be available for students' investigations. See Lab Topics 2, 5, 6, 7, 14, 15, 19, and 20 for examples.

Tips and techniques for accommodating investigative lab prep to your staff and facilities are noted throughout the text.

The Lab Exercises

This lab manual contains 20 labs on a range of topics that are typically covered in an introductory course. The lab topics are classified as investigative, traditional, or observational, but keep in mind that the differences among the categories depend as much on the types of thinking that students are asked to do as on the activities themselves.

Organization

Open-ended investigations are featured prominently in this manual. Eight of the 20 lab topics have students perform introductory exercises and then design and perform their own experiments using scientific process skills.

Traditional investigations. The elements of each experiment are predetermined by the step-by-step procedure in this type of laboratory. But where appropriate, students are asked to identify components of the experiments they perform and make predictions about the results. Most of the emphasis in these lab topics is on interpretation of the results of procedures.

Observational labs stress the theme of adaptation as students observe similarities and differences of taxonomic groups or structures within an organism.

Each of these labs is supported by:

- **Objectives** presented at the beginning of each exercise, which help students focus on the important information.
- **Key terms** printed in bold type so students can locate an explanation of any significant term used in the lab topic.
- **Background information** at the beginning of each lab topic and each exercise that give students a clear idea of the purpose of each activity they perform.

- **Questions** at the end of each activity that ask students to reflect on what they have done. A set of Questions for Review is included at the end of each lab topic.

New in the Second Edition

The second edition of Laboratory Investigations includes two new lab topics. Lab Topic 8, Chromosomes and Cell Division, covers mitosis and meiosis. In addition to pop bead simulations of these processes, students extract DNA so they can actually see the genetic material. Lab Topic 10, Forensic Application of Molecular Genetics, introduces students to the science of DNA profiling by using paper models. Its concluding exercise requires students to apply their knowledge of STR (short tandem repeat) analysis, probability, and Mendelian genetics to solve a mystery. In addition, several lab topics have been updated with new information and activities.

Special Features

Safety Considerations: Students are cautioned about safe use of equipment and hazardous substances by an

icon for general safety

and one for biohazards.

A disposal icon alerts students to use special waste containers for substances that should not be poured down the drain. Safety concerns are also pointed out in the Instructor Notes.

Keys to Success: Special notes provide tips for obtaining good results with the procedures and other reminders.

Color Insert: Eight pages of color photographs illustrate organisms discussed in the labs, especially Plant Diversity and Animal Diversity. All the photographs are cross-referenced to the text. The insert also includes color comparison charts that can be used in place of spectrophotometers for the color change reactions in Lab Topic 5 (Enzymes) and Lab Topic 6 (Cellular Respiration).

Appendices: Appendix A reviews basic measurement techniques and data presentation and interpretation. It can be used as a laboratory exercise, or for student reference. Appendix B, "Guide to Writing Scientific Reports," describes the purpose and content of each section of a formal lab report and illustrates each section with a sample student paper. It also addresses the problem of plagiarism from published sources.

Instructional Support

AIE. The Annotated Instructor's Edition is a source of practical assistance for every type of lab instructor, from first-time teaching assistant to experienced faculty member.

Features of the AIE include:

- Margin Notes

  (a) Instructor overview. The first page of each lab topic has a thumbnail sketch of its themes and activities.

  (b) Time requirements for each exercise.

  (c) Additional information, helpful hints, troubleshooting advice, and suggestions for extending or supplementing the described activities.

- Answers to student questions.

- Sample data and sketches to help prepare first-time instructors for student results.

- Instructor notes at the end of each lab topic discuss organization and presentation of the lab, including suggested content for the introduction and summary. The notes also contain further information and explanations, and suggestions for guiding students through investigative labs. More estimates of time information are also included in the notes.

- Investigative extensions: Each investigative lab topic includes a section that describes possible student investigations. In addition, several other lab topics include suggestions for instructors who want to use an investigative approach in more lab topics, to extend coverage of a topic, or to assign independent projects. See Lab Topics 3, 8, 11, 16, 17, and 18.

- Suggestions for a 2-hour lab. Because the lab topics are written to fill a 3-hour lab period, the Instructor Notes include suggestions for adapting the labs to a 2-hour period.

Supplements

Prep Guide: A thorough Preparation Guide for Laboratory Investigations in Biology accompanies the Lab Investigations manual. It provides:

- Descriptions of the materials used for each exercise, including sources, catalog numbers, recipes, and instructions.

- Suggestions for less expensive alternative materials wherever possible.

- A summary list of the materials required for each lab topic.

- A Planning Guide listing which labs require specific pieces of equipment, living organisms, and solutions.

I hope that you find this lab manual useful and easy to use. If I can provide further assistance, please e-mail, write, or call. I'd also like to hear about your experiences using these lab topics.

Jean Dickey
Biology Instruction
307 Long Hall, Clemson University
Clemson SC 29634–0325
(864) 656-3827
E-mail: dickeyj@clemson.edu

Acknowledgments

I'm grateful for the assistance of many people in the development and preparation of this laboratory manual. Barbara Nicodemus, a talented teacher, made important contributions to both the first and second editions. I've had the benefit of feedback from a great many teaching assistants and undergraduate students at Clemson University. At Clemson, I have especially appreciated the help and support of Darrell Bayles, a former graduate student, and colleagues John Cummings and Bob Kosinski, in developing the investigative lab program. From the start, Benjamin Cummings has provided excellent editorial and production teams. For the second edition, I am very grateful to have had Ginnie Simione Jutson leading the project. With a tight schedule, it was a pretty neat trick for her to keep me on track without making me feel overwhelmed. My thanks also go to Merrill Peterson at Matrix Productions, to Travis Amos, Steve Anderson, and Evelyn Dahlgren at Benjamin Cummings, to Judy and John Waller for the new artwork, and to Anna Trabucco for her copyediting. I remain indebted to former editors Edith Beard Brady, Lisa Donohoe, Korinna Sodic, and Kim Johnson for their dedication to making the concept of an investigative laboratory manual into a reality, and to Neil Campbell for his continuing support.

The Process of Scientific Inquiry

 Before coming to lab, you should read through all of Lab Topic 1.

Introduction

Scientific inquiry is a particular way of answering questions. It can't be used for all types of questions. The questions that can be answered by science must meet specific guidelines and scientific investigations must be carried out using certain rules. When an investigation is designed properly and meets these guidelines, then the results are acceptable to other scientists and are added to the body of scientific knowledge. If an investigator cannot show that his or her experiment was done according to the guidelines, then the results of that experiment will not be recognized as valid by other scientists.

The purpose of such guidelines can be understood by comparing them to sports records. For example, a new record set in a track and field event only counts if the meet was approved by the governing body that sets the guidelines. The site and equipment used are scrutinized to be sure that they are within the regulations and the athlete is tested for use of illicit substances. Only when these required conditions are met is the record certified as valid.

In this laboratory you will learn about the basic elements of scientific inquiry and how to apply this process to solving problems.

Outline

Exercise 1.1: The Black Box

Exercise 1.2: Defining a Problem

Exercise 1.3: The Elements of an Experiment

Exercise 1.4: Designing an Experiment

Exercise 1.5: Application of Scientific Inquiry

Instructor Overview
This lab topic is an introduction to scientific inquiry and is an essential foundation if the investigative approach will be used. Students discover elements of scientific problem-solving through attempting to determine the contents of a sealed container (the "black box"). The components of experimental design are further defined through examples and discussions (see the Instructor Notes at the end of the lab topic regarding team and class discussions). To synthesize and apply what they have learned, students critique a proposed experiment. The final exercise shows students how to apply their understanding of scientific inquiry to evaluating research that affects their daily lives.

It is important for students to take home the message that they can understand and do science.

Time Requirements
Exercise 1.1: 25–30 minutes
Exercise 1.2: 20–30 minutes
Exercise 1.3: 20–30 minutes
Exercise 1.4: 30 minutes
Exercise 1.5: 15–40 minutes

E X E R C I S E   1 . 1
The Black Box

Objective

After completing this exercise, you should be able to

1. Explain the scientific inquiry method, which you apply to various examples in this exercise.

The Prep Guide suggests what to put in each box. The plastic bag has in it an assortment of objects from which the items in the containers were selected. Each lab team will have different objects in its box; all the bags contain the same objects.

You will use the "black box" exercise as a model of how scientific inquiry is carried out. Each lab team has a container with one or more objects sealed inside. Each team also has an empty container of the same type and a plastic bag holding objects that might be inside the sealed container. Your task is to devise a way to find out what is in the box without opening it. The steps listed below give you some idea of how to proceed. Answer the questions to keep a record of what you did.

Procedure

1. Make observations.

    Investigate the container by any means available to you *except* opening the container.

    What are your observations?

    Students should describe the sounds they hear or other clues, such as the heaviness of the container.

    How did you make your observations?

    The most useful information is gathered by shaking the container to hear how the object sounds inside. The heaviness of the container might also be a clue.

    What other methods that are not available to you right now might be used to make observations?

    An X-ray machine, of course, would be helpful, but a balance would be a more practical suggestion. If students have looked at the possible contents, they may suggest using a magnet.

    Why is making observations an important first step in solving this problem?

    It narrows the number of possible answers. For example, if the box feels heavy, it's more likely that it contains a heavy object than a light one. The observations allow students to make a better guess.

2. Make a guess about the contents of the box.

What did you base your guess on?

Students can use their observations and their knowledge (the objects they are given as possible box contents) to make their guesses.

3. For now, you still can't open the sealed container. How can you test whether your guess is correct?

   Students can put the object they think is in the sealed box into the empty box and observe whether it feels and sounds like the sealed box. If balances are available, the weights can be compared (the lesson on using the balance is in the next lab). Magnets can also be used.

   You can weigh the containers for the students if you want them to have quantitative information for their experiments.

4. Use the method you described in step 3 to check your guess. Record your results below. Was your guess correct? How sure are you?

   Students should use whatever means they have devised as a test and write down the results (that is, whether the guess seems to be correct). They should state how confident they feel about the conclusion they draw from the results.

5. If you aren't sure you know yet what is in the box, what should you do next?

   Make another guess and test it.

6. Short of opening the box, what's the best you can do to find out what's in it?

   Keep doing experiments (testing objects) until a combination that feels and sounds right is found. Have each team member feel and listen to the objects in the two boxes and arrive at a conclusion independently. Ask other lab teams to do the same.

7. Suppose you tell your instructor what you have concluded is in the box, and he or she says that you are wrong. What are some things that could have led you to make the wrong conclusion?

   - *Inaccurate observations or subjective differences among observers.*
   - *Lack of information needed to solve the problem, including:*
     - *The objects that were supplied for testing may not include all the objects that were in the box.*
     - *The piece of thread that is in the bag is undetectable in the box by the observational methods used, so students can guess that it is there but they can't test their guess.*

8. Summarize the methods you used to solve the problem of the black box.

   1. *Observe.*
   2. *Guess.*
   3. *Test guess.*

   4. *Draw conclusion from test.*
   5. *If test doesn't confirm guess, make another guess and test again.*

Suppose the evidence best supports the hypothesis that the box contains a penny. There may be similar boxes all over the country; other scientists may also have reached the conclusion that these boxes contain pennies. That would be stronger evidence that the boxes do contain pennies, but it still would not be proof. Science accumulates knowledge through many small experiments that add support to hypotheses. When a number of related hypotheses are supported and form a bigger picture or framework, a theory may emerge.

The steps you used to determine the contents of the black box are similar to the procedure followed in one type of scientific investigation. The investigator poses a question—for example, "What is in the box?" From the question and preliminary observations, the investigator makes an educated guess (known as a hypothesis) about the answer. She then devises an experiment to test the hypothesis, performs the experiment, and draws a conclusion from its results. The hypothesis may be revised, and further experiments may be done if the results are not conclusive. Eventually the investigator reaches a point where she is confident that her conclusions are correct.

In Exercises 1.2 and 1.3 you will learn to recognize the elements of a good scientific investigation. In later laboratories you will design your own investigations.

E X E R C I S E   1 . 2
Defining a Problem

Objectives

After completing this exercise, you should be able to

1. Identify questions that can be answered through scientific inquiry and explain what characterizes a good question.

2. Identify usable hypotheses and explain what characterizes a good scientific hypothesis.

See additional discussion in Instructor Notes at the end of this lab topic.

Every scientific investigation begins with the question that the scientist wants to answer. The questions addressed by scientific inquiry are based on observations or on information gained through previous research, or on a combination of both. Just because a question can be answered doesn't mean that it can be answered *scientifically*. Discuss the following questions with your lab team and decide which of them you think can be answered by scientific inquiry.

What is in the black box?

Are serial killers evil by nature?

What is the cause of AIDS?

Why is the grass green?

What is the best recipe for chocolate chip cookies?

When will the Big Earthquake hit San Francisco?

How can the maximum yield be obtained from a peanut field?

Does watching television cause children to have shorter attention spans?

How did you decide what questions can be answered scientifically?

If an experiment can be done or data can be collected that will answer the question, then the question can be addressed scientifically.

A scientific question is usually phrased more formally as a **hypothesis,** which is simply a statement of the scientist's educated guess at the answer to the question.

A hypothesis is usable only if the question can be answered "no." If it can be answered "no," then the hypothesis can be proven false. The nature of science is such that we can prove a hypothesis false by presenting evidence from an investigation that does not support the hypothesis. But we cannot prove a hypothesis true. We can only support the hypothesis with evidence from *this particular investigation.* For example, you used hypotheses to investigate the contents of your sealed box. A reasonable hypothesis might have been, "The sealed box contains a penny and a thumbtack." This hypothesis could be proven false by doing an experiment: putting a penny and a thumbtack in a similar box and comparing the rattle it makes to the rattle of the sealed box. If the objects in the experimental box do not sound like the ones in the sealed box, then the hypothesis is proven false by the results of the experiment, and you would move on to a new hypothesis. However, if the two boxes do sound alike, then this does not prove that the sealed box actually contains a penny and a thumbtack. Rather, this investigation has supplied a piece of evidence in support of the hypothesis.

You could test almost any hypothesis you made by putting objects in the empty box. What one hypothesis could *not* be proven false by experimentation?

> *That there was a piece of thread in the box. Although there was a piece of thread in the baggie, it wasn't detectable by the method used. Students couldn't show that there wasn't thread in the sealed container.*

You may now open the sealed container. Was your final conclusion about its contents correct?

See additional discussion in Instructor Notes at the end of this lab topic.

If your conclusion has now been disproven, explain how you reached an erroneous conclusion. (You may have found that your conclusion was wrong in spite of accurate observations and careful experimentation. Conclusions reflect the best evidence available at the time.)

> *If students had known that the box might contain a safety pin, they might have reached a different conclusion. The thread was undetectable by the means of investigation available to the students, so its presence could not be scientifically tested.*

Can you think of any areas of scientific inquiry where a new technology or technique might challenge or disprove hypotheses that are already supported by experimental evidence?

See additional discussion in Instructor Notes at the end of this lab topic.

> *For example, subcellular components have been made visible by advances in microscopy. DNA technology has had many outcomes. For example, it is the basis of the recent reorganization of classification into Domains.*

See additional discussion in Instructor Notes at the end of this lab topic.

The scientific method applies only to hypotheses that can be proven false through experimentation (or through observation and comparison, a different means of hypothesis testing). It is essential to understand this in order to understand what is and is not possible to learn through science. Consider, for example, this hypothesis: More people behave immorally when there is a full moon than at any other time of the month. The phase of the moon is certainly a well-defined and measurable factor, but morality is not scientifically measurable. Thus there is no experiment that can be performed to test the hypothesis. Propose a testable hypothesis for human behavior during a full moon.

> *Examples: More murders are committed during a full moon. More people are committed to mental institutions during a full moon. More babies are born during a full moon. Any factor for which data can be collected would make a testable hypothesis.*

The criterion for judging testable hypotheses is: Can an experiment be devised that would prove the hypothesis false? In order to do an experiment, the factors that are being investigated must be measurable.

Which of the following would be useful as scientific hypotheses? Give the reason for each answer.

Plants absorb water through their leaves as well as through their roots.

> *Yes (water absorption can be measured).*

Mice require calcium for developing strong bones.

> *Yes (you can give mice different amounts of calcium in their diets and measure aspects of bone structure).*

Dogs are happy when you feed them steak.

> *No (happiness is a subjective human emotion).*

An active volcano can be prevented from erupting by throwing a virgin into it during each full moon.

> *Yes (it would never pass a Use of Human Subjects committee, but a valid experiment can be designed).*

The higher the intelligence of an animal, the more easily it can be domesticated.

> *No (animal intelligence can't be measured).*

The earth was created by an all-powerful being.

> *No (no experiment can be designed to test this).*

HIV (human immunodeficiency virus) can be transmitted by cat fleas.

> *Yes (ability of HIV to infect or to tag along on fleas can be observed and tested, as can ability of HIV carriers to infect flea host).*

EXERCISE 1.3
The Elements of an Experiment

Objectives

After completing this exercise, you should be able to

1. Define and give examples of dependent, independent, and standardized variables.
2. Identify the variables in an experiment.
3. Explain what control treatments are and why they are used.
4. Explain what replication is and why it is important.

Once a question or hypothesis has been formed, the scientist turns his attention to answering the question (that is, testing the hypothesis) through experimentation. A crucial step in designing an experiment is identifying the variables involved. **Variables** are things that may be expected to change during the course of the experiment. The investigator deliberately changes the **independent variable**. He measures the **dependent variable** to learn the effect of changing the independent variable. To eliminate the effect of anything else that might influence the dependent variable, the investigator tries to keep **standardized variables** constant.

Dependent Variables

The **dependent variable** is what the investigator measures (or counts or records). It is what the investigator thinks will vary during the experiment. For example, she may want to study peanut growth. One possible dependent variable is the height of the peanut plants. Name some other aspects of peanut growth that can be measured.

Number of peanuts, weight of the plants, leaf area, time to maturation.

All of these aspects of peanut growth can be measured and can be used as dependent variables in an experiment. There are different dependent variables possible for any experiment. The investigator can choose the one she thinks is most important, or she can choose to measure more than one dependent variable.

For this experiment, which dependent variable would the investigator choose? Probably the number or weight of peanuts, since that's the part of the crop that is marketed. The investigator might also want to measure some other things at the same time, such as time to maturation.

Independent Variables

The **independent variable** is what the investigator deliberately varies during the experiment. It is chosen because the investigator thinks it will affect the dependent variable. Name some factors that might affect the number of peanuts produced by peanut plants.

Amount of fertilizer, type of fertilizer, temperature, amount of water, day length, genetic type of the peanuts.

If you want to extend this idea beyond this laboratory, you could have students collect data (for example, some newspapers list police reports) to compile and test a hypothesis. This would also be an interesting exercise to use for students in the lecture class to demonstrate hypothesis testing.

In many cases, the investigator does not manipulate the independent variable directly. He collects data and uses the data to evaluate the hypothesis, rather than doing a direct experiment. For example, the hypothesis that more crimes are committed during a full moon can be tested scientifically. The number of crimes committed is the dependent variable and can be measured from police reports. The phase of the moon is the independent variable. The investigator cannot deliberately change the phase of the moon, but can collect data during any phase he chooses.

Although many hypotheses about biological phenomena cannot be tested by direct manipulation of the independent variable, they can be evaluated scientifically by collecting data that could prove the hypothesis false. This is an important method in the study of evolution, where the investigator is attempting to test hypotheses about events of the past.

The investigator can measure as many dependent variables as she thinks are important indicators of peanut growth. By contrast, she must choose only one independent variable to investigate in an experiment. For example, if the scientist wants to investigate the effect that the amount of fertilizer has on peanut growth, she will use different amounts of fertilizer on different plants; the independent variable is the amount of fertilizer. Why is the scientist limited to one independent variable per experiment?

She needs to know which factor is primarily affecting the dependent variable(s).

The single-factor experiment is ideal, but scientists have also had to develop ways of investigating multifactorial systems where one independent variable cannot be isolated. This is particularly a problem in ecology, where complex systems are studied and investigations are often carried out in the field, rather than in controlled laboratory conditions.

Time is frequently used as the independent variable. The investigator hypothesizes that the dependent variable will change over the course of time. For example, she may want to study the diversity of soil bacteria found during different months of the year. However, the units of time used may be anywhere from seconds to years, depending upon the system being studied.

What was the independent variable in your black box investigation?

The objects that were placed in the box to test. (This is what the experimenter can vary.)

What was (or were) the dependent variable(s)?

Sound made when the box was shaken, weight, magnetic attraction.

When the variables have been identified correctly, the examples should make sense when stated in the following format: (Dependent variable) depends on (independent variable). For example, "The adult weight of dogs depends on litter size."

Identify the dependent and independent variables in the following examples (circle the dependent variable and underline the independent variable):

Height of bean plants is recorded daily for 2 weeks.

Dependent: height; independent: time.

Guinea pigs are kept at different temperatures for 6 weeks. Percent weight gain is recorded.

Dependent: percent weight gain; independent: temperature.

The diversity of algal species is calculated for a coastal area before and after an oil spill.

Dependent: algal diversity; independent: oil spill.

Light absorption by a pigment is measured for red, blue, green, and yellow light.

Dependent: light absorption; independent: wavelength of light.

Batches of seeds are soaked in salt solutions of different concentrations, and germination is counted for each batch.

Dependent: germination; independent: concentration of salt.

An investigator hypothesizes that the adult weight of a dog is higher when it has fewer littermates.

Dependent: adult weight; independent: litter size.

Standardized Variables

A third type of variable is the **standardized variable.** Standardized variables are factors that are kept equal in all treatments, so that any changes in the dependent variable can be attributed to the changes the investigator made in the independent variable.

The term "standardized" variable replaces "controlled" variable to avoid confusion between controlled variables and control treatments.

Since the investigator's purpose is to study the effect of one particular independent variable, she must try to eliminate the possibility that other variables are influencing the outcome. This is accomplished by keeping the other variables at constant levels, in other words, by *standardizing* these variables. For example, if the scientist has chosen the amount of fertilizer as the independent variable, she wants to be sure that there are no differences in the type of fertilizer used. She would use the same formulation and same brand of fertilizer throughout the experiment. What other variables would have to be standardized in this experiment?

Amount of water, temperature, day length, genetic type of peanuts.

Predictions

A hypothesis is a formal, testable statement. The investigator devises an experiment or collects data that could prove the hypothesis false. He should also think through the possible outcomes of the experiment and make **predictions** about the effect of the independent variable on the dependent variable in each situation. This thought process will help him interpret his results. It is useful to think of a prediction as an if/then statement: *If* the hypothesis is supported, *then* the results will be . . .

You can have students practice by writing predictions for the hypotheses listed at the end of Exercise 1.2.

For example, a scientist has made the following hypothesis: Increasing the amount of fertilizer applied will increase the number of peanuts produced. He has designed an experiment in which different amounts of fertilizer are added to plots of land and the number of peanuts yielded per plot is measured.

What results would be predicted if the hypothesis is supported? (State how the dependent variable will change in relation to the independent variable.)

If using more fertilizer increases peanut yield, then the number of peanuts produced per plot will increase with the amount of fertilizer applied to the plot.

What results would be predicted if the hypothesis is proven false?

If using more fertilizer does not increase peanut yield, then the number of peanuts produced will not be directly related to the amount of fertilizer.

Levels of Treatment

Once the investigator has decided what the independent variable for an experiment should be, he must also determine how to change or vary the independent variable. The values set for the independent variable are called the **levels of treatment.** For example, an experiment measuring the effect of fertilizer on peanut yield has five treatments. In each treatment, peanuts are grown on a 100-m^2 plot of ground, and a different amount of fertilizer is applied to each plot. The levels of treatment in this experiment are set as 200 g, 400 g, 600 g, 800 g, and 1000 g fertilizer/100 m^2.

The investigator's judgment in setting levels of treatment is usually based on prior knowledge of the system. For example, if the purpose of the experiment is to investigate the effect of temperature on weight gain in guinea pigs, the scientist should have enough knowledge of guinea pigs to use appropriate temperatures. Subjecting the animals to extremely high or low temperatures can kill them and no useful data would be obtained. Likewise, the scientist attempting to determine how much fertilizer to apply to peanut fields needs to know something about the amounts typically used so he could vary the treatments around those levels.

Control Treatments

It is also necessary to include **control treatments** in an experiment. A control treatment is a treatment in which the independent variable is either eliminated or is set at a standard value. The results of the control treatment are compared to the results of the experimental treatments. In the fertilizer example, the investigator must be sure that the peanuts don't grow just as well with no fertilizer at all. The control would be a treatment in which no fertilizer is applied. An experiment on the effect of temperature on guinea pigs, however, cannot have a "no temperature" treatment. Instead, the scientist will use a standard temperature as the control and will compare weight gain at other temperatures to weight gain at the standard temperature.

For each of the following examples, tell what an appropriate control treatment would be.

1. An investigator studies the amount of alcohol produced by yeast when it is incubated with different types of sugars. Control treatment:

   A treatment with no sugar.

2. The effect of light intensity on photosynthesis is measured by collecting oxygen produced by a plant. Control treatment:

   Put a plant in the dark.

3. The effect of NutraSweet sweetener on tumor development in laboratory rats is investigated. Control treatment:

   One group of rats receives no NutraSweet sweetener.

4. Subjects are given squares of paper to taste that have been soaked in a bitter-tasting chemical. The investigator records whether each person can taste the chemical. Control treatment:

   Use some papers that do not contain the chemical.

5. A solution is made up to simulate stomach acid at pH 2. Maalox antacid is added to the solution in small amounts, and the pH is measured after each addition. Control treatment:

   Add water instead of Maalox antacid.

Replication

Another essential aspect of experimental design is **replication.** Replicating the experiment means that the scientist repeats the experiment numerous times using exactly the same conditions to see if the results are consistent. Being able to replicate a result increases our confidence in it. However, we shouldn't expect to get exactly the same answer each time, because a certain amount of variation is normal in biological systems. Replicating the experiment lets us see how much variation there is and obtain an average result from different trials.

A concept related to replication is **sample size.** It is risky to draw conclusions based upon too few samples. For instance, suppose a scientist is testing the effects of fertilizer on peanut production. He plants four peanut plants and applies a different amount of fertilizer to each plant. Two of the plants die. Can he conclude that the amounts of fertilizer used on those plants were lethal? What other factors might have affected the results?

> *No. Many things, such as insufficient water or light, or attack by parasites or insects, could have caused the plants to die.*

When you are designing experiments later in this lab course, consider sample size as an aspect of replication. Since there are no hard and fast rules to follow, seek advice from your instructor regarding the number of samples and the amount of replication that is appropriate for the type of experiment you are doing. Since the time you have to do experiments in lab is limited, inadequate replication may be a weakness of your investigations. Be sure to discuss this when you interpret your results.

Methods

After formulating a hypothesis and selecting the independent and dependent variables, the investigator must find a method to measure the dependent variable; otherwise, there is no experiment. Methods are learned by reading articles published by other scientists and by talking to other scientists who are knowledgeable in the field. For example, a scientist who is testing the effect of fertilizer on peanuts would read about peanut growth and various factors that affect it. She would learn the accepted methods for evaluating peanut yield. She would also read about different types of fertilizers and their composition, their uses on different soil types, and methods of application. The scientist might also get in touch with other scientists who study peanuts and fertilizers and learn about their work. Scientists often do this by attending conferences where other scientists present results of investigations they have completed.

In this course, methods are described in the lab manual or may be learned from your instructor.

Summary

Figure 1.1 summarizes the process of scientific investigation. The process begins and ends with the knowledge base, or what is already known. When a scientist chooses a new question to work on, he first searches the existing knowledge base to find out what information has already been published. Familiarity with the results of previous experiments as well as with the topic in general is essential for formulating a good hypothesis. After working through the rest of the process, the scientist contributes his own conclusions to the knowledge base by presentations at professional meetings and publication in scientific journals. Because each new experiment is built upon past results, the foundation of knowledge grows increasingly solid.

To summarize the elements of an experiment, have the students generate a list of things to consider when designing an experiment. A typical list includes defining the question or hypothesis; identifying dependent, independent, and standardized variables; proper control treatments; and adequate replication.

See additional discussion in Instructor Notes at the end of this lab topic.

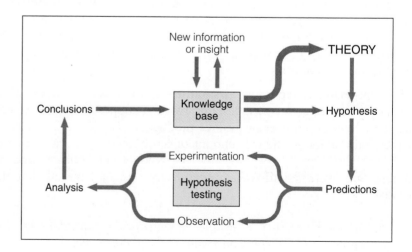

Figure 1.1.
Summary of scientific inquiry.

A theory pulls together evidence into a general explanation of some aspect of nature. It is the result of repeatedly testing a broad scope of hypotheses, not just a single one.

Scientific knowledge is thus an accumulation of evidence in support of hypotheses; it is not to be regarded as absolute truth. Hypotheses are accepted only on a trial basis. When you read about current scientific studies in the newspaper, keep in mind that the purpose of the media is to report news. In science, "news" is often preliminary results that are therefore quite tentative in nature. It is not unusual to hear that the results of one study contradict another. Some results will hold up under future scrutiny and some will not. However, this does not mean that scientific knowledge is flimsy and unreliable. All scientific knowledge falls somewhere along a continuum from tentative to certain, depending on the evidence that has been amassed. For example, it takes an average of 12 years to get a new drug approved by the FDA as researchers progress through laboratory evaluation of possible compounds, animal studies, and an escalating series of trials in humans. Even so, there are cases of drugs being recalled when new information is discovered. In a way, every medicine you take is still being tested—on you. We don't object to this because we feel confident that the knowledge base is firm, that the science is "done" to an acceptable degree.

EXERCISE 1.4
Designing an Experiment

Objectives

After completing this exercise, you should be able to

1. Given a proposed experiment, critique the experimental design.
2. Given a method for measuring a dependent variable, design an experiment.

Science is almost always a collaborative effort. Teams of scientists work together to solve problems; often these teams include scientists-in-training (graduate students). Working with others brings a variety of perspectives, knowledge bases, and experiences to bear on the problem. When scientists propose a project, they may seek funding from an agency such as the National Science Foundation. In this process, the team's proposal is reviewed by other scientists, who decide whether the problem is worth addressing, whether the proposers have the knowledge required to address it, and whether the design of the experiments is scientifically sound. When scientists have finished an investigation, they present the results to other scientists in the field by making oral presentations at conferences and by publishing articles in scientific journals.

Whether you are going to be a scientist or not, you will find many of the skills that scientists use are applicable to your own career. Most jobs require cooperative effort of some kind, just as you will collaborate with your lab team. Effective communication skills are especially important. You will almost certainly have occasion to present your work or defend your ideas to your coworkers and supervisor.

When you are asked to design an experiment in this course, your lab team will be provided with possible dependent variables and methods and procedures that you can use to measure them. You will decide what independent variable might affect these results. For example, if the topic being

Students should critique this sample proposal. As in the other exercises, it works best to have each lab team discuss the proposal among themselves for a while before having a class discussion. Most students have no difficulty picking out the major flaws in this proposal. Some major features that should be discussed are given here.

Subjects: The proposal doesn't specify that subject variables such as sex, age, and health habits will be standardized. The athletes chosen as subjects should also be similar to the nonathletes so that athletic training is the only significant difference. Selection of the type of athletes is also important, since some sports require more cardiovascular conditioning than others.

Replication: The proposal has each subject performing the test only once, so there is no replication.

Sample size: Three subjects in each group is a very small sample. A large sample size is better, but time, effort, and availability of subjects must also be considered.

Performing the test: The proposal says the subjects will run up and down the stairs, which is pretty vague. Given the instruction to run up the staircase, subjects will go at different speeds and expend different amounts of effort. Some standardized form of exercise is needed. Students may suggest using a treadmill or bicycle.

Data tables: These could also be improved. A revised version of the proposal is provided in the Instructor Notes at the end of the lab. This version is more representative of a good student proposal. Photocopy and hand it out to students if you plan on having students design experiments in later labs.

studied is the circulatory system, pulse rate and blood pressure might be the dependent variables that you measure to assess cardiovascular fitness. Your lab team would decide what factors (the independent variables) might affect a person's pulse rate and/or blood pressure and then design an experiment to test the effect of one of these factors.

Before you design your own experiments in later laboratories, you will work with your lab team in this part of the laboratory to critique a proposed experiment. This exercise is a rough draft of a proposal that could be improved. Use your knowledge of the scientific process to revise this experiment.

Sample Proposal

Hypothesis

Athletes have better cardiovascular fitness than nonathletes.

Dependent Variable(s)

Pulse rate, blood pressure.

Independent Variable

Athletic training.

Control(s)

Subjects who have had no athletic training (to have a comparison group of subjects); readings taken before exercise (to get a baseline measurement for each subject).

Replication

Three subjects will be used in each group. Each subject will perform the exercise once.

Brief Explanation of Experiment

The pulse rate and blood pressure of athletes and nonathletes will be measured. The subjects will then perform 5 minutes of aerobic exercise. The pulse rate and blood pressure of each subject will be measured again immediately after exercise.

Predictions

We think that the pulse rates and blood pressure of athletes will be lower after exercise and will return to normal rates more quickly than those of nonathletes, indicating better cardiovascular fitness of athletes.

Method

1. Recruit three athletes to be subjects. Our lab team will be the nonathlete subjects.

2. Record resting pulse rate and blood pressure for each subject.

3. All subjects will run up and down the stairs for 5 minutes.

4. Pulse rate and blood pressure of each subject will be measured immediately after the exercise.

5. Pulse rate will continue to be taken until it returns to the resting value. The time taken for each subject's pulse to return to normal will be recorded.

6. Measure and record blood pressure for each person when the resting pulse rate is reached.

Tables Designed to Collect Data

	Resting blood pressure	Resting pulse rate
Athlete 1		
Athlete 2		
Athlete 3		
Tom		
Jennifer		
Eric		

	Blood pressure after exercise	Pulse rate after exercise
Athlete 1		
Athlete 2		
Athlete 3		
Tom		
Jennifer		
Eric		

	Time to reach resting pulse	Blood pressure at resting pulse
Athlete 1		
Athlete 2		
Athlete 3		
Tom		
Jennifer		
Eric		

If there is time, have students brainstorm ideas for other independent variables that could be used with these techniques. Choose one that can be done in class, and design and perform the experiment. Directions for measuring blood pressure are given in Lab Topic 15, Circulation.

List other independent variables that might be investigated using these techniques of measuring pulse and blood pressure.

Age, gender, weight, race, stress, consumption of drugs (for example, caffeine), smoking, time of day.

E X E R C I S E   1 . 5

Application of Scientific Inquiry

Objective

After completing this exercise, you should be able to

1. Evaluate research reported in the media using the criteria provided in this exercise.

If you are not a science major, you may be asking why you should learn how scientific inquiry works. One reason is that understanding the science process gives you perspective on the scientific knowledge described in your textbook. When you read a science textbook you are only seeing the tip of the iceberg. Beneath all of that information lies a vast foundation of observation, experimentation, and analysis. Each experiment that you do in lab should give you a tiny taste of how the information in your textbook came to be there—how we know what we know.

Understanding scientific inquiry matters in a personal sense, too. You may never be a producer of scientific knowledge—a scientist—but you make decisions daily as a consumer of science. Do the vitamins and minerals added to Cap'n Crunch cereal make it a health food? Will zinc lozenges shorten my cold? Should I have the Big Mac or the grilled chicken salad? Do I want fries with that? Should I walk to class or drive my SUV? Does it matter whether I get more than 5 hours of sleep a night? Should I steer clear of genetically engineered food? Is a soft drink that has ginseng or ginkgo biloba in it worth the extra cost? Understanding the state of the research can help you make informed choices about issues that affect your health, your wallet, and your impact on our environment. In this exercise, you will learn how to evaluate the science supporting matters that affect you personally.

From Exercises 1.3 and 1.4 you know how to evaluate whether an experiment is designed properly. But accounts of scientific studies in magazines and newspapers don't always provide those details. Some other factors to look for include the following:

Tobacco companies, for example, fund research. Although their scientists may be highly qualified and do good science, one can't help but be a bit skeptical when the funding agency has a profit motive.

Where was the study carried out and who funded it? For objectivity, look for studies funded by competitive agencies such as NIH (National Institutes of Health) and NSF (National Science Foundation). These government-funded organizations require peer review of proposed experiments. That is, a group of knowledgeable scientists determines funding based on criteria such as experimental design and the qualifications of the investigators to perform the experiments. On the other hand, some funding groups have a vested interest in the outcome of research.

How large was the sample size? As you have learned, conclusions drawn from a large sample are more reliable than conclusions from a small sample. Small studies are not bad science, their results are simply more preliminary than those from large studies.

Where was the study published? The most reliable sources are peer-reviewed journals. Knowledgeable scientists critique each article submitted to determine whether it merits publication. Other types of publications do not require the same rigorous screening. For example, an industry newsletter or a Web site promoting herbal supplements will probably not offer a thorough and objective review of the research.

Procedure

Your instructor may have asked you to find a news article yourself, or she may have one to assign to you. In either case, read the article and briefly describe the following aspects of the science reported on. (Your article may not include all of these.)

Experimental Design:

Independent variable —

Dependent variable(s) —

Standardized variables —

Control treatment —

How large was the sample size?

Where was the study carried out, and who funded it?

Where was the study published?

How convinced are you by the results reported in the article? Rate your conclusion on a scale of 0 (this isn't even science!) to 3 (I would base a personal decision on these results). Explain how you arrived at your rating.

Students—and perhaps most non-scientists—put great faith in anecdotal information. If two friends lost weight on the Atkins diet or bulked up on "andro," that's all the evidence one needs. Here is a good reminder: "The plural of anecdote is not data."

There are other criteria to consider, as well. An article from the June 1997 issue of Consumer Reports, "How to Assess the Latest Medical Breakthrough", does a nice job of explaining how to interpret media reports on health issues. In addition to the items mentioned already, Consumer Reports suggests looking at the population studied (bacteria? animals? humans? specific group of humans?), the type of study (prospective or retrospective), and the real-life meaning of any statistical significance reported.

The examples are experiments that should be understood by non-biologists. However, it doesn't matter whether students completely understand the content of the experiments. The idea is for them to recognize the components of scientific investigations. An excellent supplement would be for you to clip current science news items and ask the same kinds of questions.

Questions for Review

1. A group of students hypothesizes that the amount of alcohol produced in fermentation depends on the amount of glucose supplied to the yeast. They want to use 5%, 10%, 15%, 20%, 25%, and 30% glucose solutions.

   a. What is the independent variable?

   Percentage of glucose.

   b. What is the dependent variable?

   Amount of alcohol produced.

c. What control treatment should be used?

0% glucose.

d. What variables should be standardized?

Amount and type of yeast, temperature, total volume in the reaction vial, and time allowed for fermentation.

2. Having learned the optimum sugar concentration, the students next decide to investigate whether different strains of yeast ferment glucose to produce different amounts of alcohol. Briefly explain how this experiment would be set up.

The independent variable is now type of yeast; the dependent variable is still the amount of alcohol produced. Standardized variables are the same (except type of yeast), with the addition of the percent sugar solution used.

3. A group of students wants to study the effect of temperature on bacterial growth. To get the bacteria, they leave petri dishes of nutrient agar open on a shelf. They then put the dishes in different places: an incubator (37°C), a lab room (21°C), a refrigerator (10°C), and a freezer (0°C). Bacterial growth is measured by estimating the percentage of each dish covered by bacteria at the end of a 3-day growth period.

a. What is the independent variable?

Temperature.

b. What is the dependent variable?

Growth, measured by percent cover.

c. What variables should be standardized?

Bacteria should come from the same source, the agar in the dishes should be the same, and the growth period should be the same length for all treatments.

4. A team of scientists is testing a new drug, XYZ, on AIDS patients. They expect patients to develop fewer AIDS-related illnesses when given the drug, but they don't expect XYZ to cure AIDS.

a. What hypothesis are the scientists testing?

XYZ will decrease the incidence of AIDS-related diseases in patients with AIDS.

b. What is the independent variable?

XYZ treatments.

c. What is the dependent variable?

Development of AIDS-related diseases.

d. What control treatment would be used?

A group of AIDS patients who do not receive XYZ; the control group usually receives a placebo (a pill that does not contain the drug) so subjects do not know who is getting the experimental treatment.

e. What variables should the researchers standardize?

Pharmaceutical researchers usually try to have groups of subjects who are the same sex and race or ethnic background and within the same age range. Length of time that the patients have had AIDS should also be standardized.

5. A group of students decides to investigate the loss of chlorophyll in autumn leaves. They collect green leaves and leaves that have turned color from sugar maple, sweet gum, beech, and aspen trees. Each leaf is subjected to an analysis to determine how much chlorophyll is present.

   a. What is a reasonable hypothesis for these students?

   Leaves that have changed color have less chlorophyll than green leaves.

   b. What is the independent variable?

   Leaf color. The students are actually doing four experiments at once, since they are using four different species, so tree species is also an independent variable.

   c. What is the dependent variable?

   Amount of chlorophyll.

   d. What would you advise the students about replication for this experiment?

   For example: Use ten leaves of each type and make three chlorophyll determinations for each leaf.

6. A scientist wants to study mating behavior in crickets. She hypothesizes that males that win the most male-vs.-male contests mate with the most females. She observes the crickets to obtain data. For each male, she counts the number of male-male fights he wins and the number of females he mates with.

   a. What is the independent variable?

   Number of contests won.

   b. What is the dependent variable?

   Number of females mated with.

   c. What constitutes replication in this experiment?

   The number of males observed.

The Process of Scientific Inquiry
Instructor Notes

Organization

Many instructors hesitate to use investigative labs (for which this lab serves as an introduction) because they assume that students who are working independently need more direction than they will be able to provide. We routinely use the "divide and discuss" technique for groups ranging from 22 (our typical lab size) to 80 or more (for our TA training sessions). In this technique, participants are given a problem (such as the black box exercise) or a topic to discuss with a small team (three to five members). After a period of talking with their teammates, the entire class reassembles for a general discussion of each team's findings. If you have never tried this technique, you will be amazed at how well students can work through problems on their own with only intermittent attention from the instructor. The quality of the general discussion is usually very good. Since students have had a chance to try out their ideas on each other, they are more willing to contribute to the class discussion. Creativity is enhanced by the small-group interactions, and often students can answer each other's questions without involving the instructor. The class discussion is also more thorough and to the point, since students have already worked through most of their difficulties.

Because it is very important to create a good atmosphere for cooperative learning in investigative labs, you may want to spend the first part of the lab making the students feel comfortable with you and each other. Many of our instructors have the students introduce themselves by telling one interesting fact about themselves or talk to each other briefly and then introduce each other. The instructors also tell a little about themselves (for example, their research interests, how they came to be at this university, their own experiences as freshmen). Introductions such as these help break the ice and get students used to the idea that they need to participate actively in lab. This is especially useful if the class is mostly made up of first-semester freshmen.

Exercise 1.1 (25–30 minutes)

Divide the class into teams of three and give them 15 to 20 minutes to work on the black box problem. At the end of that time, have each group briefly recap the investigative process it used and state its conclusions.

Exercise 1.2 (20–30 minutes)

Unless students have had a chance to read the lab ahead of time, the instructor should lead the class through both exercises. Get as much student participation as possible. The best device for this is to have students break up into their lab teams to discuss each series of questions before having a class discussion (for example, identifying good questions and usable hypotheses). You provide introductory explanations and guide the discussion. This technique keeps the lab from becoming a lecture punctuated by responses from a few students and provides the benefits mentioned above.

Posing Scientific Questions

In order for scientists to answer a question, they must be able to do an experiment (or a series of experiments). "Evil," for example, implies a moral judgment of a person's character. This is a human concept, a statement of our values. Science does not make value judgments, so there is no way to answer a question about them scientifically. The "best" cookie is also a human judgment. However, if "best" can be defined in measurable terms, it can be tested scientifically.

Scientists may study indicators of forthcoming geological and meteorological events, but they can't predict the future. Thus, scientists can say that conditions favor a major earthquake at a particular place, but they can't say when it will happen.

The remaining questions can be answered through scientific investigation.

The question must be narrow and well-defined so that useful experiments can be designed. How to obtain maximum yield of peanuts can be determined scientifically, but this question is really a series of questions: How many seeds per acre should be planted? How much and what kind of fertilizer should be applied? What genetic variety of peanuts gives the best yield in this climate? On what date should the planting be done? When should the crop be harvested? When experiments are performed to answer each of the individual questions, the larger, complex question can be answered.

Black Box Results

When it is time to open the containers, tell students that a new technology has been invented: the razor blade. This scientific advance allows us to study black boxes as never before and points out the tentative nature of the conclusions that are drawn from experiments. The boxes will be set up so that there is an unexpected result in some.

Developing new methodologies is critical to the advancement of scientific knowledge. An analogy with this exercise is the electron microscope, which allowed scientists to resolve images at much higher magnifications than were possible with the light microscope. With this tool, previously unknown subcellular structures such as the cytomembrane system were identified.

Scientists do sometimes find that a hypothesis that was supported by experiments was in fact false. They may find that their hypotheses were based on incomplete information: New objects are discovered. Some boxes could contain a small feather or another light object that was not among the test items in the bag. Students would have no clue that such an object might be in the sealed box. In a similar vein, one of the items in the bag is a paper clip. One of the sealed boxes may contain a safety pin, which sounds and feels like the paper clip but is not in the bag. The students will conclude that the box holds a paper clip. This is a valid conclusion given the information they have, but it turns out to be incorrect.

Exercise 1.3 (20–30 minutes)

There are many examples of medical studies that are not yet conclusive. Is it healthy, unhealthy, or a matter of no difference to take aspirin, eat oat bran, drink wine, avoid red meat, and eat broccoli? What differentiates the conclusions from these studies from, for example, studies linking smoking to lung cancer?

Recently, there has been a controversy over the effects of dioxin. An area near a dioxin-contaminated site was abandoned, at great financial and emotional cost to the residents, on the basis of studies that condemned dioxin as extremely hazardous to human health. The scientist who conducted the studies later said that his conclusions had been misinterpreted as final, rather than tentative, results, and that if people had understood the tentative nature of such conclusions, they would not have panicked as they did.

The Tentative Nature of Science

Thus, in teaching how scientific investigations are carried out, we emphasize that hypotheses are not proven to be true, but are only supported by the evidence. We want students to understand that it takes many repetitions of results to reach a reliable conclusion. But as a reult, many students misunderstand the nature of scientific knowledge as being entirely tentative. This misconception is manifested most clearly in the position that evolution is "only a theory." To address that problem, it is worthwhile to contrast the examples of inconclusive studies with examples of a more established nature. In my class, I use textbooks to demonstrate the difference. Some information has changed from one edition of the textbook to the next. On the other hand, a book published 40 years ago has the same information for many topics as the current edition of our textbook.

Even long-established dogma may be challenged. We have all learned that average human body temperature is 98.6°F. This figure was arrived at in 1868 by a scientist who measured body temperature in 25,000 individuals, taking over a million separate readings. That's a pretty thorough study. However, better methods of measuring body temperature have increased the accuracy of the readings and it appears that the average is actually 98.2° F. (*Science News* 142 (13): 196, 1992)

As a summary activity, conduct a class discussion of students' ratings. One approach would be to work through the rating numbers as a group. For example, ask all of the students who gave their articles a rating of 3 to explain what made the results convincing so that the common factors can be identified. Ask the students who gave a 2 rating what further evidence would convince them, and so on.

Exercise 1.4 (30 minutes)

Introduce this exercise by asking lab teams to brainstorm about possible experiments they can do once they know how to measure pulse and blood pressure.

If you are going to ask students to present their own proposals later in the semester, you can present the sample proposal orally to model how you expect students to present to the class.

An improved version of the sample proposal is included below. Photo-copy and hand out this revised version if you intend to require students to hand in their proposals in this form, which will be reproduced in the labs that ask students to design their own experiments.

If there is time at the end of the lab, have the class design and perform an experiment using the step test. (Student-designed investigations of pulse and blood pressure are also included in Lab Topic 15, Circulation.)

Exercise 1.5 (15–40 minutes, depending on how much is done during lab time)

Application of Scientific Inquiry

If there is enough lead time, you can assign students to bring an article from the newspaper or Web to lab with them. Alternatively, have a stack of short news articles ready to hand out. They should be very recent—no more than days or weeks old—to reinforce the idea that science is "made fresh daily." If you provide the articles, you can decide whether to have a different article for each student, or to streamline the process by having four or five different articles with multiple copies. After an introduction, you may wish to assign the article evaluation for homework and follow up with a discussion during the next lab period.

For a 2-Hour Lab

Omit Exercise 1.4. Either omit Exercise 1.5 or assign it as homework.

Revised Sample Proposal

Hypothesis

Male varsity basketball players have better cardiovascular fitness than males who are not varsity basketball players.

Dependent Variable(s)

Pulse rate, blood pressure.

Independent Variable

Athletic training.

Control

Recording pulse and blood pressure before exercise.

Replication

Each subject will perform the exercise three times.

Brief Explanation of Experiment

The pulse rate and blood pressure of ten male varsity basketball players and ten non-basketball players will be measured. The subjects will then perform 5 minutes of aerobic exercise, the "step test." The pulse rate and blood pressure of each subject will be measured again immediately after exercise. We will also record the time it takes each person's pulse rate to return to the resting value after exercise (recovery time).

Predictions

We think that the pulse rates and blood pressure of basketball players will be lower after exercise and will return to normal rates more quickly than those of nonathletes, indicating better cardiovascular fitness of athletes.

Method

1. Recruit ten varsity basketball players to be subjects. Recruit ten males who are 18–21 years old and are not athletes.
2. Record resting pulse rate and blood pressure for each subject.
3. All subjects will perform the "step test" for 5 minutes (subjects step up and down on a low platform, keeping the rate constant).
4. Pulse rate and blood pressure of each subject will be measured immediately after the exercise.
5. Pulse rate will continue to be taken until it returns to the resting value. The time taken for each subject's pulse to return to normal will be recorded.
6. Measure and record blood pressure for each person when the resting pulse rate is reached.

Table Designed to Collect Data

Subject	Resting		After exercise		Recovery time (min)
	Blood pressure	Pulse (beats/min)	Blood pressure	Pulse (beats/min)	
1					
2					
3					
4					
5					
⋮					

Subjects 1–10 are basketball players.
Subjects 11–20 are nonathletes.

Extension

Student-designed lab experiments teach the fundamentals of the science process but they are necessarily limited in scope. I have used a supplementary project in my classes to bolster the connection between the process of scientific inquiry and real-life relevance. To identify interesting topics, I had each student submit a current article from either the Web or print media pertaining to scientific research. From those topics, I chose an appropriate number to assign to groups of students. Topics have included aspartame, RU-486, benefits and costs of alcohol consumption, medicinal benefits of marijuana, biological basis of homosexuality, cholesterol reduction, creatine, vitamin C, ginkgo biloba, the Atkins diet, the effect of religion on health, herbal remedies for colds, androstenedione, life on Mars, and chocolate, among others. Finding good topics is never a problem!

Each group produced a fact sheet to be photocopied for the entire class. Students were required to include the following elements:

- *Define the issue:* What is its importance; what claims are being made?

- *Evaluate the science* that has been done using the criteria discussed in class and in lab. What do we know? How credible is the evidence?

- *Communicate your findings* to educate your peers. What's the take-home message?

- *Identify sources of information.* Each group should have at least 8 sources of information that relate either directly or indirectly to the topic. *Designate two* of your sources as "must reads" for someone who wants to know more about the topic.

Each group also made a 10-minute presentation to the class in any format it chose. Having suffered PowerPoint fatigue in their classes, most students chose creative methods such as skits, talk show parodies, game shows, and home videos to convey information.

A less time-intensive approach is to assign a single topic for which there are contradictory results and have students locate information that is pro or con. One year I put the chicken on trial for producing "lethal" eggs. Each student found a study pertaining to some effect of egg consumption on human health. The class was divided into witnesses for the defense and for the prosecution. Each side gave its evidence, including the criteria for evaluation we had discussed. As an assessment, each student wrote an essay justifying a verdict of guilty or not guilty based on the scientific evidence.

This type of exercise extends students' understanding of scientific inquiry to encompass its effect on their own experiences and daily lives.

Acknowledgment

The scientific inquiry diagram in Exercise 1.3 was based on discussions with V. Christine Minor of Clemson University.

pH and Buffers

Introduction

Molecules that are dissolved in water may separate (ionize) into charged fragments. pH is a measure of the concentration of one of those charged fragments, hydrogen ions (H^+), in solution. A substance that has a high concentration of H^+ is acidic. A substance that has a low concentration of H^+ is basic (alkaline).

The pH scale ranges from 0 (most acidic) to 14 (most basic). There is a tenfold difference between pH units. For example, a solution with a pH value of 6 has a ten-times-greater concentration of hydrogen ions than a solution with a pH value of 7. Some examples are shown in Table 2.1.

Outline

Exercise 2.1: Introduction to Acids, Bases, and pH

Exercise 2.2: Using Red Cabbage Indicator to Measure pH

    Activity A: Making a Set of Standards

    Activity B: Comparing pH of Beverages and Stomach Medicines

Exercise 2.3: Using the pH Meter to Determine Buffering Capacity

Exercise 2.4: Designing an Experiment

Exercise 2.5: Performing the Experiment and Interpreting the Results

Instructor Overview
In this lab topic, students measure the pH of several beverages and medicines using an indicator that is made in the lab from red cabbage. Students use a pH meter to determine the buffering capacity of known buffers. After completing these exercises, which demonstrate techniques for investigating pH and buffering capacity, each team of students will design its own experiment. This lab includes a list of prompts to help students organize their experimental designs (see Exercise 2.4). After students have collected data for their experiments, they answer a series of questions that leads them to interpret their results (see Exercise 2.5). Alternatively, you may require students to write lab reports.

EXERCISE 2.1
Introduction to Acids, Bases, and pH

Objectives

After completing this introductory exercise, you should be able to

1. Explain what makes a solution acidic or basic.
2. Explain the pH scale.
3. Describe the phenol red test for pH.

Preparation
If you want students to do investigations, it is a good idea to introduce the concept of pH the previous week in lab (or lecture) and ask students to bring in samples to test for their investigations.

Time Requirements
Exercise 2.1: 5–10 minutes
Exercise 2.2: 35–40 minutes
Exercise 2.3: 40–45 minutes
Exercise 2.4: 20 minutes
Exercise 2.5: 60 minutes or time remaining in lab period

An **acid** is a substance that releases or causes the release of H^+ into solution. Solutions that have pH values lower than 7 are considered to be acids. Some common acids are hydrochloric acid, acetic acid, carbonic acid, and sulfuric acid. All of these compounds contain hydrogen. When

Not all solutions have a pH, because not all substances can ionize to produce H$^+$ even if they have hydrogen in them. Ethyl alcohol, for example, doesn't ionize. The pH of alcoholic beverages is due to the water they are made with or other ingredients or impurities. Keep this in mind when students design their investigations later in the lab.

Table 2.1
pH Scale

pH	Relative strength	Examples
0	Strong acid	
1		Battery acid
2		Gastric fluid
3	Moderate acid	Orange juice
4		Tomato juice
5	Weak acid	
6		Rainwater Milk
7	Neutral	Pure water Blood
8		
9	Weak base	Baking soda Milk of Magnesia
10		
11	Moderate base	Household ammonia
12		
13	Strong base	Hair remover
14		Oven cleaner

Pour a small amount of phenol red indicator solution into a small beaker. The initial color of the solution should be red. Use a pasteur pipet to add a little HCl to the indicator and swirl to mix. (Use the 0.1N HCl provided for Exercise 2.3.) It won't take much to change the color.

the compound is dissolved in water, hydrogen ions are released, and the pH of the solution is low. Your instructor will use an indicator called **phenol red** to demonstrate the acidity of hydrochloric acid. Phenol red is red when the solution is basic and turns yellow in acidic solution.

What color is the phenol red solution initially?

Red.

What happens when hydrochloric acid is added?

It turns yellow.

To do the CO$_2$ demo, pour a fresh solution of phenol red into a clean beaker. Use a straw to blow into it while the students watch (or have a student do the blowing). The exhaled CO$_2$ will turn the indicator yellow.

A compound does not have to contain hydrogen ions itself in order to be an acid. Carbon dioxide (CO$_2$), for example, can combine with water to generate H$^+$. Your instructor will use phenol red again to demonstrate that CO$_2$ is an acid. Briefly describe this demonstration.

The phenol red solution is initially basic (red). When CO$_2$ is blown into the phenol red solution it turns yellow, indicating that the solution has become acidic.

The reaction of SO_2 (sulfur dioxide) is similar to that of CO_2. The presence of SO_2 in the atmosphere is partially responsible for acid rain.

A **base** is a substance that can remove H^+ from solution, thus lowering the concentration of H^+. Many bases ionize to produce hydroxyl ions (OH^-), which combine with H^+ to make water (H_2O). Some common bases are sodium hydroxide (NaOH), magnesium hydroxide ($Mg(OH)_2$), and potassium hydroxide (KOH).

NH_3 (ammonia) combined with water makes NH_4OH, which ionizes to $NH_4^+ + OH^-$, so ammonia is basic. Students may need to know this if they use cleaning solutions for their investigations.

Mixing an acid with a base can produce a neutral solution by combining the H^+ with the OH^- to make water (H_2O). Pure water, which ionizes to produce equal numbers of H^+ and OH^-, is neutral (pH 7). Don't expect to get a pH of 7 when you measure water in the lab, though. Tap water contains impurities, and its pH varies a great deal. Distilled water is weakly acidic; its pH is usually around 6.

It is important for organisms to maintain a constant internal pH. As you will learn in later laboratories, biological molecules, especially proteins, are sensitive to pH, and they may not function correctly when the pH is changed.

In the following exercises you will make an indicator solution to measure pH and also learn how to use a pH meter. You will determine the pH values of some common substances and investigate how buffer systems work to maintain a constant pH.

⚠ **Wear safety glasses while performing these exercises. Strong acids and strong bases are corrosive. Inform your instructor immediately if any solution is spilled or comes in contact with your skin or clothing.**

EXERCISE 2.2
Using Red Cabbage Indicator to Measure pH

Objectives

After completing this exercise, you should be able to

1. Explain what a pH indicator is used for.
2. Describe how to measure pH using red cabbage indicator.

Several methods are available for determining pH. Many of these methods rely on the ability of certain chemicals called **indicators** to change color,

To Make Cabbage Extract:
Fill a 1-liter beaker to the 500-mL mark with coarsely chopped red cabbage. Add distilled water to cover the cabbage. Bring to a boil on a hot plate, then boil for three minutes. Filter through cheesecloth into a second 1-liter beaker. Bring the volume up to 500 mL with distilled water. Allow the solution to cool, then pour it into dropping bottles to distribute to the students. Leftover extract can be used by subsequent sections, but don't keep it overnight because it begins to smell bad.

depending on the pH of the surrounding solution. Papers saturated with indicators, such as litmus paper and alkacid test paper, can also be used.

An indicator can easily be made from a solution of anthocyanins, the pigments responsible for red, blue, and purple colors in flowers, fruits, and autumn leaves. These pigments change color as the pH changes. Red cabbage is loaded with anthocyanins, so we can make a pH indicator by boiling red cabbage to extract the pigments. Your instructor will make the extract at the beginning of class.

The use of standards, a set of known quantities, is an important technique in biological research. By comparing unknowns with the standards, we can determine what we want to know about the unknowns. The color of cabbage extract depends on the pH of the solution it is in. Your set of standards will show the color of the cabbage extract at pH 2, 4, 6, 7, 8, 10, and 12. You will then determine the pH values of various substances by mixing each substance with cabbage extract and comparing its color to the standards.

Activity A: Making a Set of Standards

Your lab team should make a set of standards using the cabbage extract and solutions of known pH according to the following procedure.

Procedure

Appendix A explains the use of pipets and pi-pumps. Also, demonstrate how to use Parafilm.

1. Put seven clean test tubes in a rack and label them 2, 4, 6, 7, 8, 10, and 12.

2. Pipet 5 mL of the appropriate buffer into each tube (pH 2 buffer into the tube labeled 2, and so on).

3. Get a dropping bottle of cabbage extract from your instructor. Add 3 mL of cabbage extract to each tube.

4. Cover the tubes with Parafilm and mix well.

5. Record the color in each tube in Table 2.2.

 Save this set of standards for use throughout the lab period.

Table 2.2
Color of Standard Solutions for
Red Cabbage Indicator

pH	Color
2	*Red*
4	*Red-violet*
6	*Violet*
7	*Blue*
8	*Blue-green*
10	*Green*
12	*Green (turns yellow)*
Record both the initial and final colors at pH 12. The pigments are not stable at this pH.	

Activity B: Comparing pH of Beverages and Stomach Medicines

Look at Table 2.3 to see what aspect of pH is being investigated in this experiment, and answer the following questions.

Table 2.3
pH Values of Beverages and Medicines

Beverages	pH	Medicines	pH
White grape juice	*4 (3.4)*	Milk of Magnesia ($Mg(OH)_2$)	*10 (9.8)*
7-Up	*2 (2.4)*	Sodium bicarbonate ($NaHCO_3$)	*8 (8.25)*
White wine	*2 or 4 (3.1)*	Maalox	*8 (8.2)*
Seltzer water	*4 (3.8)*		

pH value obtained by pH meter follows value obtained with cabbage method.

What hypothesis could be tested with this experiment?

Example: pH values of beverages are the same as pH values of stomach medicines. (Any testable hypothesis that compares these two groups of substances is acceptable.)

What is the independent variable in this experiment?

Substance category (beverage and medicine).

What is the dependent variable?

pH.

What substance could be used as a control for this experiment?

Water, since they are all aqueous solutions.

Predict the outcome of the experiment in terms of your hypothesis. What results will support the hypothesis? What results will prove the hypothesis false?

Example: If the hypothesis is supported, then all the substances will be the same color when mixed with cabbage indicator (which means they have the same pH). If the hypothesis is proven false, then the substances will produce different colors. (Prediction should follow hypothesis. If students have additional background knowledge, they could go further and predict the approximate pH they expect to see for each group.)

Use the cabbage indicator method to measure the pH of the substances listed in Table 2.3.

Procedure

1. Put 2 droppersful of the solution to be tested in a clean test tube.
2. Add 1 dropperful of cabbage extract.
3. Swirl the tube gently to mix.
4. Compare the color of the solution to the colors of your cabbage indicator standards.
5. Record the pH value for each substance in Table 2.3.
6. Measure and record the pH of your control.

   Control: ___water___

   pH of control: ___~6–7___

The solutions to be tested and the cabbage indicator are in dropping bottles. Use that dropper for measurements.

Students should switch to pasteur pipets for their investigations, since their solutions will not be in dropping bottles. Use the same proportions (2 pipetsful of solution to be tested plus 1 pipetful of cabbage extract).

Was your hypothesis proven false or supported by the results? Use data to support your answer.

Example: It was proven to be false. The pH values of the beverages were 2 or 4, while the pH values of the medicines were 8 or 10.

What components of the beverages you tested might be responsible for the pH values of the beverages?

Carbonation, acids present in juice (for example, citric acid).

What components of the medicines you tested might be responsible for the pH values of the medicines?

Although the solutions will be in dropping bottles for convenience, have the original containers available so students can read the ingredients.

Maalox contains aluminum hydroxide and magnesium hydroxide; Milk of Magnesia contains magnesium hydroxide; the bicarbonate ion in sodium bicarbonate can pick up an H^+ (most students won't know this).

Explain why stomach medicines should have pH values that are much higher than normal stomach pH, which is around 2. (Hint: Why do people take these medicines?)

These medicines are taken for acid indigestion or heartburn, conditions of excess acidity in the stomach. The bases neutralize the acid.

E X E R C I S E   2 . 3
Using the pH Meter to Determine Buffering Capacity

Objectives

After completing this exercise, you should be able to

1. Define buffer, and explain why buffers are important to organisms.
2. Describe how to use a pH meter.
3. Interpret a titration curve (graph of pH versus milliliters of HCl and NaOH added) to determine whether a solution has buffering capacity and, if so, over what pH range.
4. Explain why some solutions have buffering capacity and others don't.

In order for normal physiological processes to occur, pH must remain relatively constant. An excess of H^+ or OH^- can interfere with the functioning of biological molecules, especially proteins. In our bodies, for example, blood pH is usually maintained between 7.3 and 7.5. However, blood returning to the heart contains CO_2 picked up from the tissues, which lowers the blood pH. Metabolic reactions in cells may contribute an excess of hydrogen ions. Our diets may also affect blood pH. Several buffering systems keep the pH constant.

A **buffer** is a solution whose pH resists change on addition of small amounts of either an acid or a base. To be a good buffer, a solution should have a component that acts as a base (takes H^+ out of solution) and a component that acts as an acid (puts more H^+ into solution when there is an excess of OH^-).

The buffering capacity of a solution is tested by adding small amounts of acid (for example, HCl) and base (for example, NaOH) and checking the pH after each addition. If the pH changes only slightly, the solution is a good buffer. Eventually its buffering capacity will be exhausted, however, and the pH will change dramatically.

A buffer operates in a specific pH range. The buffering systems in our blood, for example, buffer at around pH 7.4. That is, they maintain the pH at or very close to 7.4. The solutions you used to make up your standards for red cabbage indicator maintain each buffer at a certain pH. The pH 2 buffer maintains pH at 2, the pH 4 buffer maintains pH at 4, and so on. Notice that the purpose of a buffer is *not* to make the pH neutral (7).

Figure 2.1.
Titration curve for an unknown solution.

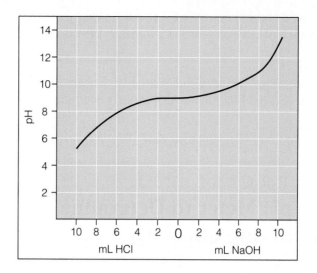

Look at Figure 2.1. Is this solution a good buffer? Explain how you know.

Yes. The y-axis shows change in pH. There is a part of the curve where change in pH is very slight when HCl and NaOH are added.

At what pH does the solution buffer?

8.5 or 9.

Figure 2.2.
Titration curve for an unknown solution.

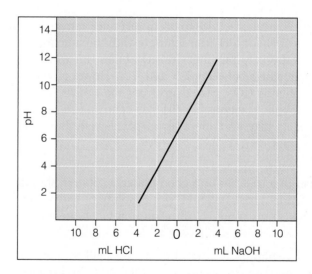

Look at Figure 2.2. Is this solution a good buffer? Explain how you know.

No. There is an immediate change in pH when HCl or NaOH is added.

You will use a pH meter to test buffering capacity in this exercise. The pH meter has a sensitive electrode that measures the H^+ concentration in solution. It can measure in tenths of pH units; some models measure hundredths of pH units.

Most of the control knobs are used only to calibrate the machine (a buffer of known pH is used to standardize the pH meter). Figure 2.3 illustrates

Figure 2.3a.
Analog pH meter.

Figure 2.3b.
Digital pH meter.

two commonly used types of pH meters: digital and analog. Your instructor can help you identify the parts you will need to use for the model of pH meter available in your laboratory.

Familiarize yourself with the pH meter in your laboratory by locating the following parts.

Controls

Readout meter: Shows the pH of the solution. On an analog meter, there are usually two scales. One shows pH and the other shows millivolts. You will use the pH scale. (If you are using a digital pH meter, only the number representing the solution's pH will be displayed.)

Demonstrate how to use the pH meters available in your lab. This is the only lab requiring the use of a pH meter, so you will save time if you or the prep staff calibrate the meters before students use them and tell them not to touch the calibration knobs. Better still, put a piece of tape over knobs that have been adjusted. When calibrating the meters, remember to use a standard buffer solution. Water does not necessarily measure pH 7.

Function selector (pH/standby switch): Use the pH position only when the electrode is immersed in the solution you want to measure. (Some digital models do not have this switch.)

Standardization knob: Used to calibrate the machine.

Temperature control: Temperature affects pH measurement, so adjusting the temperature control should be part of the calibration procedure.

Electrode: The delicate glass electrode is generally protected by a plastic sleeve. Even so, be careful not to bang the electrode on the glassware or stir bar.

If you have magnetic stir plates, demonstrate how to use them.

When you are measuring the pH of a solution, swirl it gently to assure good mixing and proper sampling by the electrode. A magnetic stir plate is a convenient way to make sure the solutions are well mixed for your experiment on buffering capacity. To use the stir plate, put a small stir bar in the beaker, and set the beaker in the middle of the stir plate. Turn the knob slowly, and the stir bar will begin to revolve in the beaker. Let it stir gently. If the bar starts to jump around, turn the knob off and then back on again more slowly.

Your instructor will assign your team one of the buffering solutions used to make the cabbage indicator standards in Exercise 2.2. You will test the buffering capacity of that solution and of water.

Buffering solution assigned to your team:

What hypothesis is being tested?

> *Example: The pH of Solution X will remain near X (its original pH) when small amounts of acid and base are added.*

What is the independent variable in this experiment?

> *Amount of acid or base added to Solution X.*

What is the dependent variable?

> *pH.*

On the axes of Figure 2.4, sketch the curve you expect to see for Solution X if your hypothesis is supported. On the same axes, sketch the curve you expect to see if your hypothesis is proven false.

Why should you determine the buffering capacity of water as part of this experiment?

> *It serves as a control.*

Procedure

1. Pour 40 mL of your assigned solution into a 100-mL beaker. Put a stir bar in the beaker and put it on the magnetic stirrer.

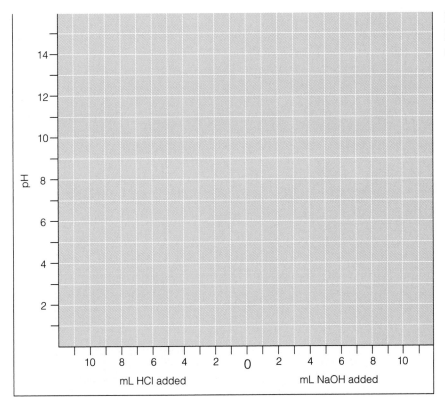

Figure 2.4.
Predicted results of buffering capacity experiment.

2. Determine the pH of the solution by following steps a–d below.

   a. Raise the electrode out of the soak beaker and rinse it with distilled water from the wash bottle.

   b. Immerse the electrode in the solution you want to measure. Swirl the beaker gently. If you are using a magnetic stir plate, make sure the bar clears the electrode before you turn it on.

   c. If your pH meter has a function switch, change it from standby to pH.

   d. Read the pH value on the readout meter or digital display.

3. Record the pH value at 0 on the x-axis of Figure 2.5 on the next page.

4. Add 1.0 mL of 0.1N HCl. N stands for normal, a measure of concentration. (If you're using a magnetic stirrer, you can leave it on with the electrode immersed throughout the procedure. If you must take the electrode out of the beaker to mix, turn the function switch to standby first.)

5. Record the new pH at "1 mL HCl added" on the x-axis of Figure 2.5.

6. Add another 1 mL of HCl and record the new pH at "2 mL HCl added" on the x-axis of Figure 2.5.

7. Continue to add 1 mL of HCl at a time and record the pH until you have added 10 mL *or* there is a significant decrease in pH, whichever comes first.

8. If the pH meter has a function switch, turn it back to standby.

9. Raise the electrode out of the solution.

1-mL pipets are used to add the HCl and NaOH. A common mistake students make is to add 0.1 mL instead of 1 mL. If the buffering curves are flat forever, that is probably the problem. You can help avoid the error by pointing out the capacity of the pipet during your introduction.

Figure 2.5.
Results of buffering capacity
experiment.

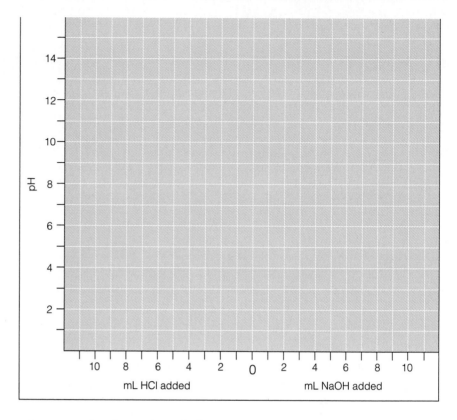

10. Rinse the electrode with distilled water, and wipe it with a cleaning tissue.

 Always rinse the electrode with distilled water after use to avoid contamination of solutions.

11. Dispose of your solution, and rinse and dry the beaker and stir bar.

12. Put another 40 mL of the same solution into the beaker and repeat the procedure using $0.1N$ NaOH instead of HCl until you have added 10 mL *or* there is a significant increase in pH, whichever comes first. Record the pH values on the "mL NaOH added" side of the x-axis of Figure 2.5.

13. When you are finished determining the buffering capacity of the solution you were assigned, repeat steps 1–12 to determine the buffering capacity of water. Use the axes in Figure 2.5 to graph the results.

 When you're done with this procedure, leave the electrode immersed in some solution (water or a buffer). It should never be allowed to dry out.

Write a descriptive title for Figure 2.5.

Example: pH changes when acid or base is added to pH 9 buffer.

Review your prediction (Figure 2.4) about the buffering capacity of the solution you were assigned. Was your prediction correct?

If the solution is a buffer, at what pH does it buffer?

Evaluate your hypothesis. Is it supported or proven false by the results?

Example: Supported. The pH of the solution changed only slightly until 6 mL of acid had been added, and it changed only slightly until 7 mL of base had been added.

Describe the buffering capacity (or lack of it) of this solution.

The solutions were made up to buffer at a certain pH. Therefore the solution must have a component that can take added H^+ out of solution and a component that can add H^+ to solution when excess OH^- are present.

Describe the buffering capacity (or lack of it) of water.

Water does not buffer. When additional H^+ are put in solution, the pH decreases immediately. When OH^- are added, the pH rises immediately. Water contains an equal number of H^+ and OH^-, so there is no component that can combine with the excess H^+ or OH^- to maintain pH.

E X E R C I S E   2 . 4
Designing an Experiment

Objective

After completing this exercise, you should be able to

1. Design an original experiment to investigate some aspect of pH or buffering capacity.

In Exercises 2.2 and 2.3 you learned a method of measuring pH and a method of determining buffering capacity. In Exercises 2.4 and 2.5, your lab team will design an experiment using one of these methods, perform your experiment, and present and interpret your results.

The following materials will be supplied for your group.

The Instructor Notes at the end of the lab topic give suggestions for helping students come up with ideas for their investigations. The notes also contain specific information about possible investigations.

Let the prep staff know what additional materials you would like to have available for the students. If possible, give students time to think about their investigations before lab and bring in their own materials for testing. You may allow students to choose the method they want to use for their investigation, or you may want to assign a method to each team in order to make the prep more predictable.

For pH Measurement Using Red Cabbage Indicator

You will already have the standards made up from Exercise 2.2.

extra test tubes

pasteur pipets and rubber bulbs

For Determining Buffering Capacity

You will use the same materials used in Exercise 2.3. Your instructor will be able to tell you what additional materials will be available.

Proposed Experiment

If you are considering an experiment using red cabbage indicator, you might want to review Table 2.1 for ideas. Also, try to think of substances whose function may be pH dependent. For example, you may recall hearing the terms pH, acid, or base used in advertisements. If you are planning an investigation of buffering capacity, consider what substances might be expected to be good buffers.

Describe your experiment below.

Question or Hypothesis

Dependent Variable

Independent Variable

Explain why you think this independent variable will affect pH or buffering capacity.

Control Treatment(s)

Replication

Brief Explanation of Experiment

Predictions (What results would support your hypothesis? What results would prove your hypothesis false?)

Method

Design a Table to Collect Your Data

List Any Additional Materials You Will Require

EXERCISE 2.5
Performing the Experiment and Interpreting the Results

Objectives

After completing this exercise, you should be able to

1. Perform the experiment your lab team designed.
2. Present and interpret the results of your experiment.

Before you do the experiment, be sure that everyone on your lab team understands the techniques that will be used. You may want to divide up the tasks before you begin work.

Be thorough in collecting data. Don't just write down numbers; record what they mean as well. Don't rely on your memory for information that you need when reporting on your experiment later! If you have any questions, doubts, or problems during the experiment, be sure to write them down, too.

Results

Before you begin to prepare your results for presentation, decide on the best format to use. Remember, you want to give the reader a clear, concise picture of what your experiment showed. Refer to the data presentation section of Appendix A (Tools for Scientific Inquiry) for help. If you are drawing graphs, use graph paper. Complete your tables and/or graphs before attempting to interpret your results.

Write a few sentences *describing* the results (don't explain why you got these results or draw conclusions yet).

Discussion

Look back at the hypothesis or question you posed in this experiment. Look at the graphs or tables of your data. Do your results support your hypothesis or prove it false? Explain your answer, using your data for support.

Did your results correspond to the prediction you made? If not, explain how your results are different from your expectations and why this might have occurred.

Describe how your data are supported by information from other sources (for example, textbooks or other lab teams working on the same problem).

If you had any problems with the procedure or questionable results, explain how they might have influenced your conclusion.

If you had an opportunity to repeat and extend this experiment to make your results more convincing, what would you do?

Summarize the conclusion you have drawn from your results.

Questions for Review

1. You have blown air from your lungs into a solution of phenol red and changed its color from red to yellow. Suggest a way to turn the color back to red.

    Add any base.

2. Give an example of two substances that, when mixed together, will produce a neutral solution.

    Any acid plus any base.

3. You measure the pH of your garden soil and find that it is 6. You measure the pH of peat moss and find that it is 4. How much greater is the concentration of hydrogen ions in peat moss than in the garden soil?

    100 times.

4. What's one ingredient that could make soft drinks acidic?

    CO_2. (But this isn't actually the main acidifier, because flat soda is just as acidic as fresh soda.)

5. Aspirin has a pH of 3. Some people who take large amounts of aspirin (for example, for arthritis) take a pill that combines aspirin with Maalox. What's the purpose of this combination?

    The acidity of aspirin alone may cause stomach upset or even ulcers. Maalox has a basic pH that neutralizes the aspirin.

6. If you want to do an experiment to measure buffering capacity, why is red cabbage indicator not a good choice of methods?

    The indicator standards can only give you pH values of 2, 4, 6, 7, 8, 10, and 12, with no values in between. This makes it impossible to detect small changes in pH.

7. On the graph in Figure 2.6, which compound is the best buffer? Explain why. Over what pH range does the compound buffer?

Solution A, which buffers at around pH 4–5, shows the least change in pH when acid or base is added, so it is the best buffer.

Figure 2.6.
Graph of buffering activity of three different compounds.

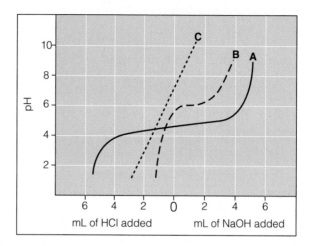

8. Considering that CO_2 is produced as a by-product of cellular metabolism, why is it important for our blood to contain buffers?

CO_2 combines with water to make carbonic acid (as explained earlier in the lab), so CO_2 production adds acid to the blood, lowering the pH. A buffer is needed to minimize the pH change.

LAB TOPIC 2

pH and Buffers
Instructor Notes

Organization

Students should work in teams of 2–4 (use whatever group size is convenient for your labs).

Exercises 2.1, 2.2, and 2.3

Introduction (5–10 minutes): Briefly review the information in Exercise 2.1 and in the introduction to Exercise 2.3. Do the phenol red demonstrations as described in Exercise 2.1.

Prep for procedures in Exercises 2.2 and 2.3 (5 minutes):

- Demonstrate pipet use.
- Demonstrate use of pH meter and stir plate.
- Review safety procedures. Everyone should wear safety glasses. Bottles of strong acids and bases should have warning labels. Tell students what to do in case strong acid or base is spilled.

Exercises 2.2 and 2.3 (60 minutes): Exercise 2.3 requires pH meters. If you don't have enough for all the teams, split up the class so that some teams do Exercise 2.2 first while the others do Exercise 2.3.

Summary of results from Exercises 2.2 and 2.3 (5–10 minutes): The purpose of this summary is to make sure students understand the methods used in Exercises 2.2 and 2.3 and to discuss any problems that arise.

Exercise 2.4 (20 minutes or less)

- You may want to tell each team which method to use for its investigation (red cabbage or buffering) so you can control the amount of equipment and materials needed, especially if the number of pH meters is limited.

- Before you tell students to do Exercise 2.4, give them some idea of extra materials that are available for their investigations (for example, beverages, shampoos, deodorants, antacids, or cleaning solutions). Writing a list on the board is helpful.

- You can tell students at the previous lab meeting to bring in ideas and materials that they can use for this experiment. A grade incentive helps. For example, you can give them a few points for bringing to lab a list of items that might have different pH values or buffering capacity.

- If your lab schedule permits, you can end the lab at this point and give students until the next lab period to design their experiments and bring in any extra materials they will need.

It may initially be difficult to get students started on designing an experiment. One useful technique is to have each team brainstorm ideas before settling on one. They should quickly write down everything they can think of that's affected by pH. (If they can't think of anything at all you can help out with questions such as "Have you ever heard about pH—acidity and alkalinity—anywhere besides biology class? Maybe in advertisements? Or in the news?") Then they can go back through their lists and consider each idea as a starting point for an experiment, eliminating those that don't lend themselves to the kinds of investigations that can be done in lab (for instance, the long-term effects of acid rain on pond life). Keep in mind that students are used to having the instructor tell them what experiment to do. Many students will delay independent thinking for as long as possible, hoping that you will eventually give up and tell them what to do. Some students believe that you have an experiment in mind and they are supposed to figure out what you want. They will keep asking if their ideas are correct. Resist the temptation to take the traditional role of the instructor; allow your students to develop their own ideas. If you see a problem with the experiment they propose, present them with the problem rather than the solution. For example, suppose students want to test the pH values of various fruits. Instead of telling them how to make juice extracts of equal concentrations, ask them how they intend to do it.

Exercise 2.5 (up to 60 minutes, or time remaining in lab period)

Since students already know how to perform the procedures from Exercises 2.2 and 2.3, this part of the lab goes very quickly.

For evaluation, students can formalize and hand in their results (tables and graphs) and answer the questions in the lab manual in Exercise 2.5 or they can write a report using the guidelines given in Appendix B.

For a 2-Hour Lab

For a 2-hour lab period, you have two options:

1. Do either Exercise 2.2 or Exercise 2.3, and then have students design their experiments using the method from the exercise they did. For example, if you only do Exercise 2.2, then all teams design an investigation on measuring pH.

2. Do Exercises 2.2 and 2.3 one week and Exercises 2.4 and 2.5 the next week.

Possible Investigations for Red Cabbage Indicator

For the cabbage indicator method, the samples must be clear or colorless to be tested. Other samples can be used if you are planning to use pH meters. Again, it is best to discuss pH with students before this lab meets and have them bring in their own samples.

Beverages

Students may investigate fruit juices (acidic, but different brands of a juice such as orange and apple vary considerably), various types of milk (all around 6.6 except buttermilk, which has a value of 5) and carbonated drinks (pH range 3–4, even after they are allowed to go flat).

Shampoo and Other Hair Care Products

The pH of hair in solution is 5.5–6.5. Shampoos below pH 5 are damaging to hair. Most shampoos are between 6.5 and 7.5, and conditioners are more basic. Hairdressers are a good source of information for students who want to investigate shampoo pH because they know quite a bit about how pH affects the hydrogen bonding and the cuticle of hair. Perming and chemical coloring involve manipulating the cuticle and hydrogen bonds, partly by changing pH. People with treated hair are advised to use more alkaline shampoos. Shampoos must be diluted with distilled water before the pH can be determined.

Hygiene Products

For example, mouthwashes (values range from pH ~4 to 8.5) can be tested. Toothpaste must be diluted with water before testing. Values for toothpastes generally range from 6.5 to 7, but Arm & Hammer, which contains baking soda, has a pH around 8.5. One team of students collected their saliva at various times of the day. Another team tested deodorants and found very acidic pHs (2.2–3.0) but no difference between products designated for men and for women.

 Do not allow students to test deodorants using the pH meter. They will damage the electrodes.

Cleaning Solutions

They are generally basic. Soaps (~6.5–7.5) are less alkaline than detergents and cleaning solutions such as Lysol and PineSol (~8.5–10.5).

Water or Soil from Various Sources

Results from water testing are better if a pH meter is used because the range of values will be small (5.5–7.5). Soil pH can be tested by soaking soil samples in distilled water and then measuring the pH of the water.

Pain Relievers

Aspirin has a low pH (~3.5), but other pain relievers such as Tylenol and Mediprin are much more basic (6–7.5). Buffered aspirin has a value of ~5. To measure the pH, weigh out equal amounts and dissolve in water.

Antacids

Students can investigate the neutralizing power of antacids by adding equal amounts of products such as Tums and Rolaids to acid.

Possible Investigations of Buffering Capacity

The buffering capacity of any substance can be investigated by using the procedure described in Exercise 2.3. Students may be interested in testing the buffering capacities of beverages such as milk, juices, and sodas in addition to medicines such as buffered aspirin, Pepto-Bismol, and Alka Seltzer.

Remind students to rinse the electrodes thoroughly after testing these substances.

Macromolecules

Introduction

Living organisms are composed of molecules that come in diverse shapes and sizes and serve a variety of purposes. Some molecules form the structure of an organism's body—for example, the cellulose that makes up the cell walls in plants, the proteins and phospholipids that comprise cell membranes, and the fibers that make up animal muscles.

There is also a wide array of molecules that perform all the functions of life. For example, enzymes catalyze the chemical reactions necessary for biological processes, neurotransmitters convey information from one brain cell to another, and visual pigments absorb light so that you can read the words on this page.

In this laboratory you will study three classes of the largest biological molecules, called **macromolecules**: carbohydrates, lipids, and proteins. A fourth class of macromolecules, the nucleic acids, will not be studied in this laboratory.

Outline

Exercise 3.1: Carbohydrates

    Activity A: Monosaccharides and Disaccharides

    Activity B: Starch

    Activity C: Hydrolysis of Carbohydrates

Exercise 3.2: Lipids

Exercise 3.3: Proteins

Exercise 3.4: Macromolecules in Food

    Activity A: Separation of Butter

    Activity B: Tests with Food

Instructor Overview
In this lab topic, students learn simple qualitative tests for detecting three of the four macromolecules: carbohydrates, lipids, and proteins. Exercises 3.1, 3.2, and 3.3 introduce the techniques by illustrating a positive test with a known substance. The I_2KI test is used to detect starch. Students also hydrolyze starch and use Benedict's reagent to identify the subunits as small sugars (the term "reducing sugar" is not used). Lipids are identified using the paper test. Biuret reagent is used to identify proteins. In Exercise 3.4, students test foods. Butter is separated (clarified) to produce a lipid fraction and a water-soluble fraction containing carbohydrates and proteins. Students use Benedict's reagent, I_2KI, biuret reagent, and the paper test to identify the types of molecules that make up each fraction. All four tests are also used to identify the components of banana, coconut, milk, peanuts, and potatoes. You may add additional foods for student testing.

Once students have learned to perform the tests and identify positive results, they can design and perform their own experiments using the same methods. Suggestions for student investigations are given in the Instructor Notes.

Time Requirements
(Some of the exercises overlap)
Exercise 3.1: 50 minutes
Exercise 3.2: 5 minutes
Exercise 3.3: 5–10 minutes
Exercise 3.4:
    Activity A: 40 minutes total.
Students should work on Activity B during waiting periods, which total 30 of the 40 minutes.
    Activity B: 50–75 minutes (10–15 minutes per sample)

EXERCISE 3.1
Carbohydrates

Objectives

After completing this exercise, you should be able to

1. Define monosaccharide, disaccharide, and polysaccharide and give examples of each.

2. Name the monosaccharide components of sucrose and starch.

3. Describe the test that indicates the presence of most small sugars.
4. Describe the test that indicates the presence of starch.
5. Define hydrolysis and give an example of the hydrolysis of carbohydrates.

Most **carbohydrates** contain only carbon (C), oxygen (O), and hydrogen (H). The simplest form of carbohydrate molecules are the **monosaccharides** ("single sugars"). One of the most important monosaccharides is glucose ($C_6H_{12}O_6$), the end product of photosynthesis in plants. It is also the molecule that is metabolized to produce another molecule, ATP, whose energy can be used for cellular work. There are many other common monosaccharides, including fructose, galactose, and ribose.

Some **disaccharides** ("double sugars") are also common. A disaccharide is simply two monosaccharides linked together. For example, maltose consists of two glucose molecules, lactose (milk sugar) consists of glucose and galactose, and sucrose (table sugar) consists of glucose and fructose. Can you discern a rule used in naming sugars?

The names generally end in -ose.

Carbohydrates are also found in the form of **polysaccharides** ("many sugars"), which are long chains of monosaccharide subunits linked together.

Starch, a polysaccharide composed of only glucose subunits, is an especially abundant component of plants. Most of the carbohydrates we eat are derived from plants. What was the last starch you ate?

Examples: potatoes, pasta (from wheat flour), bread (from wheat flour), and beans and other legumes.

Starch is the plant's way of storing the glucose it makes during photosynthesis. When you eat starch, you are consuming food reserves that the plant has stored for its own use. The starch of potatoes and root vegetables, for example, would be used the next spring for the plant's renewed growth after the winter die-back. All perennial plants (those that come up year after year, such as tulips) have some kind of food storage for overwintering. Beans, on the other hand, contain starch in the seeds. Beans are annual plants; they will die at the end of the growing season. So the seeds are stocked with starch to use when they have a chance to germinate the next spring.

Animals store glucose in **glycogen,** which is another form of polysaccharide. Although starch and glycogen are both composed of glucose subunits, the glucose molecules are bonded together in different ways, so these polysaccharides are not identical. Glucose subunits are bonded together a third way in the polysaccharide **cellulose.** While starch and glycogen are meant to be metabolized for energy, cellulose, which is the most abundant carbohydrate in the world, is a structural molecule that is designed *not* to be metabolized. Cellulose makes up the cell walls of plants and is a primary component of dietary fiber. For most animals it is completely indigestible. Those that can digest it, such as termites and cows, do so only with the assistance of organisms such as bacteria, fungi, or protistans.

Most disaccharides and polysaccharides can be broken down into their component monosaccharides by a process called **hydrolysis,** which is accomplished in organisms by digestive enzymes. This process is important in seeds. If the seed's food resource is starch, it must be able to convert the starch to glucose. The glucose is then used to generate ATP, which in turn is used to provide the growing plant embryo with energy for metabolic work. Hydrolysis of starch begins when the seed takes up water and begins to germinate.

Germination of barley seeds is part of the process of brewing beer. When the barley is germinated, the starch-to-sugar conversion begins. In the breakdown of starch, disaccharide maltose molecules are formed before the final product, glucose, is obtained. At a certain point in the germination, the barley is dried so that no further hydrolysis takes place. The maltose sugar is extracted and used in the brewing process. That's the "malt" listed on the beer can as an ingredient. The process of germinating the barley is called malting.

A chemical hydrolysis can be done in the laboratory by heating the molecules with acid in the presence of water. You will perform a chemical hydrolysis in this exercise.

 Wear safety glasses throughout the lab session.

Activity A: Monosaccharides and Disaccharides

You will use **Benedict's reagent** as a general test for small sugars (monosaccharides and disaccharides). When this reagent is mixed with a solution containing single or double sugars and then heated, a colored precipitate (solid material) forms. The precipitate may be yellow, green, orange, or red. If no monosaccharide or disaccharide is present, the reaction mixture remains clear. However, Benedict's reagent does not react with all small sugars. For example, sucrose gives a negative Benedict's reaction.

Glucose will be used in this laboratory to demonstrate a positive Benedict's test (Figure 3.1). What should be used as a negative control for this test?

Set up a tube that contains the reagent but no glucose.

Procedure

1. Make a boiling water bath by filling a beaker about half full of water and heating it on a hot plate. Put six or seven boiling chips in the beaker. You will need to use this water bath in several activities.

 Set the hot plate where it will not be in your way as you work. Be careful—it will be very hot!

Sample   Benedict's reagent

Beaker of water

Hot plate

Mix 1 dropperful of sample and 2 droppersful of Benedict's reagent

Heat 5 minutes

Cool and observe precipitate

Figure 3.1.
Benedict's test for detecting small sugars.

2.   Get two test tubes and label them 1 and 2 with a wax pencil.

 Make heavy marks so that they don't melt off in the water bath.

3.   Put 1 dropperful of glucose into Tube 1. Tube 1 is the positive control.
4.   Tube 2 is the negative control. What substance goes in it? How much should be used?

   Water. One dropperful, because that's the amount of glucose that was used.

5.   Add 2 droppersful of Benedict's reagent to each tube.
6.   Place the tubes in the boiling water bath and let them heat for 5 minutes.
7.   After 5 minutes, remove the tubes from the water bath.

 Use a test tube holder to retrieve test tubes from the boiling water.

8.   Allow the tubes to cool at room temperature for several minutes in the test tube rack while you go on to the next procedure.

9. Record your observations below.

    Tube 1 (glucose):

    There will be a brick red precipitate that settles out on cooling.

    Tube 2 (negative control):

    The solution will appear just as it did before heating: blue and clear.

In general, precipitates from this test range in color from yellow to red, depending on the concentration of sugar. The 0.1M glucose solution used in this procedure results in a brick red precipitate.

Interpretation of Results

Describe a positive Benedict's test.

> *The solution changes from clear blue to red with a precipitate.*

What are the limitations of this test?

> *It isn't quantitative (that is, you can't tell the amount of sugar present). It doesn't detect sucrose, which is a very common small sugar. It is nonspecific: You can't tell what sugars are present. Sugars would have to be extracted from any solid sample you might want to test (for example, a leaf or many food items) before the test could be performed. If a sample has an intense color, it can't be tested.*

Activity B: Starch

Starch is tested by using **iodine reagent** (I_2KI—iodine potassium iodide). A dark blue color indicates the presence of starch (Figure 3.2).

You will use a solution of potato starch to demonstrate a positive test. What negative control should be used for this test?

> *Use the reagent on water or some other sample that doesn't contain starch.*

Sample   Iodine reagent

Mix

Observe color

Figure 3.2.
The iodine test for detecting starch.

Procedure

1. Get two test tubes and label them 1 and 2.
2. Put a dropperful of starch solution in Tube 1. This is the positive control.
3. Tube 2 is the negative control. What substance goes in it? How much should be used?

    Water. One dropperful, because that's the amount of starch that was used.

4. Put 3 or 4 drops of iodine reagent into each tube.
5. Record the results below.

    Tube 1 (starch):

    Solution will be very dark, almost black.

    Tube 2 (negative control):

    Solution is straw yellow, the color of the iodine reagent.

Interpretation of Results

Describe a positive test for starch.

> *The iodine reagent turns dark blue in the presence of starch.*

What are the limitations of this test?

> *It isn't quantitative. The reaction isn't specific to starch; other polysaccharides also react. Students may notice that the iodine reacts with substances other than starch—for example, paper and fabrics.*

Activity C: Hydrolysis of Carbohydrates

As discussed earlier, disaccharides are composed of two monosaccharides linked together. Polysaccharides are long chains of monosaccharides. The bonds joining these subunits can be broken in a process called hydrolysis. In this procedure, you will hydrolyze sucrose and starch by heating them with acid.

What monosaccharides will result from the hydrolysis of sucrose?

> *Glucose and fructose.*

What monosaccharide will result from the hydrolysis of starch?

> *Glucose.*

The hydrolysis reactions will be carried out in two large test tubes. As Figure 3.3 shows, one contains sucrose and hydrochloric acid (HCl) and

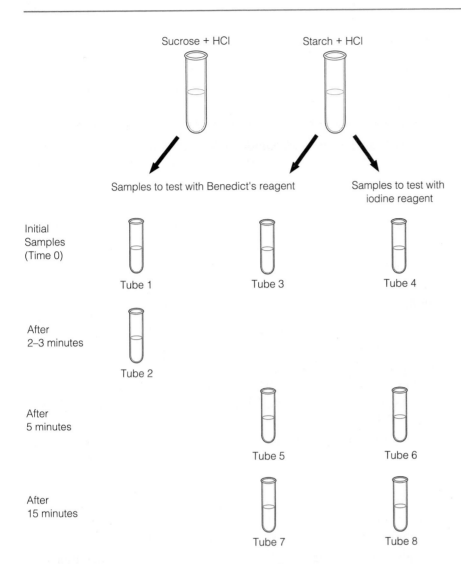

Figure 3.3.
Sampling hydrolysis products of
sucrose and starch.

the other contains starch and HCl. You will sample the sucrose tube twice:
once before the hydrolysis has begun and again after 3 minutes. You will
take six samples from the starch tube: two before the hydrolysis has been
done, two after 5 minutes of hydrolysis, and two after 15 minutes. Two
samples are needed at each time so that one can be tested for small sugars
(Benedict's test) and one can be tested for starch (iodine test).

Procedure

    Check your boiling water bath and add water if needed.

1. Get eight test tubes and label them 1 through 8. Line up the test
    tubes in order in a test tube rack.
2. Get two extra large test tubes and label them starch and sucrose. Use
    an empty beaker as a test tube holder if the test tubes don't fit in
    the rack.

Hydrolysis of Starch and Sucrose

3.   Pipet 10 mL starch solution and 5 mL 2*N* HCl into the tube labeled starch.

 HCl is a strong acid. Handle it with caution. After you use the pipet, replace it in its holder. Do not lay it down on the bench.

4.   Pipet 10 mL sucrose solution and 2 mL 2*N* HCl into the tube labeled sucrose.
5.   Swirl each tube gently to mix the contents.

Sampling

6.   Use a pasteur pipet to draw 1 pipetful of solution from the sucrose tube and put it in Tube 1.
7.   Using a *different* pasteur pipet, draw 1 pipetful of solution from the starch tube and put it in Tube 3. (Skip Tube 2 for now.)
8.   Draw an additional pipetful of solution from the starch tube and put it in Tube 4.
9.   Place the extra-large starch and sucrose tubes in your boiling water bath. Note the time:

     _____

10. After 2 or 3 minutes, draw 1 pipetful of solution from the sucrose tube and put it in Tube 2. You are now finished with the sucrose solution. You may remove it from the water bath.
11. After 5 minutes, draw 1 pipetful of solution from the starch tube and put it in Tube 5.
12. Put a second pipetful of starch solution in Tube 6.
13. Wait 10 more minutes and then repeat steps 11 and 12, putting the solution in Tubes 7 and 8.

 Do Exercise 3.2 during the waiting period.

Testing for Starch and Sugar

It is important to let the Benedict's tests from starch hydrolysis stand for a while to allow the reddish precipitate to settle. It can be seen better by holding the tube up to the light. You can also pour off the solution to see the precipitate remaining in the tube.

14. Add two droppersful of Benedict's reagent to Tubes 1, 2, 3, 5, and 7. Place these tubes in the boiling water bath for 5 minutes.
15. Add 3 or 4 drops of iodine reagent to Tubes 4, 6, and 8. Record the results in Table 3.1 on the next page.

Table 3.1

	Tube Number							
	Sucrose		Starch					
	1	2	3	4	5	6	7	8
Time (min)	0	2–3	0	0	5	5	15	15
Benedict's reagent	−	+	−		+ (slight)		+	
Iodine reagent				+ (dark blue)		− (brown)		− (yellow)

16. Remove the tubes from the water bath and wait 5 minutes for them to cool. Record the results in Table 3.1.

Interpretation of Results

Explain the results you obtained using the Benedict's test on the sucrose solution.

Initially, sucrose tests negative with Benedict's reagent. After a short period of hydrolysis, its component monosaccharides are detectable with Benedict's.

Explain the results you obtained using the iodine reagent test with starch.

Initially, the starch solution tests positive with iodine reagent. It is still positive after 5 minutes. After 10 minutes, the starch has been broken down to shorter chains (dextrins), which give the brownish color. After 20 minutes, the iodine reagent test is completely negative.

Explain the results you obtained using the Benedict's test with starch.

Initially, starch gives a negative Benedict's test. The test is still negative after 5 minutes. After 15 minutes, there should be a noticeable amount of reddish precipitate, indicating the presence of glucose (and/or maltose).

Why does hydrolysis of starch take longer than hydrolysis of sucrose?

Starch is a long, complex molecule with many glucose-to-glucose bonds. Sucrose, on the other hand, is a simple disaccharide with only one bond to break.

Hydrolysis doesn't work by clipping off one glucose at a time. Instead, the starch molecule is first broken down to shorter chains.

EXERCISE 3.2
Lipids

Objectives

After completing this exercise, you should be able to

1. Define lipid and give examples.
2. Describe the test that indicates the presence of lipids.

Lipids are compounds that contain mostly carbon and hydrogen. They are grouped together solely on the basis of their insolubility in water. The lipids we will consider in this laboratory are fats and oils, which are generally used as storage molecules in both plants and animals. You are no doubt already familiar with the fact that your body converts excess food into fat. This fat is stored in your adipose tissue until your food intake is lower than your metabolic needs, at which time the fat can be metabolized to generate ATP, whose energy can be used for cellular work. Plants, too, can store fats. Seeds are often provisioned with fats that can be metabolized by the developing embryo when germination time comes. Thus we obtain corn oil, peanut oil, sunflower oil, and others by pressing the seeds.

You will use the **paper test** (Figure 3.4) to indicate the presence of lipids in various foods. Although this test is not very sophisticated, it is quick and convenient.

Figure 3.4.
Brown paper test for lipids.

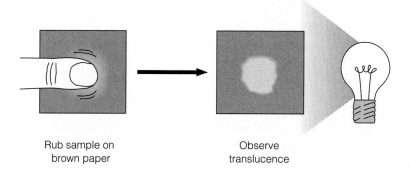

Rub sample on   Observe
brown paper   translucence

Procedure

1. Get a small square of brown paper. Write "oil" on one half and "water" on the other.

2. Put a tiny drop of salad oil on the half of the paper labeled oil. Rub it gently with your fingertip.

3. As a negative control, put a tiny drop of water on the half of the paper labeled water. Rub it gently with a different fingertip to avoid contamination.

The water may soak through the paper and appear to have the same effect as the oil until it dries, so it is important to let the paper dry thoroughly.

4. Allow the spots to dry. This may take quite a while, so go on to another exercise while you wait.

5. When the spots are dry, hold the paper up to the light.

Interpretation of Results

Describe a positive test for lipids.

> *There is a translucent spot on the paper where the oil is.*

What are the limitations of this test?

> *It is not quantitative. It is not specific: Any lipid will give a positive test. There has to be a large enough concentration of lipid in a sample to give a positive test. (For example, milk contains fats but does not give a positive test). It may be difficult to extract the lipids from a sample in order to perform the test.*

E X E R C I S E   3 . 3
Proteins

Objectives

After completing this exercise, you should be able to

1. Define protein and give examples.
2. Explain why the structure of a protein is important for its function.
3. Describe the test that indicates the presence of protein.

A **protein's** structure is determined by the amino acid subunits that make up the molecule. Although there are only 20 different naturally occurring amino acids, each protein molecule has a unique sequence. The amino acids are linked by fairly tight bonds, and the side groups that are part of the amino acids also interact with each other to help shape the molecule.

Proteins have a greater diversity of roles than either carbohydrates or lipids. The shape of a protein is key to its purpose: Proteins work by selectively binding to other molecules.

You will use **biuret reagent** as a test for proteins (Figure 3.5). This reagent, which is blue, reacts with proteins to give a light violet or lavender color.

You will use a solution of egg albumin (a protein extracted from egg whites) to demonstrate a positive biuret test. What negative control should be used for this test?

> *Test the reagent with water or some solution that doesn't contain protein.*

Procedure

1. Get two test tubes and label them 1 and 2.
2. Put two droppersful of egg albumin into Tube 1.

Figure 3.5.
Biuret test for protein.

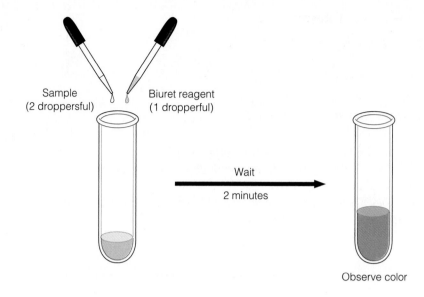

3. Tube 2 is the control. What substance goes in it? How much should be used?

Water. Two droppersful, because that's the amount of protein that was used.

4. Put 1 dropperful of biuret reagent into each tube and swirl gently to mix.

5. After 2 minutes, record the color in each tube.
Tube 1 (egg albumin):

Violet.

Tube 2 (negative control):

Blue.

Interpretation of Results

Describe a positive biuret test.

The solution changes color from blue to violet.

What are the limitations of this test?

It is not quantitative. It could not be used with colored samples. The proteins in solid samples (for example, foods) must be extracted before the test so that they will come into contact with the reagent.

EXERCISE 3.4
Macromolecules in Food

Objectives

After completing this exercise, you should be able to

1. Explain how the components of butter are distributed in two fractions by the process of clarification.

2. Interpret the results of tests that indicate the presence of sugar, starch, lipid, and protein in an unknown sample.

We metabolize food in order to release energy to produce the ATP needed for cellular work. We also break down food molecules in order to use their subunits as raw materials for synthesizing our own macromolecules. In this exercise, you will investigate certain foods to learn which macromolecules are present in each.

Activity A: Separation of Butter

Most foods are complex mixtures of substances. Butter, for example, may appear to be solid fat, but it is actually a mixture of proteins, carbohydrates, and lipids. It is an emulsion, which means that the lipids occur in very small droplets dispersed throughout the water-soluble portion. (You will learn more about emulsions in Lab Topic 14, Digestion.)

The lipid can be separated from the water-soluble, protein-containing part of the butter in a process called clarification. Butter is often clarified for use in cooking. Once the water-soluble part has been removed, the lipid that remains can be used to fry at higher temperatures than for whole butter because it is the protein in butter that scorches first. It is also the protein that spoils most readily, so clarified butter keeps longer.

Clarifying raises the smoke point of butter from 250°F to 350°F. The clarification process can be repeated to get a purer lipid fraction. Ghee, used extensively in Indian cooking, is highly clarified butter.

Procedure

1. Fill a 250-mL beaker approximately to the 75-mL mark with water.

2. Put the beaker on a hot plate and let the water come to a boil.

3. Cut approximately 1 tablespoon of butter into smaller chunks and put them in a 50-mL beaker.

4. When the water boils, remove the beaker from the hot plate with a potholder and set it on a paper towel on the lab bench. Choose a spot where it will not be disturbed.

5. Put the 50-mL beaker into the 250-mL beaker (water bath).

6. Leave the butter undisturbed for at least 15 minutes. While you wait, continue with steps 7–9.

You can save money on butter if the class pools its water-soluble fractions for the tests. Each group would need 1/2 teaspoon in step 3 and 1/2 teaspoon in step 7. After clarification, collect all the water-soluble (lower) layers in one test tube, and give an aliquot to each group. Each group should then do just one of the tests instead of all four.

A tablespoon of butter will yield about 4 mL of the water-soluble fraction.

7. Get approximately 1 teaspoon of butter from your instructor and put it into a test tube. (A teaspoon is $\frac{1}{3}$ of a tablespoon.)

8. Put the test tube in a warm water bath to melt the butter. (You can use the water bath the 50-mL beaker is in, but don't disturb the beaker.)

9. When the butter in the test tube has melted, perform the four tests that were introduced in this laboratory and record your results in Table 3.2. (The test procedures are reviewed in the next section.)

10. After 15 minutes, gently pick up the 50-mL beaker and set it in ice. Leave it undisturbed for about 10 minutes. Work on Activity B while the lipid (upper) layer solidifies.

11. The upper layer (clarified butter) should now be solid or semisolid. Remove it with a spatula and put it on a paper towel. Pat the bottom of the butter dry with a paper towel to remove any contaminants from the lower layer.

12. Place the clarified butter in a test tube and melt it in a warm water bath.

13. Perform the four tests on the melted clarified butter and record the results in Table 3.2.

14. Perform the four tests on the lower (liquid) layer of the clarified butter and record the results in Table 3.2. If you do not have enough liquid to perform all the tests, arrange to share results with another lab group.

Table 3.2

	Benedict's (sugar)	Iodine (starch)	Paper (lipid)	Biuret (protein)
Whole butter	+	–	+	(+) *(faintly positive)*
Clarified butter, upper layer	–	–	+	–
Clarified butter, lower layer	+	–	–	+

Interpretation of Results

Describe what happens to butter as a result of the clarification procedure.

> *The emulsion is broken down by heat so it is possible for the lipid to be separated from the other (water-soluble) components (lipids are defined as being insoluble in water). When the melted butter is cooled, the lipid solidifies. The end result is that the lipid and water-soluble parts (containing sugars and proteins) are physically separated from each other.*

Clarified butter lacks the "butter" flavor. What does this tell you about the molecules responsible for the taste of butter?

They are water-soluble—that is, they are not lipids. From the tests that are done in this lab they could be sugars or proteins, but of course there are also molecules that are not detected by our tests.

Some students try this procedure to clarify margarine. They warm the margarine and leave it on ice for 10 minutes as specified in the directions. They find that the margarine has resolidified but there is no liquid lower layer. What does this tell them about the margarine?

It is composed entirely of lipids, with no water-soluble components such as sugars and proteins.

Activity B: Tests with Food

Test each of the items in Table 3.3 for the presence of simple sugars, starch, lipid, and protein. Your instructor may want to modify the list. The procedures for the tests are reviewed below.

Benedict's Test (sugar)

Put 1 pasteur pipetful of sample into a test tube. Add 2 droppersful of Benedict's reagent; mix. Heat in a boiling water bath for 5 minutes. Allow to cool and observe the precipitate.

 Some samples may require extra cooling time, so don't be too hasty in recording results.

Iodine Test (starch)

Put a pipetful of sample into a test tube and add 4 or 5 drops of iodine reagent; mix.

 In some foods, the starch is still contained in granules inside the cells. You may see these dark granules suspended in the yellow solution instead of seeing the entire solution turning blue.

For solid foods we are using a water extraction, which does not extract the lipids in a sample, so testing a drop of the solution would not be a good test.

Paper Test (lipid)

If the sample is whole (for example, a peanut), rub a piece of it directly on the paper. If the sample is liquid, put a small drop on the paper.

 Remember to wait for the paper to dry before you record the results.

Biuret Test (protein)

Put 1 pipetful of sample into the test tube and add 1 dropperful of biuret reagent; mix.

 Allow at least 2 minutes for the reaction to occur. Some samples may take 5 minutes to react.

You may want to add other foods and have different groups test different foods.

Some of the foods to be tested are solids. Use a razor blade to mince approximately 1 cm³ (about the size of a pea) of the sample. Put it in a test tube with 10 mL distilled water. Put your thumb over the top of the test tube and shake it vigorously for 1 minute. Perform the tests using the liquid (except the lipid test). Record your results in Table 3.3. Be sure to rinse off the razor blade and cutting board between samples to avoid contamination.

Table 3.3

	Benedict's (sugar)	Iodine (starch)	Paper (lipid)	Biuret (protein)
Banana	+	+ *(granules)*	–	–
Coconut	–	+ *(granules slow to form)*	+	–
Milk	+	–	–	+
Peanut	–	–	+	+
Potato	(+)	+ *(granules)*	–	–

Interpretation of Results

Which results confirmed your expectations about the composition of foods?

Students will have various answers, depending on their knowledge of food. For example, they probably know that potatoes should contain starch.

Which results were unexpected?

Again, answers depend on students' knowledge. For example, they might not realize that coconuts contain a large amount of lipid or they might think that peanuts are starchy.

What factors might result in a false negative test (that is, the food does contain a molecule but the tests results are negative)?

Concentration is too low or the molecule wasn't extracted in sufficient amounts to give a positive test.

Why might a plant storage organ (such as a fruit or tuber) contain both starch and sugar?

The food reserves may be stored as starch, but some of the starch has been hydrolyzed so sugars are also present (this is what happens when fruits get sweet as they ripen).

If you have tested foods in addition to the ones listed in Table 3.3, compare the results from those tests with the results for the foods listed in Table 3.3.

For example, students may compare the molecules present in peas and peanuts (both legumes), in two fruits such as bananas and apples, or in white potatoes and sweet potatoes.

Questions for Review

1. What subunits make up
   a. Carbohydrates?

   Monosaccharides.

   b. Proteins?

   Amino acids.

2. Why is each test done initially using water as well as a known sample?

   As a negative control, to see the color of a negative test.

3. Why might a substance taste sweet, yet give a negative reaction with the Benedict's test?

   Sucrose would taste sweet but not be detectable with Benedict's reagent.

   What procedure could you use to check your answer to the previous question?

   Boil the substance with acid (hydrolyze it). Sucrose will release its subunits, which will react with Benedict's reagent.

4. You have been given an unknown solution. Describe how you would test it for the presence of
   a. Starch:

      Iodine reagent turns blue in the presence of starch.

   b. Lipid:

      Put a drop of solution on brown paper and see if there is a translucent spot when the paper dries.

   c. Sugars:

      Boil the solution with Benedict's reagent. Formation of a precipitate indicates the presence of sugar.

   d. Protein:

      Biuret reagent turns violet in the presence of protein.

5. You have tested an unknown sample with biuret and Benedict's reagents. The solution mixed with biuret reagent is blue. The solution boiled with Benedict's reagent is also blue. What does this tell you about the sample?

   It does not contain protein. It does not contain small sugars (although it possibly could contain sucrose).

6. Whole butter gives only a slightly positive test for protein (and may show no reaction at all). When the same butter is clarified, however, the liquid lower layer is definitely positive for protein. Explain why these different results might have been obtained.

   Students may hypothesize that the concentration of protein in the whole butter is too low to detect.

7. Since potatoes have starch in them, why don't they taste sweet after they are boiled?

   The starch is not hydrolyzed because boiling alone isn't sufficient; the solution must be acidified.

Acknowledgments

Procedures for the macromolecule tests were adapted from the following sources:

Armstrong, W. D., and C. W. Carr. Physiological Chemistry Laboratory Directions, 3rd ed. Minneapolis: Burgess Publishing, 1963.

Dotti, L. B., and J. M. Orten. Laboratory Instructions in Biochemistry, 8th ed. St. Louis: C. V. Mosby, 1971.

Oser, B. L., ed. *Hawk's Physiological Chemistry*, 14th ed. New York: McGraw-Hill, 1965.

LAB TOPIC 3

Macromolecules
Instructor Notes

Organization

Introduction (10–15 minutes)

Review the macromolecules that will be studied in this lab. It is important for students to understand what monomer building blocks compose each type of polymer.

Students should work in pairs or small groups. Each group should start with Exercise 3.1 and work through the rest of lab in order.

Prep for Procedures (5 minutes)

Students should have all the labware they will use for all the exercises already in their work areas. However, you might prefer to distribute the food items from a central location.

Caution students to be careful with the hot plates and razor blades. They should use test tube holders to remove test tubes from the boiling water bath and a potholder or gloves to move the beaker of hot water that is used as a water bath. Students should wear safety glasses while performing the lab exercises.

There are a few waiting periods during the procedures. Students should work on the next procedure, rather than sit around idly.

Summary (10 minutes)

Have students review results of hydrolysis and tests on food.

For a 2-Hour Lab

Exercises 3.1, 3.2, and 3.3 introduce the tests used in Exercise 3.4, so they can't be eliminated. (However, Exercise 3.1, Activity C, on hydrolysis, is expendable. Omitting it will save about 20 minutes.) Omit Exercise 3.4, Activity A. Adjust the number of food samples tested to fit the remainder of the time available (it takes 10–15 minutes to test each sample). Different foods can be assigned to different lab groups, with the results shared and discussed during the summary.

Investigative Extensions

This lab topic can be supplemented by having students use the tests on more foods or by making comparisons among different foods. The tests used in this lab are not quantitive, however, so students can't compare the amounts of different types of molecules present in different foods. Students can test a wider variety of foods for the presence or absence of sugar, starch, lipid, and protein. Students can attempt to predict the contents in some way. For example, one might hypothesize that fruits contain sugars and vegetables do not. The taste of the food might also be used as a predictor.

Possibilities for higher-level investigations have been introduced in the background material for some of the tests. For example, the conversion of starch to sugar can be investigated in germinating seeds. Storage conditions can also affect the starch content of foods. Potatoes, for example, are not stored in the refrigerator because the starch would be converted to sugar. These are slow changes, though, so the experimental treatments should be set up before the lab. It takes more than a week to get a detectable amount of sugar in a refrigerated potato. Seed germination takes one to several days, depending on the type. Therefore, there is little that students can do experimentally with these techniques in the time available in a lab period.

Using the Microscope

Introduction

The invention of the microscope was the first technological breakthrough in the study of biology. Although anatomists could investigate the basic architecture of organisms, very little could be learned about their composition before the microscope came into use. It made possible the discovery that all living organisms are composed of cells and that all cells are similar in some ways. It also enabled scientists to learn how various types of cells are different from each other. A light microscope can magnify specimens up to 1500×. Since a typical animal cell might be approximately 0.02 mm in diameter and the unaided eye can only distinguish objects as small as about 0.2 mm in diameter, the microscope is an essential tool for studying cells and subcellular components.

You will be using two types of microscopes, the compound light microscope and the stereoscopic dissecting microscope. Effective use of these microscopes is one of the fundamental skills required for studying biology, and it takes some practice to master this skill. This laboratory will show you how to use these instruments to their best advantage.

Outline

Exercise 4.1: The Compound Microscope

    Activity A: The Parts of the Compound Microscope

    Activity B: Using the Compound Microscope

    Activity C: Depth of Focus

Exercise 4.2: Observing Cells

    Activity A: Animal Cells

    Activity B: Plant Cells

Exercise 4.3: The Stereoscopic Dissecting Microscope

    Activity A: The Parts of the Dissecting Microscope

    Activity B: Using the Dissecting Microscope

Exercise 4.4: Observing an Animal Specimen

Instructor Overview
Students will need to know how to use microscopes effectively for several of the lab topics in this manual. In this lab topic, the terminology of microscopy is presented in its simplest form. Emphasis is on practical exercises that help students acquire this important skill. Students become familiar with the basic operation of a compound light microscope by using a letter "e" slide. From there, they progress to animal and plant cells. Procedures also illustrate how field of view and depth of focus change with magnification. The concept of staining is included, since students will examine prepared slides in other lab topics. The stereoscopic dissecting microscope is introduced through exercises that parallel the ones that are done with the compound light microscope so that students can see the similarities and differences in use of these two instruments.

Time Requirements
Exercise 4.1: 40–50 minutes
Exercise 4.2: 40–50 minutes
Exercise 4.3: 30–45 minutes
Exercise 4.4: 15 minutes

If the compound microscopes are not already out on the student benches, have the students get their scopes before beginning the lab.

E X E R C I S E   4 . 1
The Compound Microscope

Objectives

After completing this exercise, you should be able to

1. Determine total magnification for the lenses on your microscope.
2. Explain the difference between magnification and resolving power.
3. Identify the parts of the compound microscope and give the function of each part.
4. Describe the procedure you would follow to locate and focus on any specimen using the scanning, low-power, or high-power objective.
5. Define field of view and describe how the field of view changes with magnification.
6. Define depth of focus and explain how it changes with magnification and how to adjust for it when viewing specimens.

The compound light microscope has two sets of lenses: the eyepiece, which is near the eye, and the objective, which is near the specimen. The total magnification of the specimen is calculated by multiplying the magnification of each set of lenses:

total magnification =
magnification of objective × magnification of eyepiece

Many student-model scopes lack oil immersion lenses, so the directions given in this lab manual don't assume that students will have one. Some student scopes don't have a scanning power and the lowest power objective available is 10×. For this reason, the terms scanning, low power, and high power are used in this manual rather than low, medium, and high. Elsewhere, the high-power lens is sometimes called high-dry to indicate that no oil is needed.

The eyepiece magnification is usually fixed, and ranges from 8 times (8×) to 15×, depending upon the microscope model. The objective magnification is changeable. Compound microscopes generally have three to five different objectives. Typical magnifications are 4× (scanning power), 10× (low power), 45× (high power), and 100× (oil immersion). The magnification is engraved on each objective.

Calculate the total magnification for the lens combinations on your microscope:

Scanning power: eyepiece _____× times objective _____× = _____×

Low power: eyepiece _____× times objective _____× = _____×

High power: eyepiece _____× times objective _____× = _____×

Oil immersion: eyepiece _____× times objective _____× = _____×

Although theoretically we could put even higher powered lenses on microscopes to achieve greater magnifications, there is a practical limit imposed by the resolving power of the optical system. The **resolving power** of a microscope is its ability to distinguish two objects that are very close together as separate; it is what allows us to see detail.

When two objects cannot be distinguished as separate, the image looks blurry. Resolving power is thus a measure of clarity. Two microscopes may give the same magnification, yet the clarity of images provided by one may be better due to superior resolving power, which in turn is a function of the quality of the objective lens.

Activity A: The Parts of the Compound Microscope

Although there are numerous variations in the features of different models of compound microscopes, they are all constructed on a common plan. Figure 4.1 shows a typical model.

Figure 4.1.
The compound monocular light microscope.

Eyepiece

Arm

Revolving nosepiece

Objectives

Stage clips

Coarse focus

Stage

Fine focus

Condenser

Diaphragm

Light

Base

The microscope shown in Figure 4.1 is configured so that the arm is turned away from the user and the user reaches behind the stage to manipulate the focus knobs. Some microscopes are configured so that the eyepiece is turned around 180° and the arm is toward the user. If your microscopes fit the latter description, be sure to point out the different orientation to your students.

Locate the following parts on your microscope. The descriptions of parts include the common variations found in different models of microscopes. In the space below each item, answer the questions about the microscope you will be using.

Eyepiece

The eyepiece, also called the ocular, is the topmost lens system, and serves to magnify the image of the specimen. Some models have one eyepiece (monocular). When using a monocular scope, it is recommended that you leave the other eye open (if you can master this technique) or else cover the other eye with one hand, rather than squinting with one eye tightly shut. Binocular (two-eyepiece) models are also available.

How many eyepieces does your microscope have?

Is it monocular or binocular?

On a binocular microscope, the eyepieces can be moved closer together or farther apart to match the distance between the user's pupils. There is sometimes also a diopter ring on one eyepiece, which allows you to adjust the focus for that eye before focusing with both eyes.

Objectives

The number that is engraved next to the magnification is the numerical aperture (N. A.). The higher the N. A., the greater the resolving power of the lens. The condenser should also have an N. A. engraved on it. In order to obtain all the potential resolving power of the lens, the N. A. of the condenser must be equal to or greater than the N. A. of the lens.

The objectives magnify and resolve the image of the specimen. They are set in a revolving nosepiece so that the user can readily change lenses. Most microscopes have a ring above the lenses. Use this to turn the nose-piece rather than grabbing the lenses themselves. You should feel the lens click into place when it is engaged. This is the **working position.**

What are the magnifications of the objective lenses on your microscope?

Stage

The slide holding the specimen is placed on the stage and secured with clips. There is a hole in the center of the stage to allow light to pass through. On most microscopes, you move the slide around the stage manually in order to locate areas of interest on the specimen. Other models, however, have a movable stage. Once the slide is clipped in place, you rotate knobs to move the slide from side to side or toward or away from yourself.

Does your microscope have a movable stage?

Light

The specimen is illuminated by substage lighting. If the light source is built in, the light is focused into a beam by a **condenser** located just below the stage. Other models use an external light source; you adjust a mirror to direct the light upward through the stage. A **diaphragm** is used to control the aperture (opening), or width of the beam of light that reaches the specimen. Some microscopes allow the user to vary the aperture continually by using a lever. Other models have a wheel with fixed apertures of various sizes; you turn the wheel to select the appropriate size.

Describe the illumination system for your microscope:

Focus Knobs

Once the slide containing the specimen is mounted on the stage, you look through the eyepiece and adjust the focus using first the coarse-focus knob (to bring the specimen into view) and then the fine-focus knob (to sharpen the image). On most microscopes, turning a focus knob toward you raises the objective, and turning it away from you lowers the objective. On some microscopes, however, the stage, rather than the objectives, moves up and down.

Describe how the focus knobs on your microscope work:

Support Structure

The body of the microscope is constructed to protect the lenses and the focusing system and to provide a sturdy, vibration-free support stand.

 You should never attempt to remove the lenses or any other part of the microscope.

When working with a microscope, observe the following rules:

1. Carry the microscope with two hands. Hold the arm with one hand and use your other hand to support the base.
2. Keep all parts of the microscope dry. If you spill anything on the stage, wipe it up immediately.
3. If the lenses need to be cleaned, always use lens paper. Never use paper towels, tissues, or cleaning tissues. They will scratch the lenses.
4. When you are finished using the microscope, remove the slide and turn the lowest power objective to the working position.

Students often mistake cleaning tissues for lens paper, so be sure to point out the difference.

Activity B: Using the Compound Microscope

Before you begin using the microscope to examine unfamiliar creatures, you will first do an exercise using a recognizable object, the letter "e," to help you learn to use the microscope effectively. The keys to successful microscope work are locating the specimen, focusing on the specimen, and adjusting the light for the best contrast and resolution.

If you wear glasses, you should remove them to look through the microscope unless they correct an astigmatism.

Procedure

You may want to have students make their own "e" slides. If so, let the prep staff know that you will need the materials to do this (as listed in the Prep Guide). You will need to demonstrate the procedure for making a wet mount, which is first introduced in Exercise 4.2 in this lab. Cut out an "e" from a piece of paper. Put a drop of water on a clean slide. Pick up the "e" with forceps and place it on the drop of water. Add a coverslip as shown in Figure 4.4.

1. Turn on the light source and make sure the lowest power (scanning power, if available) objective is in the working position.

2. Place the "e" slide on the stage, centering the "e" over the hole in the stage. Secure the slide with the clips.

3. Look at the slide on the stage from the side (*not* through the eyepiece) and turn the coarse-focus knob until the objective is as close to the stage as it will go. (Most microscopes have a built-in stop to prevent you from crashing the lens into the stage, but be careful in case yours doesn't.)

4. If your microscope is binocular, adjust the distance between the eyepieces by pushing them closer together or moving them farther apart until you can comfortably see one image. If there are markings on the microscope, record the setting so you can preset it in the future:

5. Look through the eyepiece(s) and slowly move the stage and objective farther apart by turning the coarse-focus knob until you have the "e" or some part of it in focus. Move the slide to center the "e" in your **field of view** (the area of the image that you see).

If you know the width of the pointers of your microscopes, the pointer can also be used to estimate the size of specimens. See the reference manual for your microscopes.

 You may see a black line in the field of view. This is a pointer. It allows you to indicate a structure that you want someone else to view.

6. If you are using a monocular microscope or a binocular microscope that doesn't have a focusing ring for one eyepiece, use the fine-focus knob to bring the image into sharp focus.

    If you are using a binocular scope that has a focusing ring, first look only through the eyepiece that doesn't have the focusing ring. If the focusing ring is on the left eyepiece, look through the right eyepiece and cover the left eyepiece. Focus the "e" for that eye using the fine-focus knob. Then look through the eyepiece with the focusing ring, keeping the other eye covered. Adjust the focusing ring until the image is sharp. Finally, view the image with both eyes.

    How is the image of the "e" oriented compared to the image of the "e" on the slide itself?

    It is upside down.

    Look through the eyepiece and move the slide to your right.

What happens to the "e"?

It moves to the left.

What happens when you move the slide toward you?

The "e" moves away.

7. Open the diaphragm all the way (or set the disk to the largest aperture), and then slowly close it (or choose smaller and smaller apertures in the disk). You can see that the amount of light affects the image you receive. In general, you should use a smaller opening for scanning and low-power objectives and a larger opening for high-power and oil-immersion objectives. However, the type of specimen also affects how much light is needed, so it is best for you to try different settings. Begin with the largest aperture and reduce the aperture until the image has good contrast, but you are still able to see details.

8. Bring the low-power objective into the working position. If your microscope is parfocal, you should only need to adjust the focus a little using the fine-focus knob. **Parfocal** means that when an image is focused with one objective, it will also be in focus in the other objectives. Center the "e" in your field of view.
    Can you still see the entire "e"?

    Probably. It fills the field of view at 100×.

9. Now turn the high-power objective to the working position, being careful not to bang the objective on the stage clips as you turn it.

 This objective leaves very little space between the slide and the objective. You should use only the fine-focus knob when you are on high power.

Describe what you see at this magnification.

The "e" is bigger but only part of it will be seen.

How does the field of view at low power compare to the field of view at high power?

The field of view at low power is much greater.

Steps 10–15 will give you an idea of the relative fields of view that you see with your microscope.

10. Put the scanning (or lowest power) objective in the working position.

11. Turn the coarse-focus knob so that it is as close to the stage as it will go.

12. Place a transparent ruler across the stage and focus on the edge that has millimeter markings.

 If you can't bring the markings into focus, put a clean glass slide under the ruler to raise it closer to the lens.

---

13. Move the ruler so that one of the markings is at the very edge of your field of view (see Figure 4.2). Each marked interval on the ruler is 1 mm. Estimate the diameter of the field of view at this magnification.

Figure 4.2.

Alignment of ruler to measure the field of view.

The field of view varies with the objective lens and the type of eye-piece used (two eyepieces may have the same magnification yet yield different fields of view). Therefore you should perform these measurements yourself to determine the correct answers for the microscopes that your students use. Approximate ranges are given.

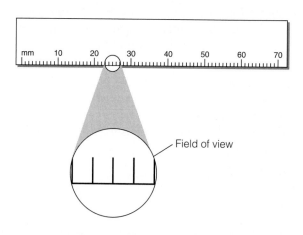

Magnification: _40×_

Diameter of field of view: _3.5–4.2 mm_

14. Change to the next higher objective and estimate the diameter of the field of view.

Magnification: _100×_

Diameter of field of view: _1.5–2.0 mm_

 If your microscope only has low-power and high-power objectives, omit step 15.

---

15. Measure the field of view again at the next higher objective.

Magnification: _450×_

Diameter of field of view: _0.3–0.4 mm_

Obviously, microscopic images are too small to be measured with a ruler. Instead of expressing measurements in millimeters, they are usually expressed as micrometers (μm). 1 μm = 1×10^{-6} m. Since 1 mm = 1×10^{-3} m, 1 mm = 1×10^{3} μm. For example, to convert 5 mm to μm:

$$5 \text{ mm } (1 \times 10^{3} \text{ μm/mm}) = 5 \times 10^{3} \text{ μm} = 5000 \text{ μm}$$

If you are not familiar with these units or working with scientific notation, see Appendix A.

Convert the field of view measurements you recorded above to micrometers:

Magnification: _40×_

Diameter of field of view: _3500–4200 μm_

Magnification: _100×_

Diameter of field of view: _1500–2000 μm_

Magnification: _450×_

Diameter of field of view: _300–420 μm_

Why is this technique for measuring the field of view not useful for an oil immersion lens?

The field of view is so small (<0.2 mm) that even if you start with a millimeter marking at one edge of the field, you won't be able to estimate the distance to the next marking.

Activity C: Depth of Focus

Although a specimen mounted on a slide appears to be completely two-dimensional, there is actually 3 to 10 μm of depth. That might not sound like a lot, but keep in mind that the specimens are magnified up to 1000×. At higher magnifications, the lens will not focus on the entire specimen at once (Figure 4.3). It is therefore useful to focus up and down through the entire depth of a specimen using the fine-focus knob, especially when you are using high power. The thickness of the specimen that can be seen in focus at any time is called the **depth of focus.** You will investigate depth of focus in the following procedure.

Figure 4.3.
Depth of focus. When a specimen is thick, only part of it can be in focus at a time.

You will also learn the technique for making a **wet mount,** which simply means that you place your specimen in a drop of water (or another liquid) on a slide and add a coverslip. The coverslip flattens the specimen, giving the slide a uniform depth for easier viewing, and protects the lens from getting wet.

Procedure

1. Get a clean microscope slide. With forceps, pull a *tiny* amount of cotton off of a cotton ball and put it on the middle of the slide.

2. Put a drop or two of *Paramecium* culture onto the cotton. (Hint: get a drop from the bottom of the container.)

3. Add a coverslip, using a dissecting needle (as shown in Figure 4.4) to lower the coverslip onto the slide. The trick is to do it slowly enough to keep any air bubbles from being trapped under the coverslip.

Figure 4.4.
Making a wet mount.

Students are very likely to get air bubbles in any wet mount they make. If a student says "What's this real round, thick-walled organism?"—it's an air bubble. Check for these pseudocritters in students' sketches, too.

4. Put the slide on the stage of the microscope and focus using scanning power. Of the cotton threads that appear in your field of view, how many are in focus?

    All of them.

5. Try different apertures with the scanning power objective. Which aperture is best for this objective?

6. Change the objective to low power. Remember, if your microscope is parfocal you should only have to adjust the fine focus a little. How many of the cotton threads are in focus now?

    When you focus on a thread with a 10× objective, you will be able to see that there are other threads that are out of focus.

    Try different apertures with the low-power objective. Which aperture is best for this objective?

7. Change the objective to high power. You may need to adjust the fine focus slightly, but don't touch the coarse focus! Now how much of the cotton is in focus?

    At high power you'll only get one thread in focus at a time.

    Use the fine-focus knob to focus all the way down to the bottom thread, then back up.

8. Try different apertures with the high-power objective. Which aperture is best for this objective?

9. Locate a *Paramecium* and watch it swim among the threads.

E X E R C I S E   4 . 2
Observing Cells

Objectives

After completing this exercise, you should be able to

1. Explain what stains are and why they are often used with biological materials.
2. Draw a cheek cell and identify its nucleus and plasma membrane.
3. Draw an *Elodea* cell and identify the chloroplasts and cell wall.

Activity A: Animal Cells

Animal cells contain numerous organelles—such as Golgi bodies, endoplasmic reticulum, lysosomes, microbodies, and mitochondria—that are too small to be seen with the compound light microscope. The only animal organelle that is reliably visible by compound light microscopy is the nucleus, which may be 4–6 µm in diameter. You can also tell where the plasma membrane is (though you can't see it, since it's less than 0.01 µm thick) because it forms a boundary around each cell.

Most animal cells are relatively colorless and appear transparent under the light microscope. We can tag certain structures by attaching pigment molecules to them in a procedure called **staining.** In this exercise you will use the stain methylene blue, which dyes the acidic molecules of the nucleus (the nucleic acids) blue. It also imparts a light blue tint to the rest of the cell.

If your microscopes have very good optics you may be able to see mitochondria; many labs on cell structures include a procedure for locating them. However, most student-model scopes that are used in general biology are not powerful enough to magnify and resolve mitochondria.

Procedure

1. Gently scrape some cheek cells from the inside of your cheek with a clean, flat toothpick.

 Discard your toothpick immediately after use in the container provided.

---

2. Spread the scrapings in the middle of a clean slide. Wait until the slide dries before proceeding.

3. Set half of a small petri dish inside half of a large petri dish. Place the slide on top of the small petri dish (see Figure 4.5).

Figure 4.5.
Arrangement of petri dishes for staining cheek cells. The small petri dish serves as a platform for the slide, and the large petri dish will catch the excess water that washes off the slide.

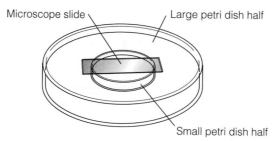

Microscope slide   Large petri dish half

Small petri dish half

4. Put several drops of methylene blue on the cheek scrapings. Wait 2 minutes for the stain to take effect.

5. *Gently* rinse the methylene blue off the slide with water from a squirt bottle. (If you rinse too vigorously, you may wash the cells off, too.)

6. Use a paper towel to blot dry the bottom of the slide. Do not wipe the top!

7. Add a small drop of water to the cheek scrapings; then add a coverslip.

8. Examine your cheek cells under the compound microscope.

Nucleus (stained blue)

Plasma membrane

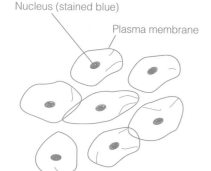

Student sketch of cheek cells.

Only oil that is made for microscopy should be used as immersion oil. Do not substitute mineral oil.

 Since the cells are still relatively transparent, adjust the aperture of the diaphragm to obtain maximum contrast.

9. Sketch the cheek cells in the margin of your lab manual and label the plasma membrane and nucleus.

If your microscope is equipped with an oil immersion lens, complete steps 10–14.

10. Be sure your slide is focused and centered at high power.

11. Move the high-power objective out of the way so that no objective is directly over the slide.

12. Put a small drop of immersion oil on the coverslip in the area you wish to view.

13. Turn the ring above the objectives so that the oil immersion lens is in the working position. When it is in place, the lens will actually be in the oil drop—this is why it's called oil immersion. View the specimen. You will need to open the diaphragm as wide as possible to get enough light. Sketch a cell in the margin of your lab manual.

What has happened to the size of your field of view?

It will be very restricted at this magnification.

14. When you are finished, wipe the oil immersion lens with lens paper to remove the oil.

Your cheek cells are a type of tissue called **squamous epithelium,** the same kind of tissue that makes up your skin and other internal and external linings of your body. The cells are loosely joined together in sheets (see Figure 4.6).

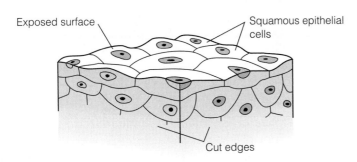

Figure 4.6.
Squamous epithelium.

The specimen you prepared showed cells lying flat on the slide, and you looked down on top of them through the microscope. Most tissues, however, don't peel off in sheets the way squamous epithelium does. For example, if you want to examine the skin tissues that lie beneath the squamous epithelium, you have to make a very thin slice from a block of tissue so that light can pass through it (see Figure 4.7). The tissue is first fixed, or killed, and then sliced and stained. When the specimen is mounted on a slide, a coverslip is fixed in place permanently. This type of slide is called a prepared slide, in contrast to the wet mounts you make yourself. You will frequently use prepared slides in this course.

Procedure

View a prepared slide of skin. How can you identify the squamous epithelium?

Squamous epithelium is seen in cross section on the outer side of the skin. Compare with Figures 4.6 and 4.7.

Figure 4.7.
Preparation of a thin section of tissue. After a block of tissue is removed from a larger specimen, it is sliced thinly enough to allow light to pass through it.

Chloroplast

Cell wall

Student sketch of *Elodea*.

Some students will be experienced in microscope use or catch on very quickly; others will have trouble just seeing the "e." Have extra specimens ready for students who finish the exercises early. Water taken from a pond or aquarium usually contains a fascinating mixture of algae, protistans, and rotifers. The Prep Guide gives instructions for making a hay infusion, which is another way to obtain a teeming culture of microorganisms.

Have students add a drop of Protoslo or methyl cellulose to the slide to slow down the organisms for observation. Be sure students don't use the immersion oil instead of Protoslo—the bottles may be very similar.

Figure 4.8.
Three-dimensional view of *Elodea* cell.

Activity B: Plant Cells

In addition to the nucleus, chloroplasts and the cell wall are visible when you look at a plant cell under the light microscope. Chloroplasts are the large (~8 μm), green, football-shaped organelles responsible for photosynthesis. The cell wall is composed of cellulose fibrils that provide strength and some rigidity. (As is the case with the plasma membrane, you cannot actually see the cell wall, but can infer its position.) There is a plasma membrane pressed up against the inside of the cell wall. Mature plant cells also have a large central vacuole, which is filled with water and water-soluble substances and may occupy 90% of the cell volume.

As an example of plant cells, you will examine *Elodea*, an aquatic plant commonly used in aquaria. The leaves are only a few cell layers thick, so a wet mount of a whole leaf allows enough light to pass through to examine the cells with light microscopy.

Procedure

1. Place one *Elodea* leaf on a slide in a drop of water.
2. Add a coverslip as shown in Figure 4.4. Try to avoid trapping air bubbles on the leaf surface.
3. Focus on the *Elodea* cells at scanning power, then at low power, and finally at high power.
4. With the high-power objective in place, move the fine-focus knob slightly up and down to see the entire thickness of a cell.

Sketch a few *Elodea* cells in the margin of your lab manual. Label the cell wall and chloroplasts. (You may not be able to see the nucleus.) In some cells you should be able to see cytoplasmic streaming as the chloroplasts move around the cell on microfilament tracks.

Try to visualize the *Elodea* cell in three dimensions. The cell is shaped like a shoebox, and inside the cell wall is the plasma membrane. A thin layer of cytoplasm lines the cell, but the largest volume is occupied by the central vacuole, which is interior to the cytoplasm. As Figure 4.8 shows, you are looking down on the top of the cell, so your view of the vacuole is blocked by the cytoplasm.

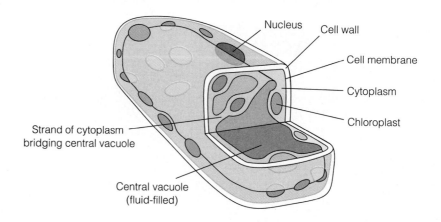

Nucleus

Cell wall

Cell membrane

Cytoplasm

Chloroplast

Strand of cytoplasm bridging central vacuole

Central vacuole (fluid-filled)

E X E R C I S E   4 . 3
The Stereoscopic Dissecting Microscope

Objectives

After completing this exercise, you should be able to

1. Explain how a dissecting scope differs from the compound micro-
    scope.
2. Identify the parts of the dissecting scope and give the function of
    each part.
3. Describe how to locate and focus on a specimen using the dissecting
    scope.
4. Identify situations in which a dissecting scope would be more useful
    than a compound light microscope.

In order to be viewed with the compound light microscope, specimens
must be thin enough for light to pass through them. They must be
mounted on glass slides; usually they are covered by a coverslip, so little
manipulation of the specimen is possible while it is being viewed. In addition, the minimum magnification is typically 40×.

The stereoscopic dissecting microscope, usually referred to as a dissecting
scope, provides a lower range of magnifications, usually between 5× and
50×, and a large working distance for manipulation of specimens. This
type of microscope is also a compound microscope, since there are two
lens systems, but in this case the objective lenses are housed inside the
scope and are changed using a knob or dial. The number and magnifications of objective lenses vary considerably from one model of microscope
to another. Some have only two settings, low power and high power, while
others can be varied continuously like a zoom lens.

All dissecting scopes have two eyepieces. In contrast to a binocular compound microscope, in which you see the same image with both eyes, in
the dissecting scope each eye sees through an independent system—there
are actually two objectives. The result is that you see depth the way you do
with unaided eyes.

Activity A: The Parts of the Dissecting Microscope

Locate the following parts on your dissecting scope (see Figure 4.9).

Eyepieces (oculars)

These lenses magnify the specimen. Notice that the distance between the
eyepieces can be adjusted to fit the distance between your eyes. Often one
eyepiece has a diopter ring on it, which allows you to adjust the focus of
the two eyepieces independently.

What is the eyepiece magnification on your dissecting scope?

Stage

Usually the specimen may be placed directly on the stage to be examined. A slide is required only for some types of specimens (for instance, if the specimen must be viewed in a drop of water). A hole in the center of the stage allows light to be transmitted from beneath the stage.

Magnification Changer

This dial or knob changes the objective lenses inside the microscope. It is marked to indicate the magnifications possible for the objectives. There may be a setting that is less than 1 (for example, 0.7). If the eyepiece is 10×, this objective would give a total magnification of 7×.

What are the objective magnifications on your dissecting scope?

What are the total magnifications for your dissecting scope?

Focus Knob

Note that there is only one type of focus knob on this kind of microscope.

Light Sources

Two types of lighting are used with dissecting scopes: transmitted light, which comes from beneath the stage, and reflected light, which shines down on the specimen from above. Transmitted light is used when the specimen is thin and transparent. Reflected light is used for specimens that are opaque. It enhances the three-dimensional quality of the image.

Describe the illumination system in your microscope.

Activity B: Using the Dissecting Microscope

Familiarize yourself with the operation of the dissecting scope by using the "e" slide again.

Procedure

1. Position the "e" slide on the stage so that the "e" is directly over the hole in the center of the stage.
2. Set the objective magnification at its lowest power.
3. Turn on the transmitted (substage) light.
4. Focus on the "e." (You may need to complete steps 5 and 6 to get the best image.)
5. While looking through the eyepieces, move them together or apart until you see a single image. If there are markings on the microscope, record the setting for your intereye distance for future reference:

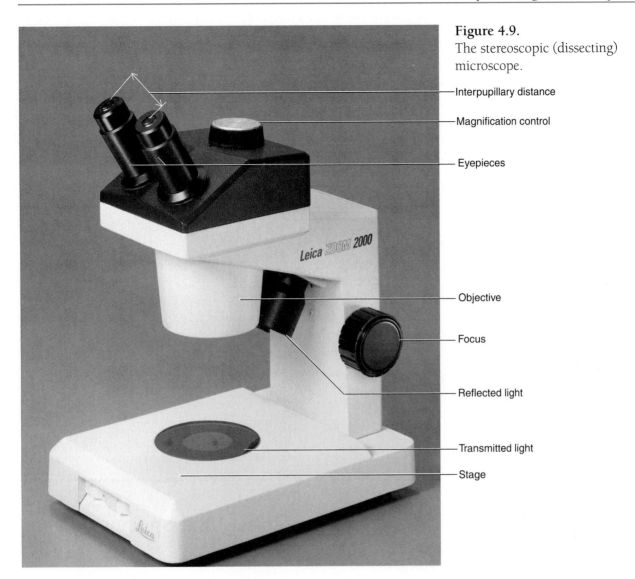

Figure 4.9.
The stereoscopic (dissecting) microscope.

— Interpupillary distance

— Magnification control

— Eyepieces

— Objective

— Focus

— Reflected light

— Transmitted light

— Stage

6.   With one eye, look through the eyepiece that doesn't have the focusing (diopter) ring. Cover the other eye. Use the focus knob to bring the image into sharp focus. Then look through the eyepiece that has the focusing ring and cover the other eye. Turn the ring until you get the sharpest focus for that eye. Then look with both eyes. The image should be in focus.

How is the image of the "e" oriented compared to the image of the "e" on the slide itself?

It is oriented the same way.

Look through the eyepiece and move the slide to your right.
What happens to the "e"?

It moves to the right.

What happens when you move the slide toward you?

The "e" moves toward you.

7. Turn the magnification knob so that you see the "e" through the entire range of magnifications available.

8. Turn off the transmitted light and turn on the reflected light. How does this change the image you see?

The image will be darker, especially if you made your own "e" slides using letters that were typewritten on paper. If the slide is permanent, you can also see the little blob of mounting medium in clear relief.

Which light source is better for this slide?

Transmitted light.

9. Next, get a photograph that has been cut out of a newspaper and place it on the stage.

10. Using reflected light, focus on the image. When you magnify the picture, you can see that it is composed of dots. The individual dots are not resolved by the unaided eye.

11. Turn off the reflected light and turn on the transmitted light. How does this change the image you see?

You can now see ink from the back of the picture coming through; this obscures the image you want to see.

E X E R C I S E 4 . 4
Observing an Animal Specimen

Objective

After completing this exercise, you should be able to

1. Determine the best illumination and magnification to use for viewing a specimen on the dissecting scope.

As you have seen, the dissecting scope is the best choice if you are working with large specimens requiring relatively little magnification. In this exercise, you will examine a hydra, an animal related to sea anemones and corals. The hydra is commonly found in ponds and streams and is a favorite laboratory organism for introductory biology.

Procedure

Students may need help locating the living hydra (step 4) if they have changed to a high magnification. Also, they may not realize that they will have to raise the objective quite a bit to focus on a specimen with this much depth.

1. Get a prepared slide of a hydra and place it on the stage of the dissecting scope.

2. Use a magnification that allows you to see the entire hydra and focus on it.

Which type of light gives you the best image?

Transmitted.

3. Sketch the hydra in the margin of your lab manual.

4. Get a dish containing living hydra. Place it on the stage and focus on one hydra.

    Which type of light gives you the best image?

    Reflected, although you may want to turn on both light sources.

    Describe how the image of the living hydra is different from the image of the prepared slide.

    The hydra was flattened in order to make the prepared slide, so that's pretty much a two-dimensional image. You could use a compound scope to view such a slide. The living hydra demonstrates that you can see a three-dimensional image on a dissecting scope.

5. Dip a piece of thread in the juice from liver. While you are looking at the hydra under the dissecting scope, place the thread in the dish near the hydra. Describe the animal's behavior.

    It will wave its tentacles as if searching for prey. (The glutathione in the liver is a feeding stimulus to the hydra.)

Your instructor will provide other specimens for viewing under the dissecting scope.

Testis

Student sketch of male hydra.

Insect specimens are great for this, but just about anything is interesting to look at under the dissecting scope.

Questions for Review

1. If the eyepiece on a microscope has a magnification of 10×, what is the total magnification with a 10× objective?

    100×.

    What is the total magnification with a 45× objective?

    450×.

2. A microscope gives a total magnification of 1500×, but the image is too blurry to be useful. What might be the problem with this microscope?

    The resolution is not good enough at 1500×.

3. Why is it important to center a specimen on low power before attempting to focus on it at high power?

    The field of view is smaller at high power. If the specimen is not centered, the user may not be able to locate the area of interest.

4. A student focuses on a specimen at low power and carefully centers it before changing to high power. At high power, however, he doesn't see the part of the specimen he was interested in. What might be the problem?

The higher the magnification of the objective, the less depth the user sees. The student should adjust the fine-focus knob up and down to see the entire depth of the specimen.

5. Inspired by her biology lab, a student decides to make a closer study of the food she eats. She uses a razor blade to make a very thin section from a raw potato and mounts it in a drop of water on a slide. To her disappointment, she can barely make out the cells under the microscope. What might she do to improve her results?

She could use some kind of stain so that the cells and organelles will show up. Or the problem could be that her slice is too thick to allow light to pass through, so she would have to make a thinner slice.

6. Compare the image obtained using a compound light microscope with the image obtained using a stereoscopic dissecting microscope.

The compound scope gives a relatively flat image from relatively flat specimens (specimens must be thin enough for light to pass through). The image is highly magnified compared to the image from the dissecting scope. The dissecting scope allows the user to see in stereo as we do with our normal binocular vision; in other words, it permits depth perception. Everything looks flat under the compound scope.

Using the Microscope
Instructor Notes

Organization

No introduction of background information is necessary.

Prep for Procedures (10 minutes)

Team size depends on how many microscopes are available. Ideally, each student should learn to use the compound microscope by himself or herself.

- If the number of dissecting scopes is limited, divide the class so that some students begin with Exercise 4.3. They will be finished with the dissecting scopes before the rest of the class is ready to use them.

- Review and demonstrate the rules for handling the microscopes. Show students what lens paper looks like, and tell them to let you know if they have mechanical problems with their scopes.

During the Laboratory Period

Circulate among the students and make sure they are actually seeing something through the microscope. A good clue is to look at their sketches. Don't wait for students to ask for help. Since they may not know what they are supposed to see, they may not realize that they aren't seeing it. You may want to ask, "Have you found it? Let me take a look."

Summary

Have students come up to the board or overhead projector and draw the organisms they saw. Give students a chance to ask any questions they couldn't answer in the exercises.

For a 2-Hour Lab

Omit Exercises 4.2 and 4.4. They are the most interesting parts of the lab, but it is essential that students be able to use microscopes effectively.

Enzymes

Introduction

Biological processes depend on molecular catalysts called **enzymes** to speed up the chemical reactions that are necessary for cells to function. It takes energy to initiate some chemical reactions. Enzymes work by lowering that amount of energy so it's easier for the reaction to get started. Without enzymes the reactions would take place far too slowly to support life. The basic components of an enzyme-catalyzed reaction are the substrate(s), the product(s), and the enzyme itself.

The **substrate** is the reactant molecule that is changed by the enzyme. Each enzyme has a specific substrate or type of substrate. For example, the substrate sucrose is acted upon by the enzyme sucrase. In the reaction shown in Figure 5.1, a disaccharide, sucrose, is broken down into its component monosaccharides glucose and fructose, which are the **products** of this reaction. The enzyme itself is neither changed nor destroyed during the reaction.

Enzymes speed up the synthesis of biological molecules as well as the breakdown, but synthesis is usually more complex than breakdown and requires a series of reactions. When sucrose is synthesized from glucose and fructose, for example, there are six steps to the process. Each step is catalyzed by a different enzyme.

There are many chemical reactions that are common to most organisms, so the enzymes that catalyze these reactions are also common. For example, the process by which energy is harvested from glucose molecules is nearly universal, so almost all organisms have common enzymes that are used in this common pathway However, the specific properties and behaviors of enzymes may be different in different organisms. Even within the same organism, the version of an enzyme found in one organ may be slightly different from the version found in a different organ.

In addition to differences in the enzyme molecules themselves, the environment in which an enzyme works is an important influence on the reaction rate.

Instructor Overview
As an introduction to enzymes, students are asked to predict the effects of enzyme and substrate concentration as well as physical factors, such as pH and temperature, on the reaction rate of enzymes. These predictions may be tested later when students design their own experiments. Students then learn a method for investigating catecholase, an enzyme that catalyzes the conversion of the colorless compound catechol to benzoquinone, which is reddish-brown. Directions are given for measuring the color change using a spectrophotometer and for estimating color intensity by comparing samples to a color standard provided. Once students have learned to measure product formation in this way, each team will design its own experiment. This lab topic prompts students to organize their experiments (see Exercise 5.3), collect data, and answer a series of questions to interpret their results (see Exercise 5.4). Alternatively, you may require students to write lab reports.

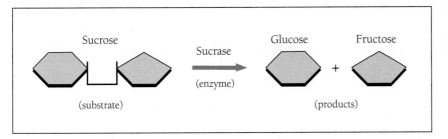

Figure 5.1.
Breakdown of sucrose by sucrase into glucose + fructose.

Preparation
Turn on the Spec 20s at the beginning of lab so they will warm up.

Time Requirements
Exercise 5.1: 30 minutes
Exercise 5.2: 30 minutes
Exercise 5.3: 15–30 minutes
Exercise 5.4: 40 minutes or time remaining in lab period

In this lab topic you will learn a method for investigating catecholase, an enzyme that is found in many plants. Your lab team will then design and perform your own experiment using this method.

Outline

Exercise 5.1: Factors Affecting Reaction Rate

   Activity A: Quantity of the Reactants

   Activity B: Physical Factors

Exercise 5.2: The Catecholase-Catalyzed Reaction

Exercise 5.3: Designing an Experiment

Exercise 5.4: Performing the Experiment and Interpreting the Results

---

EXERCISE 5.1
Factors Affecting Reaction Rate

Objectives

After completing this exercise, you should be able to

1. Make predictions regarding the effects that the amounts of enzyme and substrate present will have on the rate of an enzyme-catalyzed reaction.

2. Make predictions regarding the effects that physical factors such as pH, salt concentration, and temperature will have on the rate of an enzyme-catalyzed reaction.

Activity A: Quantity of the Reactants

Enzyme molecules do not undergo permanent changes during a reaction. After "turning over" one substrate molecule into product, the enzyme is free to engage with another substrate molecule and repeat the process. Enzymes work at a steady pace, turning over substrate into product for as long as substrate is available.

If substrate is abundant, what should happen to the reaction rate (amount of product formed/unit time) when **more enzyme** molecules are added to the reaction mixture? Sketch your prediction on the axes in Figure 5.2. Remember to graph the independent variable on the horizontal axis and the dependent variable on the vertical axis. Label both axes.

Figure 5.2.
Effect of enzyme concentration on reaction rate.

Assuming that there is plenty of substrate, product formation is proportional to the amount of enzyme. Note for student-designed investigations: If substrate is limiting, the substrate may all be converted to product. Students' results will then show that higher enzyme concentrations result in the same amount of product formation. Even when substrate is not limiting, all the substrate will eventually be converted to product if the reaction is allowed to run long enough.

If the number of enzyme molecules is constant and the number of substrate molecules is low, what do you expect to happen to the reaction rate when **more substrate** is added to the reaction mixture? Sketch your prediction on the axes in Figure 5.3 and label both axes.

Figure 5.3.
Effect of substrate concentration on reaction rate.

At a certain point (known as Vmax, or maximum velocity, in enzyme kinetics) all enzyme molecules are occupied, and an increase in substrate does not result in an increase in product. In other words, the enzyme molecules are turning substrate into product as fast as they can.

Activity B: Physical Factors

Most enzymes are protein molecules. Recall that proteins are composed of a string of amino acids linked together with relatively tight bonds. The amino acids have different chemical groups attached to them. Some of these groups attract or repel each other, and some are hydrophobic or hydrophilic. As a result of these interactions, the protein molecule is folded into a three-dimensional shape, which is closely related to the protein's function. In the case of enzymes, the shape of the protein molecule is what enables it to establish the close association with the substrate molecule(s) that is the prerequisite for making the reaction happen. The particular part of the enzyme molecule where the substrate fits in is called the **active site.** Thus anything that alters the shape of the enzyme molecule, especially at the active site, may prevent it from functioning correctly.

The attractions and repulsions between chemical groups on the amino acid chain are strongly affected by **pH,** which is a measure of the concentration of hydrogen ions in solution (Lab Topic 2, pH and Buffers). Each enzyme has a pH at which it is most perfectly shaped; it therefore functions best at that pH. If the pH is too low (acidic) or too high (basic), the enzyme molecule loses its shape and thus its ability to catalyze reactions.

How do you think reaction rate will change in solutions of different pH? Sketch your idea on the axes shown in Figure 5.4 and label both axes.

Figure 5.4.
Effect of pH on reaction rate.

Curve would be shifted depending on pH optimum of the enzyme. If students numbered the x-axis, they would probably draw their curves showing maximum activity at pH 7. However, most enzymes have an optimum pH that is related to the pH of their environment and that is not necessarily 7. For instance, digestive enzymes in the stomach have a low pH optimum, while digestive enzymes in the small intestine function best in somewhat basic conditions. A few enzymes, such as papain, do not show a pH optimum at all.

Reaction rate

pH

Salt in the enzyme's environment can also cause distortion of its shape by changing the interactions of the chemical groups on the amino acids. Just as with pH, there is an optimum salt concentration at which the enzyme is shaped perfectly to engage the substrate.

Temperature also has a profound effect on enzyme reactions. Since the enzyme molecules must be in actual contact with substrate molecules for the enzyme to catalyze the reaction, anything that increases the number of collisions between enzyme and substrate is expected to increase the reaction rate. Sketch this relationship on the axes in Figure 5.5 and label both axes.

Figure 5.5.
Effect of temperature on reaction rate.

Students will probably draw their curves something like this with the information given so far.

But heat energy has another effect in addition to speeding up the movement of molecules. It can also disrupt the delicate attractions between the chemical side groups of amino acids, causing the enzyme to change its shape and perhaps even be denatured. So when the temperature is too high, the enzyme can't catalyze the reaction. Therefore, like pH and salt concentration, enzymes work best in an optimum temperature range. Adding this information to the graph you just drew, sketch how temperature is related to reaction rate on the axes in Figure 5.6.

Figure 5.6.
Effect of temperature on reaction rate.

Students should now draw a closer approximation of a curve that shows temperature dependence. Reaction rate initially increases as the number of collisions between molecules increases, and then drops off as the enzyme is denatured.

EXERCISE 5.2

The Catecholase-Catalyzed Reaction

Objectives

After completing this exercise, you should be able to

1. Name the substrate, enzyme, and product in the experiment in Exercise 5.2.

2. Explain how the catecholase reaction is measured in the experiment in Exercise 5.2.

To prepare the potato extract: Put 80 g of potato (skins and all) in a blender with 500 mL of distilled water. Blend for 2 minutes at the highest speed. Strain the solution through four thicknesses of cheese-cloth. Immediately distribute the extract to vials and stopper tightly.

Figure 5.7.
Catecholase reaction. The colorless substrate is converted to a reddish-brown product.

In order to study how various factors affect a particular enzyme, we need to be able to measure either the disappearance of substrate or the appearance of product. In this laboratory you will study the enzyme **catecholase**, which catalyzes a reaction in which **catechol**, the substrate, becomes the product **benzoquinone**. Benzoquinone is a reddish-brown color, so we can easily determine how much benzoquinone has been formed. In fact, you are already familiar with this reaction: You are observing it when you see the cut surface of an apple or potato turn brown. A good source of catecholase is an extract made from potatoes.

Catechol (colorless substrate) + O_2 →[Catecholase (enzyme)] Benzoquinone (reddish-brown product)

Since we can use color to visualize product formation, we need a means of measuring how much color change happens during the reaction. One way to do this is to use an instrument called a spectrophotometer. A common model, and the one referred to in this lab, is the Spectronic 20 (Spec 20— see Figure 5.8). You may have a different model in your laboratory.

Figure 5.8.
Spec 20.

Read absorbance of sample (bottom scale)

Sample holder

Wavelength knob

Percent Transmittance

Absorbance

Set absorbance to infinity

Set absorbance to 0

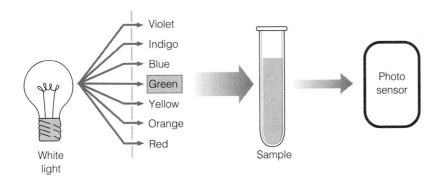

Figure 5.9.
Diagram of how a Spec 20 works. Like a prism, a spectrophotometer divides white light into its component wavelengths. In this diagram, green light has been selected to shine through the sample.

The Spec 20 measures catecholase activity by measuring the color change in reaction mixtures. As shown in Figure 5.9, this instrument shines light through a sample of reactants in a test tube and measures the amount of light that penetrates through the tube. This tells us how much light was absorbed by the sample, which in turn is a measure of how much product has been formed. The more of the reddish-brown product that has been made, the more light will be absorbed.

If spectrophotometers are not available, you may use a color chart to determine how much reaction has occurred. Compare your samples with Plate 1 and record the number for color intensity that matches the sample most closely. The more intense the color, the more benzoquinone has been formed.

On the axes in Figure 5.10, sketch how absorbance or color intensity (y-axis) varies in relation to product formation (x-axis) for the catecholase reaction. Label both axes.

Spectrophotometers differ in the width of the band of wavelengths that can be selected. Spec 20s use a diffraction grating that results in bands about 20 nm wide. Better instruments provide narrower band widths. The Spec 20 is more than adequate for the experiments in this lab manual. In fact, colorimeters (which use a limited number of colored filters rather than a diffraction grating) will also give satisfactory results.

Figure 5.10.
Relationship of product formation and absorbance.

Absorbance is not strictly linear, but is close enough for the purposes of this experiment. Students are asked to draw this graph so that they will better understand the results that they get (that is, higher absorbance = more product).

In order to make the reaction occur, the mixture must contain both substrate and enzyme. Name the specific components for this reaction.

Substrate:

Catechol.

Enzyme:

Catecholase (from potato extract).

You will set up three sample tubes to illustrate how this method works. If you are using a Spec 20 or other spectrophotometer, you must first zero the instrument using a blank. (If you are using the color chart method, you may skip the rest of this paragraph.) The blank is a type of control. You will notice that the potato extract itself has color; it already absorbs some light. That amount of light must be subtracted from the amount of light absorbed by the product of the reaction. This procedure is analogous to taring a balance. The blank contains enzyme (potato extract) but no substrate, so very little reaction will take place. (There will be some reaction, though, because the potato extract itself contains substrate.)

Look at Table 5.1 to see how the experiment has been designed and answer the following questions.

Table 5.1
Reaction Mixtures for Catecholase Experiment

Tube	Water (mL)	Catechol (mL)	Extract (mL)
1	3	2	1
2	3.5	2	0.5
3	4	2	0

What hypothesis is being tested?

Example: Increasing the amount of enzyme will increase the absorbance or color intensity. (Any testable hypothesis is acceptable.)

What is the independent variable in this experiment?

Amount of enzyme.

What is the dependent variable?

Absorbance or color intensity.

Is there a control for this experiment?

Yes. Tube 3 contains no enzyme.

Predict the outcome of your experiment in terms of your hypothesis. Explain what results would support your hypothesis and what results would prove your hypothesis false.

> *Example: If the hypothesis is supported, then the absorbance (or color intensity) in Tube 1 will be higher than the absorbance (or color intensity) in Tube 2 at the end of the experiment. Any other results would prove the hypothesis to be false.*

Procedure

Zeroing the Spec 20 (skip steps 1–9 if you are using the color chart)

1. Get a clean test tube (either a small test tube or a special Spec tube) and use a wax pencil to label it B (for blank). If the tube does not already have a short vertical mark at the top, draw one with wax pencil. When you put the tube into the sample holder, the mark should always face front.

2. Measure 1 mL potato extract into Tube B.

3. Add 5 mL distilled water to Tube B.

4. Cover Tube B tightly with Parafilm and invert it to mix the contents.

5. Set the wavelength knob on the Spec 20 to 540 nm (see Figure 5.8).

6. Using the knob on the left (Figure 5.8), set the absorbance reading (bottom scale) to infinity.

7. Wipe Tube B with a cleaning tissue and insert it into the sample holder of the Spec 20 with the vertical mark facing front.

8. Using the knob on the right, set the absorbance reading to 0.

9. Set Tube B aside. You will need to rezero the Spec 20 later.

The Spec 20 is now zeroed and ready to measure the tubes in which the catecholase reaction has occurred.

Measuring Color Change in Samples

10. Use a pipet to measure the water into Tubes 1, 2, and 3 (see Table 5.1).

11. Use a different pipet to add 2 mL catechol to each tube.

12. Use a third pipet to add 1 mL potato extract to Tube 1. Add 0.5 mL potato extract to Tube 2. Do not add any potato extract to Tube 3.

13. Place Parafilm tightly over the top of each tube and invert the tubes to mix their contents.

14. Shake each tube gently at 1 minute intervals to keep the reactants well mixed.

You may want to have everyone zero the Specs at the same time following your instructions.

Tubes made to use in spectrophotometers usually have a vertical white line at the top of the tube. We just use small test tubes, which are relatively inexpensive, in our Spec 20s and have students draw a line with wax pencil. Inserting the tube in the same orientation for each measurement nullifies any effect from differential absorbance by the glass. Of course, if only one reading is to be made from a sample (as in Tubes 1, 2, and 3 in this demonstration), this isn't necessary.

Spec 20 Method

15. After 3 minutes, use the blank to rezero the Spec 20 (follow steps 6–8).

16. Insert each tube in turn into the sample holder and read the absorbance (bottom scale). Record the absorbance reading for each tube in Table 5.2.

Color Chart Method

15. After 5 minutes, compare each tube to the catecholase chart on Color Plate 1. Observe the intensity of the colors (the intensity is more important than the actual color). In Table 5.2 record the number of the color intensity that most closely matches the color intensity of each tube.

Collect data from all the lab teams to get more replication.

Table 5.2

Results of Catecholase Experiment (absorbance or color intensity is recorded after 3 minutes)

	Tube		
	1	**2**	**3**
Absorbance/color intensity	*0.24/3*	*0.09/1*	*0/0*

Was your hypothesis proven false or supported by the results? Use data to support your answer.

> *Example: Supported. Absorbance in Tube 2 was 0.09, while absorbance in Tube 1 was 0.24. More enzyme resulted in higher absorbance readings.*

Predict the color change of a tube containing 3.75 mL water, 2 mL catechol, and 0.25 mL of potato extract. Explain how you derived this prediction from your data.

> *Example: Less enzyme should result in less color change, so the absorbance reading would be somewhere between 0 and 0.09. On the color intensity scale, the result will probably be the same as for 0.50 mL enzyme, since the differentiation among samples is much more gross.*

Why is it necessary to add the potato extract to each tube last, after the water and catechol are already measured?

> *The reaction will begin as soon as the enzyme is added. If you put all the ingredients in Tube 1 first, the reaction will be occurring while you set up Tube 2.*

Why was a different amount of water added to each tube?

> *The total volume in each tube must be equal.*

EXERCISE 5.3

Designing an Experiment

Objective

After completing this exercise, you should be able to

1. Design an original experiment to investigate some aspect of enzyme activity.

In Exercise 5.2 you learned a method of measuring the reaction as the enzyme catecholase converts catechol to benzoquinone. In Exercises 5.3 and 5.4 your lab team will design an experiment using this method, perform your experiment, and present and interpret your results. You may want to review Exercise 5.1 to help you decide on an independent variable for your investigation.

The following materials will be supplied for your group.

50 mL potato extract	30 Parafilm squares
2 1-mL pipets	30 test tubes
3 5-mL pipets	test tube rack
pi-pump for 1-mL pipet	50 mL of 0.05% catechol
pi-pump for 5-mL pipet	bottle of distilled water
wax marker	

Your instructor will be able to tell you what additional materials will be available.

Describe your experiment below.

Question or Hypothesis

Dependent Variable

Independent Variable

Exercise 5.1 should provide plenty of hints to students about the sorts of investigations they could do. The Instructor Notes at the end of this lab topic give you more specific information.

Let the prep staff know what additional materials you would like to have available for the students. For example, you might want to have a series of pH buffers, water baths, ice, hot plates, and other sources of catecholase. Let students know (perhaps by a list on the board) what materials they can use.

Explain why you think this independent variable will affect catecholase activity.

Control Treatment(s)

Replication

Brief Explanation of Experiment

Predictions

What results would support your hypothesis? What results would prove it false?

Method

Include the levels of treatment you plan to use. It might be helpful to make a table like Table 5.1, showing the contents of each reaction tube.

Design a Table to Collect Your Data

List Any Additional Materials You Will Require

E X E R C I S E   5 . 4

Performing the Experiment and Interpreting the Results

Objectives

After completing this exercise, your should be able to

1. Perform the experiment your lab team designed.
2. Present and interpret the results of your experiment.

Before you do the experiment, be sure that everyone on your lab team understands the techniques that will be used. You may want to divide up the tasks before you begin work. Since it is important to measure the volumes of reactants accurately, you may want to ask your instructor to review pipet use with you.

Be thorough in collecting data. Don't just write down numbers; record what they mean as well. Don't rely on your memory for information that you will need when reporting on your experiment later! If you have any questions, doubts, or problems during the experiment, be sure to write them down, too.

Results

Before you begin to prepare your results for presentation, decide on the best format to use. Remember, you want to give the reader a clear, concise picture of what your experiment showed. Refer to the data presentation exercise in Appendix A (Tools for Scientific Inquiry) for help. If you are drawing graphs, use graph paper. Complete your tables and/or graphs before attempting to interpret your results.

Write a few sentences *describing* the results (don't explain why you got these results or draw conclusions yet).

Discussion

Look back at the hypothesis or question you posed in this experiment. Look at the graphs or tables of your data. Do your results support your hypothesis or prove it to be false? Use your data to support your answer.

Did your results correspond to the prediction you made? If not, explain how your results are different from your expectations and why this might have occurred.

Describe how your data are supported by information from other sources (for example, textbooks or other lab teams working on a similar problem).

If you had any problems with the procedure or questionable results, explain how they might have influenced your conclusion.

If you had an opportunity to repeat and extend this experiment to make your results more convincing, what would you do?

Summarize the conclusion you have drawn from your results.

Questions for Review

1. A freshly cut potato turns brown when left standing. Why do mashed potatoes stay white?

   Catecholase, the enzyme that causes the browning, is denatured when the potatoes are cooked.

2. The three graphs in Figure 5.11 represent three different enzymes that are found in a unicellular pond organism. Assuming that these enzymes must function in order for the organism to grow, what water pH is best for this organism? Explain your answer.

   pH 5–6. Although two of the enzymes function in broader pH ranges, the third enzyme limits the organism to a narrower pH range.

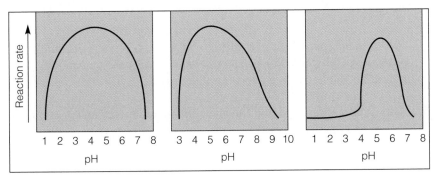

Figure 5.11.
Reaction rate versus pH of water for three enzymes.

3. Lemon juice, which has a pH of about 3, can be sprinkled on freshly cut fruit to keep it from turning brown. Propose a hypothesis to explain this observation.

   Based on the lab, students are most likely to answer that the acidic pH denatures the enzyme catecholase, which is responsible for browning. This is a reasonable hypothesis.

   How could you test your hypothesis?

   Test the pH of lemon juice. Then put an acid with the same pH (such as acetic or hydrochloric acid) on fresh fruit and observe whether the browning reaction occurs.

   Propose an alternative hypothesis. (Hint: Lemon juice is a complex substance.)

   Some other component of lemon juice either interferes with the enzyme's function or prevents the formation of benzoquinone by some other means (notice that O_2 is required for the reaction).

4. You hypothesize that reaction rate is proportional to enzyme concentration and design an experiment to test your hypothesis. Fill in the blanks below to complete the design.

	Water	Substrate	Enzyme
Tube A	20 drops	*10 drops*	*0 drops*
Tube B	15 drops	10 drops	5 drops
Tube C	*10 drops*	10 drops	*10 drops*
Tube D	*0 drops*	*10 drops*	20 drops

   Explain your answers:
   Water:

   Each tube must have an equal total volume of solution.

   Substrate:

   Substrate must be kept constant when the independent variable is enzyme concentration.

Enzyme:

Tube A is the control (no enzyme). It makes sense to vary the independent variable in a consistent fashion (in this case by doubling it), rather than choosing values haphazardly.

5. A student team is studying the effect of substrate concentration on the rate of an enzymatic reaction. They supply each reaction mixture with the same amount of enzyme. Predict what the results will be and explain why.

    At higher substrate concentrations, enzyme will be limiting, so the reaction rate will stay the same. At a certain concentration of substrate, all the enzyme is working as fast as it can to turn substrate into product. No additional substrate can be used, because the enzyme is all "busy."

LAB TOPIC 5

Enzymes
Instructor Notes

Organization

Students should work in teams of 2–4 (use whatever group size is convenient for your labs).

The time estimate for Exercise 5.2 assumes that students have gained some familiarity with laboratory procedures by performing labs earlier in the course. If you think your students may require significantly longer than 30 minutes to carry out this procedure, you may want to save lab time by asking them to do Exercise 5.1 as a homework assignment before coming to lab. Students will also be able to design their investigations more readily if they have had at least one previous investigative lab, so plan your time for Exercise 5.3 accordingly. Data collection goes very quickly once students have designed an experiment; an hour would be ample time for Exercise 5.4.

Exercise 5.1 (30 minutes)

Let students work in their lab teams to read and make the predictions in Exercise 5.1. When most groups have finished, call on students to draw the graphs and explain their predictions. Summarize the expected effect of each factor on enzyme activity.

Exercise 5.2 (30 minutes)

Very briefly explain the catecholase experiment and how product formation will be measured. If you are using spectrophotometers, you may want to go through the procedure for their use at this time (see Procedure in Exercise 5.2). If you are using the color intensity scale, point out the color chart (Plate 1). Also review pipet use.

To help students understand how the Spec 20 works, have them cut a strip of white paper and put it in a Spec 20 tube. Put the tube in the sample holder, leaving the door open so you can look in. Turn the tube so the light shines on the paper, reflecting the color of light that would pass through the sample. Rotate the wavelength knob; you can see the colors of the spectrum change. When the wavelength knob is set for this experiment (540 nm), green light is seen shining onto the paper.

Students should work with their lab teams to complete the exercise. Have groups pool their data when everyone has finished. The important lesson of this experiment is how the catecholase reaction is being measured.

Exercise 5.3 (15–30 minutes)

Give students a list of extra materials that will be available for their investigations (for example, pH buffers, water baths, alternative sources of catecholase).

Encourage all lab teams to brainstorm a list of ideas for experiments before they settle on one.

If your lab schedule permits, you can end the lab topic at this point and give students until the next lab period to design their experiments and bring in any extra materials they will need.

If students have designed an experiment on their own in a previous lab topic, this part of the lab topic goes very quickly. If not, it may initially be difficult to get students started on designing an experiment. One useful technique is to have each group brainstorm ideas before settling on one. Students should quickly write down everything they can think of that could affect enzyme activity. Reviewing Exercise 5.1 will help. Then they can go back through their list and consider each idea as a starting point for an experiment and decide which independent variable they are interested in investigating. Keep in mind that students are used to having the instructor tell them what experiment to do. Unless there are constraints on the materials available, let each lab team make this decision on its own. After selecting an independent variable, each team should fill out a proposal form to complete their experimental design. If you see a problem with the experiment they propose, present them with the problem rather than the solution. For example, suppose students want to compare catecholase activity from three different varieties of potato. Instead of telling them how to make extracts of equal concentrations, ask them how they intend to do it. The Possible Investigations section of the Instructor Notes gives some information on how to investigate a variety of independent variables.

Exercise 5.4 (40 minutes or time remaining in lab period)

The enzyme assay itself can be done very quickly once the experiment is organized. Students should have plenty of time to perform replications; they may also want to revise their original design (for example, they may need to use additional substrate concentrations in order to find the point of enzyme saturation).

For evaluation, students can formalize and hand in their results (tables and graphs) and answer the questions in the lab manual in Exercise 5.4 or they can write a formal report.

For a 2-Hour Lab

Ask students to complete Exercise 5.1 before coming to lab. Briefly review their predictions at the start of lab, and then proceed with Exercises 5.2, 5.3, and 5.4. If your students are not familiar enough with lab procedures and experimental design to complete Exercises 5.2 and 5.3 in an hour, you may want to make the investigative part of the lab more structured. For example, you can assign an independent variable to each team.

Alternatively, you can do Exercises 5.1 and 5.2 one week and Exercises 5.3 and 5.4 the next week.

Possible Investigations

Concentration of Enzyme or Substrate

Concentration of enzyme should be varied around 1 mL potato extract. Concentration of substrate should be varied around 1–2 mL catechol (0.05%). Students who are designing this type of experiment need to make up a table specifying the amount of water, substrate, and extract to be placed in each tube. Each tube must have the same volume of solution. These experiments can be performed without any extra prep, as long as there is plenty of potato extract and catechol available.

pH

pH can be varied by using 1–2 mL potato extract, 1–2 mL catechol, and 4 mL buffer in place of the water used in other experiments. You can either have a range of buffers available in the lab for students to choose from or have the prep staff make them up on request. (Note: At pH 12, the catechol is oxidized and turns light brown before the enzyme is added.)

Temperature

Use 1–2 mL potato extract, 1–2 mL catechol, and 4 mL water. You probably won't have enough water baths for students to use them for a range of temperatures. Instead, have students make up their own water baths using ice, a hot plate, and tap water. They will also need a thermometer.

Salt

Use 1–2 mL potato extract, 1–2 mL catechol, and 4 mL salt solution. Students can suggest a range of salt concentrations they would like to try. You might tell them, as a reference point, that physiological saline is 0.9%.

Source of Enzyme

Both apples and potatoes were mentioned as having the browning reaction. Students can compare potatoes with apples or try different varieties of potatoes or apples. Other fruits or vegetables that don't show a browning reaction can also be used.

Students doing this type of experiment will need a blender to make their own extracts, but no additional solutions are needed. Be sure you give students the recipe for the extract: 40 g tissue/250 mL distilled water. Blend on high speed for 2 minutes; then filter through cheesecloth. Smaller amounts can be made as long as there is enough liquid to cover the blades of the blender. You may allow students to request the additional vegetables or fruits from the prep room or ask them to bring their own (that's what we do).

Pretreatment

Students may hypothesize that treating the potato in some fashion before extracting catecholase will affect the reaction rate of the enzyme. For example, temperature or the pH or salt concentration of soaking solutions could produce differences in the enzymes that are later extracted. Students will have to decide how long the treatment should last. In this case, potatoes should be divided so that the different treatments are given to parts of the same potato. Discard the cut outer layer before making the extract.

Students doing this type of experiment will need a blender to make their own extracts (see Source of Enzyme). They may need additional solutions or a means of creating and maintaining temperature differences (which can simply be a freezer, refrigerator, incubator, or warming oven), and they may need access to these things before the lab day so as to give their treatments time to take effect.

Inhibition by Ascorbic Acid

The introduction to this lab topic does not mention enzyme inhibitors. Because of the reaction being studied, however, there is one inhibitor that might occur to students: ascorbic acid. When sliced apples are used in a recipe, for example, the directions often say to sprinkle lemon juice over the cut surfaces to prevent browning. There is even a product, Fruit Fresh, which is sold for this purpose (its main ingredient is ascorbic acid).

Students doing this type of experiment will only need the source of ascorbic acid (citrus fruit, Fruit Fresh, or a solution of ascorbic acid) in addition to the basic materials.

Cellular Respiration

Introduction

In the process of photosynthesis, plants capture light energy and use it to make the monosaccharide glucose. The energy from the light is packaged into the chemical bonds of the glucose molecule. This glucose can be converted into a number of other molecules, which are either used or stored in the plant. For example, glucose made in the leaves is changed into sucrose, a disaccharide, before it is transported to other parts of the plant. Glucose may also be stored as starch, which is a long chain of glucose molecules. When animals eat plants, the starch and sucrose can be changed back into glucose for use in metabolism. Animals also store excess glucose in long chain molecules (glycogen) or by converting it into fat. Starch and glycogen can in turn be reconverted into glucose when energy is needed by the plant or animal.

The energy contained in the glucose molecule cannot be used directly by the cell. That energy must first be repackaged in ATP (adenosine triphosphate), the molecule that provides the energy for most cellular work. The process of **cellular respiration** transfers the energy stored in glucose bonds to bonds in ATP so that it can be used more easily by the cell. Each glucose molecule can generate as many as 38 ATP molecules through cellular respiration.

In this lab topic we will consider two methods by which cells make ATP using the energy that comes from breaking down the chemical bonds in glucose. These two methods, called **metabolic pathways,** are alcoholic fermentation and aerobic respiration.

Outline

Exercise 6.1: Alcoholic Fermentation

Exercise 6.2: Aerobic Respiration

Exercise 6.3: Designing an Experiment

Exercise 6.4: Performing the Experiment and Interpreting the Results

Instructor Overview
In this lab topic students learn two methods of investigating the processes of cellular respiration. Alcoholic fermentation in yeast is measured by collecting CO_2. Aerobic respiration is measured in a suspension of lima bean mitochondria by following the color change of DCPIP, an electron acceptor. Directions are given for measuring the color change using a spectrophotometer and for estimating color intensity by comparing samples to a color standard provided. Students perform a demonstration exercise with each of these methods, then each student team designs its own experiment. This lab topic prompts students to organize their experiments (see Exercise 6.3), collect data, and answer a series of questions to interpret their results (see Exercise 6.4). Alternatively, you may require students to write lab reports.

Time Requirements
Exercise 6.1: 30 minutes (Data collection for Exercise 6.1 can overlap with introduction to Exercise 6.2)
Exercise 6.2: 40–50 minutes
Exercise 6.3: 15–30 minutes
Exercise 6.4: 40 minutes or time remaining in lab period

EXERCISE 6.1
Alcoholic Fermentation

Objectives

After completing this exercise, you should be able to

1. Describe alcoholic fermentation, naming the substrate and products.
2. Explain how alcoholic fermentation can be measured.
3. List factors that can affect the rate of alcoholic fermentation.

Alcoholic fermentation is a metabolic pathway used primarily by yeasts and some bacteria. In fermentation, glucose is broken down into ethyl alcohol (ethanol) and carbon dioxide (Figure 6.1). In the process, some of the energy that had been stored in the glucose bonds is used to form high energy bonds in ATP.

Figure 6.1.
Substrate and products of alcoholic fermentation.

The arrows in Figure 6.1 indicate that a series of enzyme-catalyzed reactions is needed to complete the conversion of glucose (the substrate) into alcohol, carbon dioxide, and ATP (the products). Therefore, any of the factors that affect enzyme activity can affect the rate at which fermentation occurs.

Name some factors that affect enzyme activity.

pH, temperature, salt concentration (ionic strength), amounts of enzyme and substrate used.

Although glucose is the substrate at the beginning of the pathway, keep in mind that other molecules can be changed into glucose and used in cellular respiration. For example, starch or glycogen (storage molecules) can be broken down into their many component glucose molecules. Sucrose (table sugar) can be broken down into its two component sugars, glucose and fructose. The fructose can then be converted to glucose by another enzymatic process.

In the procedure you will use, fermentation is performed by yeast, a single-celled fungus. That is, yeast contains the cellular machinery, including the **enzymes,** which is capable of breaking down glucose by alcoholic fermentation. Corn syrup, which contains sucrose and fructose, will be used as the **substrate.** You will measure the rate of alcoholic fermentation by collecting carbon dioxide (CO_2), which is one of the **products,** at intervals after fermentation has begun.

Figure 6.2.
Apparatus for measuring CO_2 production in alcoholic fermentation.

Figure 6.2 shows the setup you will use to collect the CO_2. Fermentation will take place in the test tube on the right, which contains the yeast and corn syrup, the fermentation solution. The test tube is capped with a rubber stopper to prevent CO_2 from escaping. Plastic tubing leads from the fermentation tube to the CO_2 collection tube, which is upside down and contains water. As CO_2 is produced by fermentation, it goes through the tubing into the collection test tube, where it displaces the water. The displacement (in mm) is recorded as a measure of the amount of CO_2 produced.

Procedure

1. You will need to have three fermentation setups, so you should have six large test tubes, three pieces of plastic tubing that have been inserted into rubber stoppers, and three beakers (400 or 600 mL). Using a wax pencil, label three of the test tubes 1, 2, and 3 and set them aside. Assemble the setups one at a time following steps 2–7 (see Figure 6.3 on the next page).

2. Fill a tub or sink with *hot* water (50°–60°C).

3. Insert the end of the plastic tubing into one of the test tubes. This tube will be the CO_2 collection tube. Submerge the collection tube and plastic tubing in the tub of hot water (Figure 6.3a).

4. Submerge the beaker. Place the collection tube in the beaker in an inverted position (Figure 6.3b).

5. Bring the beaker out of the water. One end of the plastic tubing should still be inserted in the collection tube. Hold up the other end of the tubing (the one with the rubber stopper on it) so that the water won't be siphoned out.

You should make enough yeast suspension for the class. Stir one package of Rapid Rise yeast into 50 mL warm water. If you are not using Rapid Rise, use two packages of yeast in 50 mL warm water. Divide up the suspension among the groups. Each group will only use ~5 mL for this exercise.

Leftover yeast suspension can be saved for successive lab sections throughout the day, but it should not be kept overnight.

The suggested temperature range gives the best results. If your tap water doesn't get this hot, the experiment will still work but it will take longer to get results.

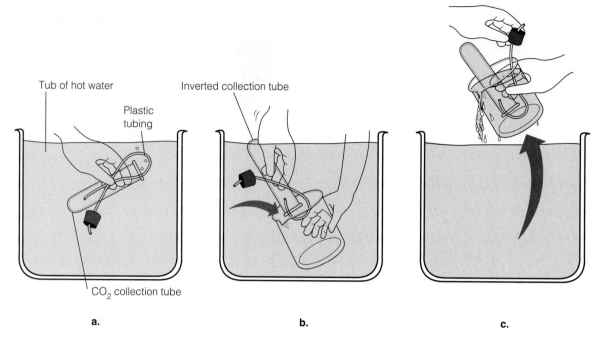

Tub of hot water

Plastic
tubing

Inverted collection tube

CO_2 collection tube

a.

b.

c.

Figure 6.3.
Setting up the CO_2 collection tube.

6. Pour some water out of the beaker so that the water level is at least 1 cm from the top of the beaker (Figure 6.3c).

7. Check the tubing for kinks. If the CO_2 can't get through the tubing, you'll have to start over.

 Assemble all three setups before proceeding to mix the fermentation solutions.

8. Mix the fermentation solutions for Tubes 1, 2, and 3 according to Table 6.1.

Table 6.1
Contents of Fermentation Tubes

	Fermentation Tube (volume in mL)		
	1	2	3
Water	4	3	1
Yeast suspension	0	1	3
Corn syrup	3	3	3

9. Swirl each test tube gently to mix the reactants. Place one test tube in each beaker.

10. Put the rubber stoppers in the fermentation tubes. This will force most of the water out of the tubing.

11. After the air bubbles from inserting the stopper have cleared the tubing (half a minute to a minute), mark the water level on each collection tube with a wax pencil. This marks the baseline for your experiment.

 If the water level is all the way to the top of the collection tube, where the tube is curved, you should wait until it has descended to the part of the tube where the sides are straight before you mark the level.

12. At 5-minute intervals, measure (in mm) the distance from the baseline mark to the water level. Continue taking data for at least 20 minutes. Record your data in Table 6.2.

After the experiment has been set up, one student in each group can be assigned to collect data while you introduce Exercise 6.2.

Table 6.2
Results of Fermentation Experiment

Tube	Minutes						
	0	**5**	**10**	**15**	**20**	**25**	**30**
1 (3 mL corn syrup)	*0 mm*	*0 mm*	*0 mm*	*0 mm*	*0 mm*		
2 (3 mL corn syrup + 1 mL yeast suspension)	*0 mm*	*0 mm*	*5 mm*	*9 mm*	*17 mm*		
3 (3 mL corn syrup + 3 mL yeast suspension)	*0 mm*	*10 mm*	*24 mm*	*41 mm*	*60 mm*		

Experiment was terminated after 20 minutes when water level in Tube 3 became too low to measure.

While you are waiting to collect data, look again at Table 6.1 to see how the experiment was designed.

What is the independent variable?

> *The number of organisms carrying out fermentation or the amount of enzymes (the yeast contains the enzymes).*

What is the dependent variable?

> *The amount of CO_2 produced (the amount of alcoholic fermentation).*

What hypothesis is being tested?

Example: The greater the amount of yeast, the more fermentation will occur. (Any hypothesis about the effect of varying the amount of yeast or enzymes on the rate of fermentation is acceptable.)

Predict the outcome of the experiment in terms of your hypothesis. Explain what results would support your hypothesis.

Example: If the hypothesis is supported, then Tube 3 will have the highest rate of fermentation because it contains the most yeast.

The instructor will collect data from all the student teams. Average the data and, in Figure 6.4, plot the average water displacement for each interval against time. This shows the reaction rate (CO_2 produced/time). The steeper the slope of the line, the faster the reaction rate. Be sure to label the axes completely.

Figure 6.4.
Rate of CO_2 production by yeast fermentation.

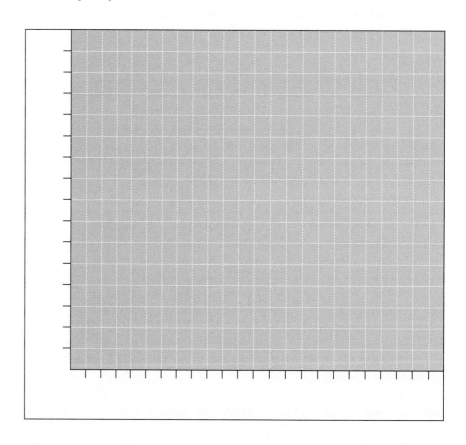

Was your hypothesis proven false or supported by the results? Use data to support your answer.

Example: Supported. Tube 3 had the most yeast and it had the greatest amount of CO_2 produced as well as the highest rate of fermentation. (Evaluation of the hypothesis depends on the hypothesis stated.)

Which fermentation tube was the control?

Tube 1, which contained no yeast.

Why were different amounts of water added to each fermentation solution?

To keep the volume constant in all tubes.

What are some other independent variables that could affect alcoholic fermentation?

Concentration of substrate, different types of substrate, temperature, pH, salt concentration (ionic strength) of the fermentation solution.

EXERCISE 6.2
Aerobic Respiration

Objectives

After completing this exercise, you should be able to

1. Name the three phases of aerobic respiration.
2. Explain how aerobic respiration can be measured using DCPIP and the succinate $\rightarrow$ fumarate reaction.
3. List factors that can affect the rate of aerobic respiration.

Most organisms, including humans, produce most of their ATP by using the process of **aerobic respiration** rather than by fermentation. Aerobic respiration begins with glucose, the same substrate that is used in fermentation. However, aerobic respiration extracts much more of the energy stored in the bonds of each glucose molecule. A single glucose molecule can yield as many as 38 ATP molecules through aerobic respiration, compared with a yield of two ATP per glucose molecule from alcoholic fermentation.

Like fermentation, aerobic respiration begins with the metabolic pathway (series of enzyme-catalyzed reactions) called **glycolysis.** In aerobic respiration, however, pyruvate (which is the product of glycolysis) enters another metabolic pathway called the **Krebs cycle.** Some of the reactions in the Krebs cycle release protons and electrons, which then enter the **electron transport system.** Most of the ATP molecules that are gained in aerobic respiration are produced by the electron transport system.

Glycolysis takes place in the cytoplasm, but the Krebs cycle and electron transport occur within a specific organelle, the **mitochondrion.** Mitochondria are found in all eukaryotic organisms (organisms whose cells have a true nucleus and membrane-bound organelles). You will use a suspension made from ground-up lima beans to study aerobic respiration. The suspension contains mitochondria, which, under our experimental conditions, will continue to carry out aerobic respiration as if they were in intact

The mitochondrial suspension should be prepared right before lab. One batch is enough for at least four lab sections. If it is kept on ice, the suspension will be usable for two successive lab sections (about 5–6 hours), but it shouldn't be kept longer than that. Consult with the prep staff about whether they will prepare the suspension or expect you to do it. The procedure takes about 20 minutes. See Instructor Notes and Prep Guide for procedure.

cells. Sucrose has been added to the mitochondrial suspension as a source of glucose for respiration, just as corn syrup was used in alcoholic fermentation.

The rate of aerobic respiration can be measured by studying the activity of one particular enzyme in the Krebs cycle. This enzyme catalyzes a reaction that converts succinate, which is the substrate for the reaction, to fumarate (Figure 6.5).

Figure 6.5.
Breakdown of glucose by glycolysis and the Krebs cycle. The enzyme-catalyzed conversion of succinate to fumarate is one step in the Krebs cycle.

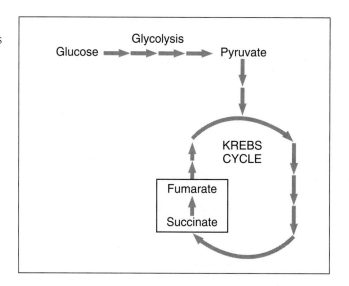

As the enzymatic reaction occurs, protons and electrons are released from the succinate molecule. In the living mitochondrion, these protons and electrons are carried to the electron transport system and used to make ATP. In order to investigate aerobic respiration, you will add a blue compound, DCPIP (di-chlorophenol-indophenol) to the reaction mixture. DCPIP intercepts the protons and electrons before they get to the electron transport system. When a DCPIP molecule picks up a proton and electron, its blue color disappears and it becomes colorless. We can use this color change as a measure of the enzymatic reaction rate, which in turn is a measure of the rate of aerobic respiration. As succinate is converted to fumarate (and as aerobic respiration proceeds), the solution in the test tube gradually turns from blue to colorless. The reaction is summarized in Figure 6.6.

Figure 6.6.
Color change of DCPIP caused by the conversion of succinate to fumarate.

The color change that DCPIP undergoes thus provides a way to visualize aerobic respiration as it occurs. We now need a method of measuring how much color change occurs. An instrument called a spectrophotometer can be used. A common model (the one referred to in this lab topic) is the Spectronic 20 (Spec 20). You may have a different model in your laboratory.

If spectrophotometers are not available, you can use a color chart to determine how much color change has taken place. Compare your samples with Plate 2 and record the number for the color intensity that matches the sample most closely.

As shown in Figure 6.7, the Spec 20 shines light through a sample of reactants in a tube and measures the amount of light that penetrates through the tube. In Lab Topic 5 (Enzymes), you recorded data from the absorbance scale: The data showed how much light had been absorbed by each sample. In this experiment you will record data from the % transmittance scale.

Spectrophotometers differ in the width of the band of wavelengths that can be selected. Spec 20s use a diffraction grating that yields bands about 20 nm wide. Better instruments provide narrower band widths. The Spec 20 is more than adequate for the experiments in this lab manual. In fact, colorimeters, which use a limited number of colored filters rather than a diffraction grating, will also give satisfactory results.

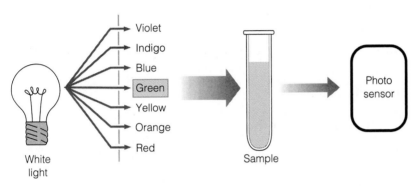

Figure 6.7.
Diagrammatic representation of how a Spec 20 works. Like a prism, a spectrophotometer divides light into its component wavelengths.

Your data will show how much light has been transmitted by the sample. The more intense the color of the sample, the less light it transmits. Percent transmittance measures how much of the blue color has disappeared. The blue DCPIP becomes colorless as aerobic respiration proceeds, so the higher the transmittance reading the more mitochondrial activity has been observed.

On the axes in Figure 6.8, sketch the relationship between % transmittance or color intensity and mitochondrial activity (aerobic respiration). Label both axes.

Figure 6.8.
Relationship of % transmittance or color intensity to mitochondrial activity.

Table 6.3

Reaction Mixtures for Mitochondrial Activity Experiment

	Tube (volume in mL)		
	1	2	3
Buffer	4.4	4.3	4.2
DCPIP	0.3	0.3	0.3
Mitochondrial suspension	0.3	0.3	0.3
Succinate	0	0.1	0.2

Before you begin the procedure, look at Table 6.3 to see how the experiment will be set up.

What is the role of each of the following components of the reaction mixture?

Lima bean extract:

Contains mitochondria.

Succinate:

Substrate for the reaction we are assaying.

DCPIP:

Indicator that provides a measure of the reaction rate.

Buffer:

Provides a favorable pH environment for the enzymes.

Why should there be a different amount of buffer in each tube?

We have to keep the total volume the same in each tube to keep the proportion of DCPIP constant so that color changes are comparable.

What is the independent variable?

The amount of succinate (substrate).

What is the dependent variable?

% transmittance or color intensity.

What hypothesis is being tested?

Example: The greater the amount of substrate present, the greater the rate of aerobic respiration. (Any hypothesis about the effect of varying the amount of succinate or substrate on aerobic respiration is acceptable.)

Predict the outcome of the experiment in terms of your hypothesis. Explain what results would support your hypothesis.

Example: If the hypothesis is supported, then Tube 3 will show the greatest color change (disappearance of blue color) because it contains the most substrate.

Procedure

Zeroing the Spec 20 (skip steps 1–7 if you are using the color chart)

When you look at the mitochondrial suspension, you will notice that it is cloudy. So even though it is colorless, transmittance of light through the sample will already be less than 100% before any DCPIP is added. If you are using a Spec 20, you will therefore prepare a "blank" tube containing mitochondrial suspension but no DCPIP and use that tube to set the transmittance at 100%. This is called zeroing the machine.

If the students did not use Spec 20s in Lab Topic 5, you may want to have everyone zero the Spec 20s at the same time following your instructions.

1. Get a clean test tube (either a small test tube or a special Spec 20 tube) and use a wax pencil to label it B. When you put the tube into the sample holder, it should always be inserted with the B facing front.
2. Measure 4.6 mL buffer, 0.3 mL mitochondrial suspension, and 0.1 mL succinate into Tube B.
3. Cover the Tube B tightly with Parafilm and invert it to mix the reactants thoroughly.
4. Set the wavelength of the Spec 20 to 600 nm (Figure 6.9).
5. Use the knob on the left to set % transmittance (top scale) to 0.
6. Wipe the blank tube (B) with a cleaning tissue and insert it into the sample holder.
7. Use the knob on the right to set % transmittance to 100%.
8. Set Tube B aside. You will need to rezero the Spec 20 later.

We just use small test tubes, which are relatively inexpensive, in our Spec 20s. Inserting the tube in the same orientation for each measurement nullifies any effect from differential absorbance by the glass.

The Spec 20 is now zeroed and ready to measure sample tubes for the experiment.

Read transmittance of sample
(top scale)

Sample holder

Wavelength knob

Set transmittance
to 0

Set transmittance
to 100%

Figure 6.9.
Spec 20.

✳ *The initial color in each tube should be blue, and the color should remain visibly blue in Tube 2 for at least 15 minutes. If you are using the Spec 20, initial transmittance readings may be as high as 30%. If initial readings are any higher or if the blue color is very light, you will have no results for the experiment; the color change will be complete in less than 5 minutes. The usual cause of this problem is that the DCPIP solution is too weak. The DCPIP should be freshly made for the labs (make it up the week before the lab will be performed), and it should be kept in brown bottles. The DCPIP solution should appear very dark (midnight) blue when drawn up in a pipet. If the solution is fresh and still appears to be too weak, try stirring it to remix the DCPIP. Also, be sure that the DCPIP solution was made exactly as specified in the Prep Guide.*

Preparing Reaction Mixtures for the Experiment

9. Get three Spec 20 tubes or small test tubes and use a wax pencil to label them 1, 2, and 3.

10. Measure the buffer, DCPIP, and mitochondrial suspension into Tubes 1, 2, and 3 as specified in Table 6.3, but it is important not to add the succinate until the other ingredients are in the tubes.

11. As quickly as possible, add the succinate to each tube.

12. Cover each tube tightly with Parafilm and invert to mix.

Spec 20 Method

13. As you complete each tube, insert it into the sample holder of the Spec 20 with its number facing to the front and record the % transmittance in Table 6.4 at time 0.

14. After 5 minutes, use the blank to rezero the Spec 20 (steps 5–7). Invert each tube to remix its contents and read and record the % transmittance.

15. Continue to record the % transmittance in the tubes at 5-minute intervals for at least 20 minutes or until Tube 3 is nearly colorless (% transmittance is near 100%). You should rezero the Spec 20 before each reading and always insert the tubes with the number facing the front.

Color Chart Method

13. Compare each tube to the right chart on Color Plate 1. Observe the intensities of the colors. In Table 6.4, record the number of the color that most closely matches the color intensity in each tube at time 0. At the beginning of the experiment all the tubes should be the same color.

Table 6.4
Results of Mitochondrial Activity Experiment
(% transmittance or color intensity)

Tube	Minutes						
	0	5	10	15	20	25	30
1 (0.3 mL mitochondrial suspension)	5	6	7	8	8	8	8
2 (0.3 mL mitochondrial suspension + 0.1 mL succinate)	7	9	12	15	19	24	30
3 (0.3 mL mitochondrial suspension + 0.2 mL succinate)	7	11	16	21	27	34	43

Look for a similarity in color intensity. The exact color does not have to match.

14. After 5 minutes, shake each tube gently to mix its contents and compare it to Color Plate 1. In Table 6.4, record the number of the color that most closely matches the color intensity in each tube.

15. Continue to record the color in the tubes at 5-minute intervals for 20 minutes or until Tube 3 is colorless (0).

The instructor will collect data from all the student teams. Average the data and plot the average % transmittance (or color intensity) for each interval against time in Figure 6.10. This shows the reaction rate for aerobic respiration. The steeper the slope of the line, the faster the reaction rate. Be sure to label the axes completely.

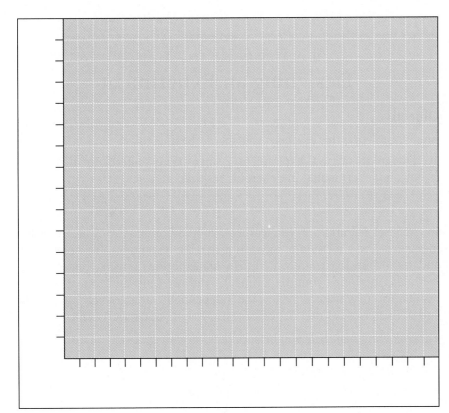

Figure 6.10.
Rate of aerobic respiration in mitochondria.

If results show similar curves for Tubes 2 and 3, it is because the enzyme is already saturated with substrate in Tube 2 (0.1 mL). If this is the case, adding more substrate (Tube 3) doesn't increase the reaction rate because all the enzyme molecules are busy turning over substrate into product as fast as they can. This substrate concentration is called Vmax, which students may know from studying enzymes. Results almost always show a difference in the two curves, but a poor yield of mitochondria (a low concentration of enzyme) could lead to Tubes 2 and 3 giving similar curves. An unexpected result can be as interesting as the one you expect!

Was the hypothesis proven false or supported by the results? Use data to support your answer.

Example: Supported. Tube 3 had the most succinate, so it should show the greatest rise in % transmittance (but see margin note this page). Evaluation of the hypothesis depends upon the hypothesis stated.

Which reaction tube was the control?

Tube 1, to which no succinate was added.

Why should the succinate be added to the reaction tubes last?

Because it is the substrate, and as soon as the substrate is added the reaction will begin.

What are some other independent variables that could affect aerobic respiration?

Concentration of succinate, volume of mitochondrial suspension, temperature, pH, salt concentration (ionic strength) of the mixture, source of mitochondria.

 When you design your own experiment to investigate mitochondrial activity using the Spec 20, be sure to include a blank tube for zeroing the instrument.

E X E R C I S E   6 . 3

Designing an Experiment

Objective

After completing this exercise, you should be able to

1. Design an original experiment to investigate some factor that affects alcoholic fermentation or aerobic respiration.

The Instructor Notes at the end of this lab topic give you some specific information about the sorts of investigations that students could do. Let the prep staff know what additional materials you would like to have available for the students. For example, you might want to have a series of pH buffers, water baths, ice, hot plates, and other substrates for fermentation. Let students know (perhaps by a list on the board) what materials they can use.

In Exercises 6.1 and 6.2, you learned a method of measuring alcoholic fermentation and a method of measuring aerobic respiration. In Exercises 6.3 and 6.4, your lab team will design an experiment using one of these methods, perform your experiment, and present and interpret your results. You may want to review the independent variables you listed at the end of Exercises 6.1 and 6.2 before you design your experiment.

The following materials will be supplied for your group.

For alcoholic fermentation:

10 fermentation kits	pi-pump
1 package yeast	50 mL corn syrup (diluted 1:1)
150-mL beaker	bottle of water
glass stir rod	wax marker
3 5-mL pipets	ruler

For aerobic respiration:

20 test tubes	pi-pump for 5-mL pipet
test tube rack	20 Parafilm squares
3 1-mL pipets	25 mL DCPIP
1 5-mL pipet	25 mL succinate
pi-pump for 1-mL pipet	100 mL buffer solution

Your instructor will be able to tell you what additional materials will be available.

Describe your experiment below.

Question or Hypothesis

Dependent Variable

Independent Variable

Explain why you think this independent variable will affect alcoholic fermentation or aerobic respiration.

Control(s)

Replication

Brief Explanation of Experiment

Predictions

What results would support your hypothesis? What results would prove your hypothesis false?

Method

Include the levels of treatment you plan to use. It might be helpful to make a table showing the contents of each reaction vessel.

Design a Table to Collect Your Data

List Any Additional Materials You Will Require

E X E R C I S E   6 . 4
Performing the Experiment and Interpreting the Results

Objectives

After completing this exercise, you should be able to

1. Perform the experiment your lab team designed.
2. Present and interpret the results of your experiment.

Before you do the experiment, be sure that everyone on your lab team understands the techniques that will be used. You may want to divide up the tasks before you begin work. Since it is important to measure the volumes of reactants accurately, you may want to ask your instructor to review pipet use with you.

Be thorough in collecting data. Don't just write down numbers; record what they mean as well. Don't rely on your memory for information you will need when reporting on your experiment later! If you have any questions, doubts, or problems during the experiment, be sure to write them down, too.

Results

Before you begin to prepare your results for presentation, decide on the best format to use. Remember, you want to give the reader a clear, concise picture of what your experiment showed. Refer to the data presentation section of Appendix A (Tools for Scientific Inquiry) for help. If you are drawing graphs, use graph paper. Complete your tables and/or graphs before attempting to interpret your results.

Write a few sentences *describing* the results (don't explain why you got these results or draw conclusions yet).

Discussion

Look back at the hypothesis or question you posed in this experiment. Look at the graphs or tables of your data. Do your results support your hypothesis or prove it false? Use your data to support your answer.

Did your results correspond to the prediction you made? If not, explain how your results are different from your expectations and why this might have occurred.

Describe how your data are supported by information from other sources (for example, textbooks or other lab teams working on a similar problem).

If you had any problems with the procedure or questionable results, explain how they might have influenced your conclusion.

If you had an opportunity to repeat and extend this experiment to make your results more convincing, what would you do?

Summarize the conclusion you have drawn from your results.

Questions for Review

1. What is the importance of cellular respiration to living organisms?

    It is the process that converts stored chemical energy into a form of energy that is usable for metabolic work.

2. Yeast is used to make bread rise. How might this work?

    Fermentation by the yeast cells produces CO_2 gas that is trapped in the dough, causing it to rise.

3. Which of the following tubes would you expect to produce the highest initial rate of CO_2 production? Explain your answer.
    Tube 1: 5 mL corn syrup, 10 mL yeast, 5 mL water
    Tube 2: 10 mL corn syrup, 10 mL yeast, 0 mL water
    Tube 3: 0 mL corn syrup, 10 mL yeast, 10 mL water

    Tube 2 has the highest amount of substrate (corn syrup), so it should have the highest rate of fermentation.

4. Suppose you do an experiment on aerobic respiration using lima bean mitochondria as described in Exercise 6.2 and the results show a final % transmittance reading of 42% in Tube 1 and 78% in Tube 2. Both tubes had an initial reading of 15%. In which tube did the greatest amount of aerobic respiration occur? Explain your answer in terms of the changes that take place in DCPIP.

    In Tube 2. The higher the transmittance reading, the more DCPIP has been changed from blue to colorless. DCPIP undergoes this color change as a result of picking up the protons and electrons generated from a Krebs cycle reaction, so the more the color changes, the more aerobic respiration has occurred.

Acknowledgment

The procedure used to assay mitochondrial activity was based on a procedure from Succinic Acid Dehydrogenase Activity of Plant Mitochondria in F. Witham, D. Blaydes, and R. Devlin, *Exercises in Plant Physiology*, Boston: Prindle, Weber & Schmidt, 1971.

LAB TOPIC 6

Cellular Respiration Instructor Notes

 If you need to prepare the mitochondrial suspension, allow 20–30 minutes before class to do it. If you are using Spec 20s, turn them on at the beginning of the lab period.

Organization

Students should work in groups of 2–4. Use whatever group size is convenient for your labs. The time estimates for this laboratory assume that students have gained some familiarity with laboratory procedures by performing labs earlier in the course. If you think your students may require significantly longer than an hour to complete Exercises 6.1 and 6.2 (even with overlapping data collections), you may want to use only one procedure rather than both. Alternatively, have half the students perform Exercise 6.1 and half perform Exercise 6.2. Students will also be able to design their investigations more readily if they have had at least one previous investigative lab, so plan your time for Exercise 6.3 accordingly. Data collection goes very quickly once students have designed an experiment, so an hour should be ample time for Exercise 6.4.

Exercise 6.1 (30 minutes)

Introduction (5 minutes)

Give a very brief introduction to alcoholic fermentation, emphasizing the general equation (glucose $\rightarrow$ ethanol + CO_2) and the involvement of enzymes in the pathway. Be sure students understand the components of the experiment: Corn syrup is the substrate, the yeast contains the enzymes, and CO_2 is the product we measure.

Prep for Procedures (5 minutes)

You may want to review pipet use.
Demonstrate how to assemble the fermentation setup.

Procedure (15–20 minutes)

Once all the lab teams have their experiments running, have each team designate one person to continue the data collection while you introduce Exercise 6.2.

Exercise 6.2 (40–50 minutes)

Introduction (10 minutes)

Give a brief introduction to aerobic respiration, emphasizing that students will assay one enzyme in the Krebs cycle as a measure of respiration.

Explain how the color change of DCPIP is used to measure respiration.

Explaining the Spec 20

Explain how the Spec 20 is used to monitor color change in DCPIP. If you did Lab Topic 5, be sure students understand that they are using % transmittance for this exercise rather than absorbance. If you are using Color Plate 1 rather than spectrophotometry, briefly explain to students how it should be used.

To help students understand how the Spec 20 works, have them cut a strip of white paper and put it in a Spec 20 tube. Put the tube in the sample holder, leaving the door open so you can look in. Turn the tube so the light shines on the paper, reflecting the color of light that would pass through the sample. Rotate the wavelength knob; you can see the colors of the spectrum change. When the wavelength knob is set for this experiment (600 nm), orange light is seen shining onto the paper.

Prep for Procedures (10 minutes)

Have each team make up its blank tube and use it to zero the Spec 20 (steps 1–7 of the exercise).

Procedure (20–30 minutes)

Have the lab teams do steps 8–14 of the exercise. It takes only a few minutes to set up the additional 3 tubes; the rest of the time is for obtaining results. Discuss the results of Exercise 6.1 and possible investigations while waiting.

Summary of Results from Exercises 6.1 and 6.2 (5–10 minutes)

The purpose of this summary is to make sure students understand the methods used in Exercises 6.1 and 6.2 and to discuss any problems that arose.

Procedure for Preparing Mitochondrial Suspension

Weigh out 50 g dried lima beans and soak them in water overnight before the suspension is made. Put the beans in plenty of water (a 250-mL beaker works well for 50 g), since they will take up water, and keep them refrigerated.

All solutions and containers should be kept ice cold! Place the blender cup and glassware you will use in the freezer several hours before you begin the procedure, and keep the solutions refrigerated.

1. Place 50 g dried lima beans (soaked overnight) in blender with 100 mL sucrose-phosphate buffer. Blend at high speed for 2 minutes.

2. Set a 600-mL beaker in an ice bucket. Strain the homogenate through cheesecloth into this beaker. You may have to change the cheesecloth once or twice as it becomes clogged with particles. Squeeze the cheesecloth gently to get the filtrate through.

3. Divide the filtrate equally among the centrifuge tubes (there will be ~10 mL per tube).

4. Centrifuge at $200 \times g$ (or at a low setting) for 1 minute. Do not decant.

5. Increase the speed and centrifuge at $1300 \times g$ (or at a medium-high setting) for 10 minutes.

6. Decant the supernatant from each tube into a beaker and put the beaker in an ice bucket. This suspension contains mitochondria. Supply it to the labs in small test tubes. Each group will use only about 1 mL, so put out two or three test tubes, which can be shared by groups. The mitochondrial suspension should be kept on ice in the lab. (Note: Many models of tabletop centrifuges do not indicate $\times g$ settings. You can obtain this information from the manufacturer. If you can't determine what the settings mean, use a low speed for the first run, then increase to a medium-to-high speed. Keep the pellets. If you centrifuged at too high a speed, the mitochondria will be in the pellet, which can then be resuspended in sucrose-phosphate buffer.)

Exercise 6.3 (15–30 minutes)

You may want to tell each team which method to use for its investigation (fermentation or aerobic respiration) so you can control the amount of equipment and materials needed.

Before you tell students to do Exercise 6.3, give them some idea of extra materials that are available for their investigations (for example, pH buffers, water baths, and different types of sugars). Writing a list on the board is helpful.

You can tell students during the previous lab meeting or in lecture to bring in ideas for this experiment. A grade incentive helps. For example, you can give them a few points for bringing to lab a list of factors that affect cellular respiration.

If your lab schedule permits, you could end the lab at this point and give students until the next lab period to design their experiments and bring in any extra materials they will need.

If students have designed an experiment on their own in a previous lab topic, this part of the lab goes very quickly. If not, it may initially be difficult to get students started on designing an experiment. One useful technique is to have each group brainstorm ideas before settling on one. They should quickly write down everything they can think of that could affect cellular respiration. Reviewing the independent variables they listed at the end of Exercises 6.1 and 6.2 will help. Then they can go back through their list and consider each idea as a starting point for an experiment and decide which independent variable they are interested in investigating. Keep in mind that students are used to having the instructor tell them what experiment to do. Unless there are constraints on the materials available, let each lab team make this decision on its own. After selecting an independent variable, each team should fill out a proposal form to complete their experimental design. If you see a problem with the experiment they propose, present them with the problem rather than the solution. For example, suppose students want to use different fruits as substrates for fermentation. Instead of telling them how to make juice extracts of equal concentrations, ask them how they intend to do it. The Possible Investigations section of the Instructor Notes gives some information on how to investigate a variety of independent variables.

If your lab schedule permits, you can end the lab at this point, giving students until the next lab period to design their experiment and bring in any extra materials they will need. This allows a greater range of creativity. For example, if students are interested in testing the fermentation of different fruit juices, you are not likely to have the materials on hand.

Exercise 6.4 (40 minutes or time remaining in lab period)

Once the teams have set up their experiments, data collection goes very quickly. For evaluation, students can formalize and hand in their results (tables and graphs) and answer the questions in the lab manual in Exercise 6.4 or they can write a formal report.

For a 2-Hour Lab

For a 2-hour lab period, you have two options:

1. Do *either* Exercise 6.1 *or* Exercise 6.2. Then have students design their experiments using the method from the exercise they did. For example, if you only do Exercise 6.1, then all groups design an investigation on alcoholic fermentation.

2. Do Exercises 6.1 and 6.2 one week and Exercises 6.3 and 6.4 the next week.

Possible Investigations for Alcoholic Fermentation

Concentration of Yeast (enzymes) or Substrate (corn syrup)

Vary the amounts used around 2.5 mL. Students who are designing this type of experiment should make a table that shows the amount of water, yeast, and corn syrup to be placed in each tube. These experiments can be performed without any extra prep as long as there is an ample supply of yeast and diluted corn syrup. Results are generally good, although with high substrate concentrations there may be so much CO_2 produced that it is difficult to measure.

Different Substrates

Type of substrate is a popular variable because students know that alcoholic beverages are made from different substrates. Our students have compared a wide variety of sugar sources, including fruit juices, fruit purees, different varieties of grapes, sports drinks (for example, Gatorade, 10K), different types of syrup (for example, maple, sorghum, and molasses), "lite" and regular syrups, and miscellaneous beverages (for example, milk, chocolate milk, KoolAid, and soft drinks). One surprising result has been that artificial sweeteners such as Sweet'n Low are readily fermented because they contain dextrose. Students may also compare sugars such as fructose, ribose, maltose, and lactose, as well as starch and sorbitol.

If students want to investigate the traditional substrates for alcohol fermentation such as corn and barley, they will need some extra information. Seeds contain stored food, usually starch, for the developing embryo. In the germination process, the seed first imbibes water, activating a series of steps that results in enzymatic conversion of the starch to usable sugar. When beer is made, for example, the barley is first germinated (in a process called malting) in order to provide the sugar substrate for fermentation.

No additional prep will be required if you have students bring their own substrate samples. If they are using chemicals such as lactose, however, you need to prepare solutions for them. They may also request additional pipets for measuring the substrates.

Different Types of Yeast

Baker's yeast is used in the basic procedure. There are also different varieties of brewer's yeast (for example, beer yeasts for ales or lagers and wine yeasts), which can be obtained from homebrewers supply stores. The brewer's yeast that is sold in nutrition stores, however, is not living yeast. It is the residue of dead yeast cells that remains after fermentation has taken place.

Inhibition

Since ethanol is known to be toxic to cells, students might want to investigate how much alcohol buildup can be tolerated before fermentation ceases. This problem should be explored by adding ethanol to the reaction vials.

Osmotic Effects

High concentrations of sugar create a hypertonic medium that cannot be tolerated by microorganisms; thus, sugar can act as a preservative. Students can investigate this effect on yeast by using highly concentrated sugar sources (for example, honey, which is 75% sugar).

pH

Effects of pH can be investigated by using buffers in the reaction vessel in place of water. These experiments generally work well, with yeast showing little activity at low and high pH values. Extra prep is required to supply the buffer solutions (see the Prep Guide for Lab Topic 2, pH and Buffers, for buffer recipes).

Temperature

For temperature experiments, first fill the beaker of the fermentation kit with water of the correct temperature; then leave the kit immersed in the water to maintain temperature. Water baths that maintain a constant temperature are ideal, but dishtubs can be used if necessary. Have ice and a hot plate available so students can mix the temperatures they want to use.

Possible Investigations for Aerobic Respiration

You or your prep staff should make the mitochondrial suspension right before lab and distribute it to the students. One recipe should be enough for an entire lab section to carry out the investigations. The procedure is given in Exercise 6.2 of this lab and is also in the Prep Guide.

Concentration of Mitochondrial Suspension (enzyme) or Substrate

Higher concentrations than are used in the basic procedure can be obtained by adding more of the solutions given, but students should also dilute the mitochondrial suspension or extract to get lower concentrations. Students who are designing this type of experiment need to design a table that specifies the amount of buffer, substrate, and suspension to be placed in each tube. Each tube must have the same volume of solution. These experiments can be performed without any extra prep, as long as there is plenty of mitochondrial suspension and succinate available. Students also need measuring glassware to make the dilutions.

Students may propose varying the amount of DCPIP as an experiment. The DCPIP is an indicator for mitochondrial activity (the subject of interest) and must remain constant.

Inhibition

Students who have a more sophisticated understanding of enzymes may want to investigate the effects of malonate or oxaloacetate, which are competitive inhibitors of succinate dehydrogenase. Mercuric chloride (100 μg/mL) is also an inhibitor: Mercury binds to the enzyme's active site. However, it is toxic and must be handled with care. It also causes flocculence, which may cause % transmittance readings to increase somewhat during the course of the experiment.

Mitochondria from Different Sources

Mitochondrial suspensions can be made from other dried beans; cauliflower also works well. Avoid using any green (chloroplast-containing) tissue, however, since DCPIP is also reduced by photosynthesis.

pH

pH can be varied by using 0.3 mL mitochondrial suspension and 0.1 mL succinate. The pH 7.2 buffer is replaced by the buffers that are to be tested. You can either have a range of buffers available in the lab for students to choose from or have the prep staff make them up on request. Buffer recipes are in the Prep Guide for Lab Topic 2.

Temperature

Use 0.3 mL mitochondrial suspension, 0.1 mL succinate, and 4.6 mL buffer. You probably won't have enough water baths for students to use them for a range of temperatures. Instead, have students make their own water baths using ice, a hot plate, and tap water. They will also need a thermometer.

Photosynthesis

Introduction

The process of photosynthesis can be summarized in the equation below.

$$\text{Chloroplasts}$$
$$6\ CO_2 + 12\ H_2O \rightarrow C_6H_{12}O_6 + 6\ H_2O + 6\ O_2$$
$$\text{Light}$$

This simple equation conceals a multitude of complex processes, which are divided into two phases. In the first phase, known as the light-dependent reactions, the pigment chlorophyll absorbs light energy, which is then invested in the high-energy chemical bonds of ATP and NADPH molecules. The second phase of photosynthesis, known as carbon fixation, does not require light, since the energy has already been captured. In this light-independent phase, ATP and NADPH, along with CO_2, are used to make the carbohydrate glucose, which is a six-carbon monosaccharide (see Lab Topic 3, Macromolecules). So the result of photosynthesis is that a small amount of light energy is absorbed and then stored in the chemical bonds of glucose. The entire process takes place within the chloroplasts of plants. The chloroplast contains the chlorophyll, the enzymes, and the other molecules that are required for photosynthesis. (The prokaryotic Cyanobacteria, which can also photosynthesize, do not have their photosynthetic apparatus enclosed in an organelle.)

As you learned in Lab Topic 6 (Cellular Respiration), virtually all of the energy used by organisms on the earth, with the exception of a few chemosynthetic organisms, is made available by photosynthesis. In the process of cellular respiration, the energy stored in the chemical bonds of carbohydrates is released and repackaged in ATP, the usable "energy currency" of the cell.

In this lab topic you will learn a method of measuring the rate of photosynthesis. You will then use this method to design your own investigation of photosynthesis.

Outline

Exercise 7.1: Measuring Photosynthetic Rate in Spinach Leaf Disks

Exercise 7.2: Designing an Experiment

Exercise 7.3: Performing the Experiment and Interpreting the Results

Instructor Overview
In this lab topic students learn a method of investigating photosynthesis by measuring oxygen evolution in spinach leaf disks. The disks, which are cut from spinach leaves with a cork borer, are infiltrated under vacuum with a bicarbonate solution. Since they no longer have gases in their intercellular spaces, the disks sink. As photosynthesis progresses, the evolved oxygen makes the disks light enough to float. Students first complete a demonstration exercise to learn the technique. Then each team of students designs its own experiment. This lab prompts students to organize their experimental designs (see Exercise 7.2), collect data, and answer a series of questions to interpret their results (see Exercise 7.3). Alternatively, you may require students to write lab reports.

Time Requirements
Exercise 7.1: 40–50 minutes (20–30 minutes to set up, 20 minutes to obtain results)
Exercise 7.2: 15–30 minutes (can be done while waiting for results from Exercise 7.1)
Exercise 7.3: 60–90 minutes

E X E R C I S E   7 . 1
Measuring Photosynthetic Rate in Spinach Leaf Disks

Objectives

After completing this exercise, you should be able to

1. Explain how the experiment measures the occurrence of photosynthesis in spinach.

2. Interpret the results of the experiment.

The light-dependent reactions of photosynthesis are summarized in the equation:

Chloroplasts

$$H_2O + ADP + \circledP + NADP^+ \rightarrow \frac{1}{2}O_2 + ATP + NADPH + H$$

Light

Light energy is used to split water molecules. Some of that energy is captured by adding P (phosphate) to ADP to make ATP and by adding hydrogen to $NADP^+$. O_2 is released as a by-product. The rate of these reactions can be estimated by measuring O_2 production in disks cut from spinach leaves.

Leaf tissue is riddled with gas-filled intercellular spaces (see Figure 7.1a), so they float when they are placed in solution (Figure 7.1b). But when leaf disks in a solution are subjected to a vacuum, the gases in the leaves are pulled out and the spaces are filled by the liquid. Since fluids are heavier than gas, replacing the gas causes the leaf disks to sink to the bottom of the flask (Figure 7.1c). Then as O_2 is produced by the light reactions of photosynthesis, it diffuses into the intercellular spaces and replaces the liquid with gas. When enough O_2 has accumulated, each leaf disk will regain its buoyancy and turn on edge or float to the surface (Figure 7.1d). You will use this technique as a method of measuring photosynthesis. Sodium bicarbonate ($NaHCO_3$), which supplies CO_2 for photosynthesis, will be the solution used to infiltrate the leaf disks.

Procedure

1. Attach the lamp to the support stand so that the lamp is approximately 25 cm from the base of the stand.

2. Fill a large (1- or 2-L) beaker with cold water to act as a heat filter for the dish you will place under the lamp. The lamps produce a lot of heat, which can affect the rate of photosynthesis. Set the beaker aside for the moment.

Figure 7.1.
Leaf disk method for measuring photosynthetic rate.

a. Spinach leaf disks. Leaf tissue contains air spaces.

Bicarbonate solution

b. Leaf disks float in bicarbonate solution.

To vacuum line

Light

O_2 bubbles on leaf edges

Bicarbonate solution

c. Under vacuum, gas in leaf disks is replaced by bicarbonate solution. Disks sink.

d. During photosynthesis, O_2 accumulates in leaf disks. They float again.

3. Pour 0.2% $NaHCO_3$ solution into three petri dishes so they are about 2/3 full. Pour approximately 100 mL of 0.2% $NaHCO_3$ solution into a 250-mL flask. The exact amount is not important, so you can use the markings on the flask to measure.

4. Get several spinach leaves and cut 40–50 disks with a cork borer or other circular cutting instrument. (See Figure 7.2 on the next page.) Do not include the large veins. Use a paper towel or Styrofoam board to cut on. As the disks are cut, put them in the flask with the bicarbonate solution.

Each group should cut and infiltrate all the disks they will need for the experiment at the same time. Leaves can be folded or stacked so that many disks can be cut at the same time.

 Stack three or four leaves together so you can cut through them all at the same time. But don't cut so many leaves that the edges of the disks are ragged.

5. Use a water aspirator or other vacuum source, as directed by your instructor, to sink the disks.

Figure 7.2.
Cutting disks from spinach leaves.

Cork borer
(Use like a cookie cutter)

Spinach leaves
(3 or 4 stacked together)

Hole in spinach leaves
where disks have
been removed

Poke a rod through
the cork borer to
remove disks

Safety Note: Inspect your flask for cracks before using.
Flawed glassware may implode under vacuum.
Wear safety goggles while using the aspirator.

a. Attach the vacuum tubing to the sidearm of the flask or the glass
tube in the rubber stopper. Put the rubber stopper firmly in the
mouth of the flask and press the tape securely over the hole in the
stopper.

 Hold the flask steady while you apply the vacuum.

*In a strong vacuum the solution will
appear to boil. Even in a weak
vacuum, however, you should be
able to see bubbles emerging from
the edges of the disk. Judge the
vacuum strength before telling
students how long to aspirate the
leaves. The tissue will be damaged
if it is subjected to vacuum too
long. Release the vacuum frequently
and swirl the flask to see if the
disks sink.*

b. Turn on the water faucets all the way. After several seconds, you
should see bubbles coming out of the edges of the disks. Leave the
disks under vacuum for approximately 15–20 seconds; then release
the vacuum by peeling back the tape.

c. Swirl the flask and wait to see whether the disks sink. Remember, the leaves will continue to float as long as they are under vacuum.

d. You will probably have to apply the vacuum two or three times. It is not necessary to sink all the disks. You will damage the tissue if you overaspirate.

 Keep the disks away from bright light!

6. Pour the contents of the flask into a large dish. *Discard any disks that are floating.* With forceps, gently transfer 10–15 disks to each petri dish. Put the lids on the petri dishes. Put one dish in a dark place (drawer or cabinet); put another one under the lamp. Set the large beaker of water on top of the petri dish; then turn on the lamp. Place the third dish on a bench where it will receive only room light.

7. Wait 20 minutes for photosynthesis to occur. While the experiment is running, answer the questions below and work on Exercise 7.2.

8. After 20 minutes, count the number of disks that are either floating or turned on edge in each petri dish. Record your data in Table 7.1.

Table 7.1
Results of Subjecting Spinach Leaf
Disks to Different Light Conditions

Light	# Disks Floating	% Disks Floating
Dark	*0/12*	*0*
Room light	*7/13*	*53.8*
Under lamp	*11/13*	*84.6*

If the disks in the bright light treatment don't float, the tissue was probably damaged during aspiration. Discard the disks and start over. It is possible that the room light will be too dim to cause enough photosynthesis in this amount of time to refloat the disks. That is a valid result. It is common for students to have a few disks floating in treatments in which no photosynthesis is expected (such as darkness). This is usually because the disks were not thoroughly infiltrated or had air bubbles adhering to them. Also, not all of the disks will necessarily rise even in the most effective treatment (such as bright light). If left alone, the disks would eventually rise unless the tissue was damaged.

What hypothesis is being tested with this experiment?

Example: Brighter light causes more photosynthesis.

Predict what the results will be if your hypothesis is supported.

Example: If more disks are found floating in the dish that is put under the lamp than in the dish that is only exposed to room light, then the hypothesis is supported.

What is the independent variable in this experiment?

Intensity of light or type of light.

It would be useful to have student teams pool their results before discussing the outcome of the experiment.

What are some other independent variables that could affect photosynthesis?

Quality (color or wavelength) of light, leaf color, addition of pollutants to the water, type of plant, age of leaf, pH of solution, amount of bicarbonate (CO_2).

What is the dependent variable?

O_2 evolution (as measured by disk flotation).

What is the control treatment?

Disks placed in the dark.

Was your hypothesis supported or proven false by the results? Explain.

Example: Supported. The greatest percentage of floating disks was found in the dish kept under the bright light. Disks that were kept in the dark did not float.

E X E R C I S E 7 . 2
Designing an Experiment

Objective
After completing this exercise, you should be able to

1. Design an original experiment to investigate some factor that affects the production of oxygen in photosynthesis.

The Instructor Notes at the end of this lab topic give you some specific information about the kinds of investigations that students could do. Let the prep staff know what additional materials you would like to have available for the students. For example, you might want to have on hand a series of pH buffers, other types of leaves besides spinach, colored filters, or different concentrations of bicarbonate solutions.

In Exercise 7.1 you learned a method of measuring photosynthetic rate in spinach leaf disks. In Exercises 7.2 and 7.3 your lab team will design an experiment using this method, perform your experiment, and present and interpret your results. Review the factors that could affect photosynthesis that you listed in Exercise 7.1 to get some ideas for possible independent variables.

The basic materials that are needed to perform an investigation will be supplied for you. Your instructor will tell you what additional materials will be available.

fresh spinach	6 petri dishes
0.2% sodium bicarbonate (1 L)	2 reflector lamps
250-mL flask with 2-hole rubber stopper	2 support stands for lamps
	Two 1-L or 2-L beakers
vacuum source	culture dish
#3 cork borer or other cutting instrument	forceps
glass stir rod	cutting board

Describe your experiment below.

Question or Hypothesis

Dependent Variable

Independent Variable

Explain why you think this independent variable will affect photo-synthesis.

Control Treatment(s)

Replication

Brief Explanation of Experiment

Predictions

What results would support your hypothesis? What results would prove your hypothesis false?

Method

Include the levels of treatment you plan to use.

Design a Table to Collect Your Data

List Any Additional Materials You Will Require

E X E R C I S E   7 . 3
Performing the Experiment and Interpreting the Results

Objectives

After completing this exercise, you should be able to

1. Perform the experiment your lab team designed.
2. Present and interpret the results of your experiment.

Before you do the experiment, be sure that everyone on your lab team understands the techniques that will be used. You may want to divide up the tasks before you begin work.

Be thorough in collecting data. Don't just write down numbers; record what they mean as well. Don't rely on your memory for information that you will need when reporting on your experiment later! If you have any questions, doubts, or problems during the experiment, be sure to write them down, too.

Results

Before you begin to prepare your results for presentation, decide on the best format to use. Remember, you want to give the reader a clear, concise picture of what your experiment showed. Refer to the data presentation section of Appendix A (Tools for Scientific Inquiry) for help. If you are drawing graphs, use graph paper. Complete your tables and/or graphs before attempting to interpret your results.

Write a few sentences *describing* the results (don't explain why you got these results or draw conclusions yet).

Discussion

Look back at the hypothesis or question you posed in this experiment. Look at the graphs or tables of your data. Do your results support your hypothesis or prove it to be false? Explain your answer, using your data for support.

Did your results correspond to the prediction you made? If not, explain how your results are different from your expectations and why this might have occurred.

Describe how your data are supported by information from other sources (for example, textbooks or other lab teams working on a similar problem).

If you had any problems with the procedure or questionable results, explain how they might have influenced your conclusion.

If you had an opportunity to repeat and extend this experiment to make your results more convincing, what would you do?

Summarize the conclusion you have drawn from your results.

Questions for Review

1. What is the function of each of the following in photosynthesis?
    Light:

    Supplies energy that will be changed into a usable form.

CO_2:

Carbon source for the light-independent reactions (provides the carbon for the glucose molecule).

Chlorophyll:

Pigment that absorbs the light energy.

2. In the leaf disk method, spinach tissue is infiltrated with bicarbonate solution so that it sinks. Why do the disks float again after being exposed to light?

    The leaf tissue photosynthesized when exposed to light. O_2, a by-product of photosynthesis, accumulated in the intercellular spaces of the leaf disks, replacing the liquid and making the leaves more buoyant.

3. Reflotation of leaf disks by O_2 is one method of measuring photosynthesis. Review the equations for photosynthesis and suggest at least one other method that could be used (assume that you have the means of measuring any substance you choose).

    Accumulation of carbohydrate could be used as a measure of photosynthesis. Another possibility is monitoring the disappearance of CO_2.

4. A team of students performs an experiment to determine whether different wavelengths of light are equally effective at producing photosynthesis. Their results are shown in Table 7.2. What conclusion would you draw from these results?

Table 7.2
Effects of Wavelength on Photosynthesis

Light	# Disks Floating	% Disks Floating
Dark	1/30	3
White light	25/30	83
Blue light	24/30	80
Yellow light	3/30	10
Green light	2/30	7
Red light	25/30	83

Blue and red light are effective at causing photosynthesis, but green and yellow light are not. White light contains all wavelengths, so it serves as a positive control.

Acknowledgments

The leaf disk technique was modified from Witham, F., D. Blaydes, and R. Devlin. *Experiments in Plant Physiology.* New York: Van Nostrand Reinhold, 1971. Those authors acknowledge the following paper for the original technique: Wickliff, J. L., and R. M. Chasson. "Measurement of Photosynthesis in Plant Tissues Using Bicarbonate Solutions." *Bioscience* 14:32–33, 1964.

LAB TOPIC 7

Photosynthesis Instructor Notes

Organization

Students should work in teams of 2–4 (use whatever team size is convenient for your labs).

The time estimates for this laboratory assume that students have gained some familiarity with laboratory procedures by performing labs earlier in the course. Students will also be able to design their investigations more readily if they have had at least one previous investigative lab. If you think your students may require significantly longer than an hour to complete Exercises 7.1 (setting up the spinach disk demonstration exercise) and 7.2 (designing an original investigation), you may want to assign an independent variable to each team rather than letting them select their own.

Exercise 7.1 (40–50 minutes)

Introduction (10 minutes)

Review the summary equations for photosynthesis. Explain how the disk infiltration method works. Students must understand that it is O_2, the by-product of photosynthesis, that causes the reflotation of the disks.

Prep for Procedures (5 minutes)

Show students how to use vacuum device.

Procedure (total of ~40 minutes)

Students can work on Exercise 7.2 while waiting for results.

Summary of Results (5 minutes)

Have students pool their data. Briefly review the results.

Exercise 7.2 (15–30 minutes)

Procedure

Before you tell students to do Exercise 7.2, give them some idea of extra materials that are available for their investigations (for example, leaves from different types of plants, colored filters, pH buffers, or different concentrations of sodium bicarbonate solution). Writing a list on the board is helpful. See Possible Investigations Using Leaf Disks below for additional suggestions.

You can tell students during the previous lab meeting to bring in ideas and materials that they might use for this experiment. A grade incentive helps. For example, you can give them a few points for bringing to lab a list of factors that affect photosynthesis or some background information on a particular aspect of photosynthesis, such as the effect of the wavelength of light.

If your lab schedule permits, you can end the lab at this point and give students until the next lab period to design their experiments and bring in any extra materials they will need.

It may initially be difficult to get students started on designing an experiment. One useful technique is to have each team brainstorm ideas before settling on one. They should quickly write down everything they can think of that might affect photosynthesis. If they can't think of anything at all, you can help out by asking them to review the components of the general summary equation for photosynthesis and identify the factors involved. Follow up with questions such as, "Chloroplasts contain the enzymes that are needed for photosynthesis. What can affect enzymes?" If they don't know, go back a step further and remind them that enzymes are proteins whose function depends on shape. Or pursue the requirement for light. What aspects of light could be important?

Once students have a list of broad ideas, they can go back through it and consider each idea as a starting point for an experiment, eliminating those that don't lend themselves to the kinds of investigations that can be done in lab.

Keep in mind that students are used to having the instructor tell them what experiment to do. Many students will delay independent thinking for as long as possible, hoping that you will eventually give up and tell them what to do. Some students believe that you have an experiment in mind and they are supposed to figure out what you want. They will constantly ask if their ideas are correct. Resist the temptation to take the traditional role of the instructor; allow your students to develop their own ideas.

Teams should work independently on performing their investigations.

For evaluation, students can formalize and hand in their results and answer the questions in the lab manual for Exercise 7.3 or they can write a formal report.

For a 2-Hour Lab

Ask students to have some ideas about factors that affect photosynthesis when they come to lab. You can demonstrate the technique for infiltrating leaf disks used in Exercise 7.1 rather than having students perform the exercise. This takes about 15–20 minutes.

Alternatively, have students do Exercises 7.1 and 7.2 one week and Exercise 7.3 the following week.

Possible Investigations Using Leaf Disks

There is a second method for collecting data using this technique. Instead of counting how many disks have floated after a period of time, watch the disks and time how long it takes each one to float, and then record the average time for each treatment. This allows the investigator to distinguish the effects of treatments in which all the disks will eventually rise.

Unless the experiment involves varying light intensity, students should use the maximum intensity for all treatments to get the best results. Maximum intensity is obtained by putting the lamp an inch or two above the beaker of water that is used as a heat shield. Measure the lamp-to-dish distance for one setup and use that for all setups.

 Be careful not to dip a hot bulb in a beaker of water!

Carbon Dioxide

Prep

Have students set their levels of treatment as % $NaHCO_3$. The fact that 0.2% $NaHCO_3$ is used in the basic technique should be a clue to appropriate levels. Leaf disks should be infiltrated with the % $NaHCO_3$ solution that is to be tested.

Students may want to measure the pH of each solution since $NaHCO_3$ is basic and unbuffered solutions are used. This could be of interest in discussing the results. Osmotic strength of the solution might also be a factor.

Results

Bicarbonate solutions that have been used by our students ranged from 0.05% to 5% $NaHCO_3$. Generally, all the disks in treatments containing any amount of $NaHCO_3$ floated in the time allowed. Students probably would have gotten better results by recording the average time it took the disks to rise in each treatment, as described above.

Different Leaves

Prep

The experiment is set up just like the basic technique. Students may want to try different types of plants, different leaves within a plant, or differently pigmented areas of variegated leaves. Keep in mind that spinach is used in the basic technique because it has a high rate of photosynthesis. Other leaf tissue may require much more time to accumulate enough O_2 to float.

Results

Our students have tried *Dieffenbachia*, collard greens, *Coleus*, African violet, croton, holly, ivy, water lily, philodendron, blue harefoot fern, dogwood, poplar, and maple with reasonably good results, but most are slower than spinach (*Dieffenbachia* tissue took an hour to rise). Oak leaves have given consistently poor results. Red and green cabbage, turnip greens, and lettuce have been tried by several groups with inconsistent results. Lettuce is practically impossible to infiltrate. Several groups have tried leaves collected outdoors in October to compare leaves that have turned with leaves that are still green. Some groups have gotten very poor results with this (that is, even the green disks didn't rise), while some have done quite well. Sweet gum and red tip (*Photinia*) are two that have given good results. Croton has given good results in experiments on variegated leaves.

Light Intensity

Prep

One way to do this is to use layers of gauze (or greenhouse shadecloth, if you have a handy source) to get different light intensities. Another is to use different wattages of light bulbs. Or the lamp can be placed at different distances from the disks. Keep the treatments well isolated from other light sources. If possible, provide students with a light meter so they can actually measure incident light.

Results

There is a decline in photosynthesis as available light decreases, regardless of the method used to change the light intensity.

Light Quality

Prep

It is necessary to use a filtering material that transmits only the wavelengths that are being tested. Green cellophane, for example, will transmit blue as well as green. The results then show that photosynthesis takes place in "green" light. Reds and blues from different sources are usually safe, but green is unlikely to have a narrow enough band pass and yellow almost always transmits many other wavelengths. The Prep Guide gives information for acetates that are available in red, blue, and green from Edmund Scientific. If you want to find your own sources, you can easily check samples in the Spec 20.

Lamps must be placed at different distances from the disks in order to equalize light intensity. If possible, you should measure the light intensity in order to determine the distances to use.

A related idea is to use different types of light bulbs—for example, incandescent, cool-white fluorescent, gro-lamps, yellow bug lights, and black lights. It is helpful to know the emission spectra for these different types of lights.

Results
This experiment works very well if the proper filters are used. Disks float in red and blue light and remain sunken in green light. White light and darkness are used as controls.

pH

Prep
Students should infiltrate the disks with the pH solutions they are testing. Make up pH buffer solutions at the desired pH values, and then add 2 g $NaHCO_3$/L. Recheck the pH.

Results
Our students have tested photosynthetic activity at pH 2–12. There is general agreement that pH 2, 4, 10, and 12 produce little activity. The optimal value is variously reported as 6, 7, and 8.

Chromosomes and Cell Division

 Before you begin the activities in this lab topic, you should be able to

1. Explain the role of DNA in genetics.

2. Explain how the eukaryotic genome is organized.

3. Define the terms chromosome and gamete.

4. Distinguish the general purpose of mitosis from the purpose of meiosis.

Instructor Overview
In this lab topic students will extract DNA from wheat germ to get concrete experience with the genetic material. Then they will look at a karyotype diagram as an introduction to examining the processes of cell division. In the remainder of the lab activities, students survey the phases of mitosis and meiosis by using pop bead models and slides showing cell division in organisms.

Introduction

The genetic information of all organisms is found in deoxyribonucleic acid (DNA), which consists of varying sequences of the four nucleotides adenine, thymine, guanine, and cytosine. Discrete sections of DNA are called genes, and they are the "blueprints" that make organisms. In the cells of eukaryotes such as wheat and humans, DNA is packaged with proteins. Most of the time this DNA-protein complex is found in a threadlike form called chromatin. Recall Lab Topic 4 when you looked at your cheek cells under the microscope. The most prominent feature of the cells was the nucleus, but you needed to use the high-power objective to see it clearly. If the DNA molecules in one of those microscopic nuclei were unwound, attached together, and pulled out like a string, it would be two meters long! It fits inside the nuclei because it is extremely thin. In this lab topic, you will extract DNA from wheat cells to see this famous molecule for yourself.

During cell division, the chromatin is elaborately wound up into the coiled structures called chromosomes. When cells give rise to new cells, there must be a way for each new cell to receive an exact copy of all of the chromosomes. In this lab topic you will examine how the process of **mitosis** accomplishes this.

Genetic information must also be passed from parents to offspring, and in this lab topic you will see how the process of **meiosis** distributes chromosomes into gametes for the purpose of sexual reproduction.

Outline

Exercise 8.1: The Genetic Material

      Activity A: Extracting DNA from Cells

      Activity B: Karyotypes

Time Requirements
(see Instructor Notes at the end of the lab topic for additional information)
Exercise 8.1:
   Activity A: 15–20 minutes to set up; 15-minute wait period for results
   Activity B: 15 minutes

Exercise 8.2: The Process of Mitosis
        Activity A: Pop Bead Simulation of Mitosis
        Activity B: Viewing Mitosis in Onion Root Tip Cells
Exercise 8.3: The Process of Meiosis
        Activity A: Pop Bead Simulation of Meiosis
        Activity B: Viewing Meiosis in Organisms

EXERCISE 8.1
The Genetic Material

Objectives

After completing this exercise, you should be able to

1. Describe how DNA can be extracted from cells.
2. Explain what a karyotype shows.
3. Define the following terms: homologous chromosomes, centromere, diploid, and haploid.

This is a quick and very dirty procedure whose purpose is to give students a bit of concrete experience with the real thing before using the chromosome models.

Activity A: Extracting DNA from Cells

DNA is found in almost every cell of every organism. You've heard about it, you've seen diagrams of its structure, but have you ever actually seen DNA molecules? In this activity you will extract DNA from the embryonic cells of wheat seeds, also known as wheat germ. After softening the cell walls of the wheat germ in warm water, detergent is used to break up the membranes. The DNA is then separated from the rest of the cell contents.

Procedure

The temperature of the water bath must be between 50 and 60 degrees C. If a water bath is not available, you may be able to use tap water. Be sure students check the temperature carefully. Fifty-five degrees is optimum.

Bamboo skewers are used to stir the slurry and to spool the DNA. You may wish to have the pointed ends cut off before they are placed in the lab.

If students do get their mixtures foamy by overly vigorous stirring, they should remove the foam with a pasteur pipet.

Cold is essential. The alcohol should be kept in a freezer, if available, or in ice buckets.

1. Weigh out 1 g of wheat germ.
2. Put the wheat germ in a large (50-mL) test tube.
3. Dip warm water from the water bath and measure 20 mL in a graduated cylinder. Pour the water into the test tube with the wheat germ.
4. Using a wooden stick, stir the wheat germ *gently* for 3 minutes.
5. Get a pasteur pipetful of detergent and release it into the test tube with the wheat germ.
6. Stir the wheat germ mixture gently for 5 minutes, being careful not to create any foam.

  Stir slowly to avoid making bubbles!

---

7. Measure 15 mL of *cold* alcohol in a graduated cylinder.

8. Tilt the test tube containing wheat germ at a 45-degree angle and *very slowly* pour the alcohol from the graduated cylinder down the side of the wheat germ tube. The alcohol should just trickle down the side and come to rest on the top of the water so that it forms a separate layer.

 Pour very slowly, taking care not to let the layers mix.

9. Once you have poured in all the alcohol, place the test tube in its rack and do not move it for at least 15 minutes. The DNA will begin precipitating out immediately between the two layers of liquid.

10. After 15 minutes, the DNA should be floating on the top of the test tube. Use the wooden stick to spool it like cotton candy.

Describe what the DNA looks like.

> *It is white and stringy, a little fluffy.*

As the wheat germ slurry settles, the cell parts fall to the bottom and there is a yellow aqueous layer above that. The cold alcohol forms a layer on top. You will see strings of DNA rising from the interface as it floats from the water into the alcohol and precipitates.

The precipitated DNA in the alcohol layer is only part of the total DNA from the wheat cells. Much of the DNA is still in the water below. You can bring this DNA into the alcohol, where it will precipitate. Tilt the test tube slightly and insert the wooden stick into the yellowish water fraction, then pull the stick up into the alcohol. You should see more DNA come up with it. Be careful not to stir the layers together, though! When it reaches the alcohol, the DNA will precipitate in its stringy form and you can spool it, too.

Most protocols for DNA extraction use an enzyme to digest proteins. That step was omitted in this procedure to get faster results. How would your results be different if you used a protein-digesting enzyme?

> *The DNA would not still be wound together with its associated proteins—it would be purer.*

Activity B: Karyotypes

Although most of the DNA in a prokaryotic organism such as bacteria is carried on a single chromosome, in eukaryotes (animals, plants, fungi, and protists) the DNA is divided among numerous chromosomes. Each eukaryotic species has a characteristic number of chromosomes, but that number does not indicate the complexity of the organism. For example, redwood trees have 66 chromosomes, gypsy moths have 62, yeast have 32, and catfish have 58. We humans have 46 chromosomes in each body cell. Guppies also have 46 chromosomes, but there are many differences in the genetic "blueprints" that make a guppy and the genetic "blueprints" that make a human.

The chromosomes do not account for a cell's entire complement of DNA. Prokaryotes also have genes on plasmids, while eukaryotes have mitochondrial DNA.

We can study an organism's chromosomes by making a karyotype of them. To make a karyotype, the nucleus of the cell is ruptured to isolate the chromosomes and then a microscopic digital image is made. A technician uses

Figure 8.1.

Karyotype of human male.

a computer program to arrange the images of the chromosomes in **homologous pairs** (Figure 8.1). The pairs are matched by comparing physical characteristics such as length and staining pattern. Homologous chromosomes have these physical similarities because they are genetically similar: they carry the same genes in the same order.

Figure 8.1 shows a finished karyotype of a human male. In humans there are 22 homologous pairs of autosomes (chromosomes that are not concerned with sex determination) plus a pair of sex chromosomes. Females have a homologous pair of X chromosomes, but the male's X and Y, shown here, don't match.

The karyotype, as well as other pictures of chromosomes that you have seen in books, shows chromosomes from cells that are undergoing mitosis. Thus the chromosomes are distinct, thickly coiled structures rather than the thready chromatin form. You may also be able to see that the chromosomes in Figure 8.1 are doubled, which makes each one look like a slender "X." Before a new nucleus can be made, the DNA in each chromosome must replicate (make an exact copy) so that each new cell will have exactly the same genetic material as the original cell. After replication, the chromosomes remain attached at a place called the **centromere.** The point of attachment is not always in the center of the chromosomes. The position of the centromere is another characteristic that technicians use to match homologous pairs.

How many *homologous pairs* of chromosomes do the following organisms have?

redwood tree—*33*

gypsy moth—*31*

catfish—*28*

guppy—*23*

A cell that contains the correct number of homologous pairs for its species is said to be **diploid.** A cell with only one member of each homologous pair is said to be **haploid.**

Almost all of the cells in our bodies are diploid. What cells are haploid, and why?

> *Gametes (eggs and sperm) are haploid. The sperm will fertilize the egg to make a new individual, and if the gametes were diploid, the zygote would contain double the correct number of chromosomes.*

Suppose the karyotype shown in Figure 8.1 was taken from a cheek cell. What cell division process would produce a new cheek cell?

> *Mitosis*

Describe what the karyotype of a new cheek cell from this same individual would look like. That is, how many chromosomes would it have? Would it be haploid or diploid?

> *It would be a diploid cell that has 23 pairs of chromosomes for a total of 46. The sex chromosomes would be XY.*

Suppose we made a karyotype of a gamete from this individual. What would it look like? How many chromosomes would it have? Would it be haploid or diploid?

> *It would be a haploid sperm (since the karyotype came from a male) that has 23 chromosomes, one member of each homologous pair. It would either have an X or a Y chromosome.*

E X E R C I S E 8 . 2
The Process of Mitosis

Objectives

After completing this exercise, you should be able to

1. Explain the purpose and location of mitosis in organisms.
2. Describe and recognize the phases of mitosis.

The union of a haploid egg with a haploid sperm produces a diploid zygote. The zygote has a nucleus that holds the chromosomes, which in turn contain the entire blueprint to make the organism. From that single cell, the process of **mitosis** produces the many (sometimes trillions of) diploid cells that make up the body of a multicellular organism. Therefore each of the organism's cells contains an exact copy of the same genetic material.

Because cells must constantly be replaced, mitosis also produces new diploid cells throughout the life of an organism. For example, your entire epidermis is replaced every 15 to 30 days, and your body makes approximately 200 billion new red blood cells every day. That's a lot of mitosis!

In the following activities you will examine the steps in mitosis by using models of chromosomes. You will then identify these steps in onion cells.

Activity A: Pop Bead Simulation of Mitosis

Give each pair of students three sets of chromosomes. If available, use two different colors to represent maternally and paternally derived members of the homologous pairs. Make the homologous pairs distinguishable from each other by length and/or position of the centromere (magnetic connector). If you have sufficient materials, you can simply supply each pair of students with a baggie of pop beads and magnetic "centromeres" and ask them to assemble three homologous pairs of chromosomes.

You will use pop beads to represent chromosomes and follow them through the processes of cell division. Your instructor will give you a set of "chromosomes" to work with.

Before you begin, take stock of your chromosomes. How many chromosomes are there?

> *6*

How many homologous pairs of chromosomes?

> *3*

How can you tell which chromosomes are homologous?

> *By the physical characteristics (length, position of centromere) that they have in common.*

How can you distinguish the members of the homologous pair *from each other*?

> *They are different colors.*

What differences between the chromosomes are represented by the different colors? That is, what is different about the maternal and paternal chromosomes?

> *They are genetically different. Although each chromosome carries genes for the same traits as its homologue, there may be different versions of the genes (different nucleotide sequences).*

Procedure

In the space below, sketch your chromosomes. You shouldn't take the time to draw each pop bead, but do be sure that homologous pairs can be identified in your drawing. An example is shown below.

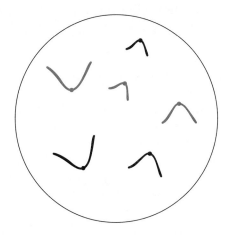

Interphase

Before cell division begins, each chromosome must replicate. This event happens while the chromosomes are still in their long, stringy, uncoiled state, so when interphase is viewed under the microscope, the chromosomes are not even identifiable.

Using more pop beads, make an exact copy of each of your chromosomes. At this point the chromosome is doubled, with an attachment point called the centromere. Each strand of the doubled chromosome is called a chromatid. Because they are identical, chromatids that are attached to each other are called sister chromatids. Sketch the chromosomes below.

Our pop bead model does not represent this stage well, since the pop bead chromosomes are already short and thick. You might ask students to offer suggestions for a better model of interphase, which would be something that is much longer and thinner such as plastic string or licorice whips.

Chromosomes should be duplicated exactly, using the same length and color of pop bead.

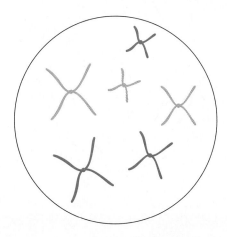

How many chromosomes are there now? (As long as the chromatids are attached, the structure is considered one chromosome.)

Still 6

How many homologous pairs of chromosomes?

Still 3

How many chromatids?

12

Besides replication of the chromosomes, what else should happen during interphase? (Hint: The entire cell is going to divide in half, not just the nucleus.)

The cell grows, new cell parts and proteins are made; the cell needs to have enough material to provision an entire new cell.

After interphase, cell division is a matter of sorting out the cell contents into two new cells. How the chromosomes are sorted is of special interest. By taking a "snapshot" of the process at several points, we can understand the mechanism of this division. Traditionally, the "snapshot" phases of mitosis are called prophase, metaphase, anaphase, and telophase, but keep in mind that cell division is a continuous process.

Prophase
It is during prophase that the chromosomes coil more tightly and become visible as distinct structures. Since we already modeled short, thick chromosomes during interphase, there are no changes to our pop bead model.

By the end of prophase, the nuclear envelope disintegrates, enabling the chromosomes to be snared by protein fibers that have grown out of two centrosomes that formed in the cytoplasm. Together these protein fibers form the spindle, an apparatus that will eventually allow the sister chromatids to separate from each other.

If you wish, you can have students model the centrosomes and spindle fibers by using the small plastic pieces supplied in some chromosome model kits and string. Or you can make your own centrosomes by cutting plastic straws into short pieces. In my experience, having to manipulate the centrosomes and string along with the pop beads gets too complicated, and students lose their focus on the chromosomes.

Metaphase
By metaphase the two centrosomes have moved to opposite ends of the cell with their protein fibers attached, and many of these protein fibers are also attached to the centromere region of a chromosome. One sister chromatid of each chromosome is connected to each pole by this spindle apparatus.

The distinctive characteristic of metaphase is the alignment of the chromosomes in a plane through the center of the cell. (Only the centromeres are aligned.) Place your chromosome models in this position and sketch metaphase in the circle below.

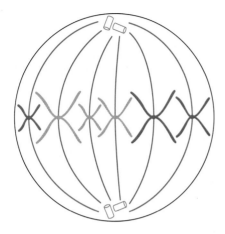

Based on what you know about the outcome of mitosis, can you guess what's going to happen next?

> *Since sister chromatids are attached to centrosomes at opposite ends of the cell, perhaps the chromatids will now separate and go to opposite sides of the cell.*

Anaphase
During anaphase the centromeres joining the sister chromatids separate and the chromatids—now called chromosomes—move along the spindle fibers to opposite poles. The number of chromosomes in the cell is briefly double as the cell is prepared for the final phase of division.

Place your chromosome models in the anaphase positions and sketch them in the circle below.

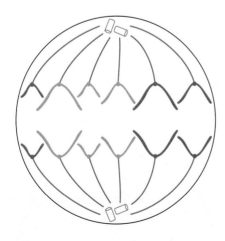

Telophase and Cytokinesis
Now that the chromatids have separated, the division of chromosomes is complete. All that remains is to reconstruct the nuclear envelopes and divide the cytoplasm between the two new cells. These processes will be finished by the end of telophase, and the chromosomes will resume their typical long, stringy form.

Place your chromosomes in the telophase positions and sketch them below.

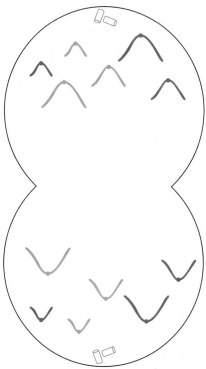

As the chromosomes go through telophase, the division of the cytoplasm, called **cytokinesis,** is also completed. The daughter cells separate from each other. Draw the chromosomes in the daughter cells below.

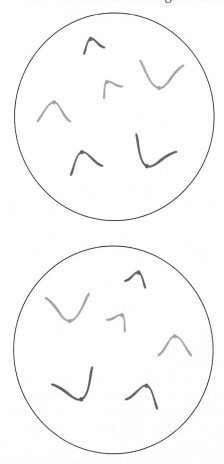

How many chromosomes are there in each new cell?

6

How many homologous pairs of chromosomes in each new cell?

3

What is the overall result of mitosis?

Two cells that are genetically identical.

After seeing the phases of mitosis, you may be wondering why cell division is such an elaborate process. Once interphase is complete, why not cut to the chase—telophase? Mitosis has to be carefully choreographed so that each daughter cell has the same number and type of chromosomes—that is, so the division of chromosomes is absolutely equal. Did you ever share a bag of M&Ms with a sibling or friend by carefully counting out and dividing one color at a time? The purpose was to be sure that the distribution was exactly equal, and mitosis has the same purpose regarding chromosomes. On the other hand, you were probably willing to divide a Snickers bar down the middle, without bothering to count every last peanut. You made the assumption that both halves were more or less equal. That's how the cytoplasm and included organelles are divided.

Activity B: Viewing Mitosis in Onion Root Tip Cells

Plants have regions of cell division where growth occurs. One such region is the root tip, whose growth enables the roots to elongate and reach through the soil. Since these tips are fine, threadlike tissues, it is easy to press them flat on a slide to view them under a microscope. Mitosis is concentrated in one particular region of the root tip (see Figure 8.2). After the new cells are produced, they will enlarge and mature into different cell types in the root. To produce these root tips, small white onions were placed in water. After several days of root growth, the root tips were cut off and preserved. They were then placed in a staining solution that makes their chromosomes visible.

Procedure

1. Use forceps to pick a root tip out of the vial and place it on a microscope slide.
2. Add a small drop of water to the slide.
3. Place a coverslip over the root tip.
4. Use the eraser end of a pencil to press gently on the coverslip. This should spread the tissues apart. Go easy at first—you don't want to crack the coverslip, and you can always press again if necessary.
5. Place the root tip slide on the stage of your microscope and find the pointed end of the root using scanning power.

Have an onion set in the lab to show students growing root tips.

Figure 8.2.
Onion root tip.

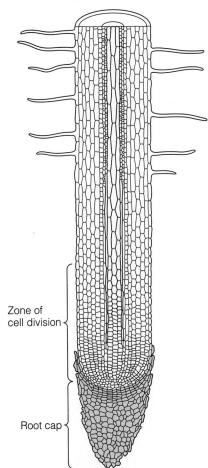

Zone of cell division

Root cap

Onion root tips have been used for decades for observation of mitotic figures. However, bulbs such as hyacinth and narcissus have reportedly been used successfully as well. Prepared onion root tip slides are available from biological supply catalogs if you wish to save prep time and lab time.

Dropping the pencil eraser onto the coverslip from about an inch above the coverslip a few times is also a good technique.

6. Turn to low power to locate the region of cell division. If the cells do not look clear, you have not squashed the root tip enough. When it is properly squashed you will be able to see a single layer of cells.

7. Once you have located the area of cell division and are able to see cells clearly, position one cell in the center of the lens and turn to high power. You should now be able to see the chromosomes. Identify which (if any) stage of mitosis the cell is in and draw it in the appropriate box below.

8. Move the slide around to find cells in different phases of mitosis. You may need to switch back and forth between high and low power to do this.

Keep in mind that mitosis is a continuous process. Some cells will be between the "snapshot" phases.

Interphase	Prophase
Metaphase	Anaphase
Telophase	Cytokinesis

EXERCISE 8.3
The Process of Meiosis

Objectives

After completing this exercise, you should be able to

1. Explain the purpose and location of meiosis in organisms.
2. Describe and recognize the phases of meiosis.
3. Explain how the process of meiosis results in genetic variation of off-spring.

Mitosis produces an exact copy of a cell and is an essential process for growth and repair in multicellular organisms. But multicellular organisms also need a special type of cell division to produce gametes (eggs and sperm), and that process is called meiosis. Unlike mitosis, which occurs throughout the body, meiosis occurs only in specialized reproductive tissues.

One outcome of meiosis is reduction of chromosome number. Look again at Figure 8.1, the human karyotype, which shows the homologous pairs of a diploid cell. Gametes made from this cell will be *haploid* with 23 chromosomes. How should 23 chromosomes be selected to go into each gamete?

A gamete should have one member of each homologous pair.

Enormous genetic variation of gametes is another outcome of meiosis. On the karyotype in Figure 8.1, randomly label each chromosome as maternal (M) or paternal (P). Then randomly choose one member of each pair to put in a gamete (or flip a coin to choose) and write your selections (M or P) below for each chromosome.

— —
 1 2 3 4 5 6 7 8 9 10 11 12 13 14 15 16 17 18 19 20 21 22 sex

Again, each chromosome is a DNA molecule that encodes information and each segment of information is called a gene. So distributed on the 46 chromosomes in the karyotype in Figure 8.1 are the 30,000 to 40,000 genes that make up the human genome, the set of instructions that produces a human being. In the next lab topic you will consider the distribution of genes during meiosis, but for now we are going to examine how the chromosomes themselves are sorted into gametes.

Compare the chromosome composition of all the gametes. Ask all the students who wrote "M" in the blank for chromosome 1 to stand. Then choose one student and see if anyone else had the same combination (before selecting a student, check quickly to be sure he or she didn't write all Ms or all Ps—there are likely to be a couple of students who took this shortcut). Ask what his or her next letter is, M or P, and have everyone whose letter doesn't match sit down. Before long, it will be clear that no one else chose exactly the same chromosomes for the gamete.

Activity A: Pop Bead Simulation of Meiosis

To see how the process of meiosis results in haploid gametes that differ from each other genetically, you will take a cell through the steps of meiosis as you did for mitosis. The mechanisms of mitosis, such as the centrosomes and spindle fibers, are also used in meiosis, and the same names—prophase, metaphase, anaphase, telophase—are given to the "snapshot" phases. But meiosis will result in four cells rather than two, so two divisions take place. The two divisions are designated meiosis I and meiosis II. Also, while many different types of cells can undergo mitosis, only specific cells can undergo meiosis. In animals, meiosis is carried out in the cells that produce gametes. In plants, meiosis produces spores (Lab Topic 12).

Where in the human body does meiosis occur?

> *In the ovaries and testes.*

Procedure

Begin with the same three homologous pairs of chromosomes that you used for mitosis.

Interphase

Before meiosis begins, the chromosomes must replicate. Make an exact copy of each of your chromosomes, then draw the interphase nucleus in the cell below.

Prophase I

As in mitosis, the chromosomes become visible and the nuclear envelope disintegrates during this phase. But something very different happens, too. Homologous pairs associate closely with each other, even to the point of entwining their "arms." Each pair of pairs is called a **tetrad.** Group your chromosomes into tetrads.

While they are in close contact, the chromatids of homologous chromosomes can exchange pieces of DNA. The size of these exchanged segments varies. You can simulate this process, called crossing over, by popping a few beads off of one chromatid and exchanging them for the same number of beads on a chromatid from its homologue.

What is the result of crossing over in a homologous pair?

None of the chromatids are genetically alike after crossing over.

Sketch prophase I below.

Be sure that students are crossing over homologues rather than sister chromatids so they will see the genetic variation.

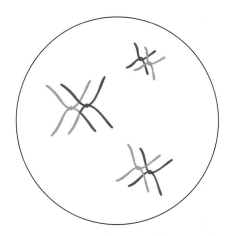

Metaphase I

Recall that metaphase of mitosis is when the chromosomes are attached to the spindle apparatus and aligned along the center of the cell. Metaphase I in meiosis is similar, but each tetrad stays together. Line up your tetrads and sketch them below.

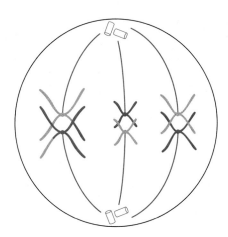

Anaphase I

In anaphase I, homologous pairs are separated from each other and move to opposite poles, but sister chromatids remain attached. How does this differ from anaphase of mitosis?

In mitosis, sister chromatids detach at this point.

Move your chromosomes to simulate anaphase I and sketch it below.

Telophase I and Cytokinesis
Once the homologous pairs of chromosomes are at opposite poles, the cytoplasm divides to form two daughter cells. Move your chromosomes into the two new daughter cells that will be formed when cytokinesis is completed. Make a sketch below.

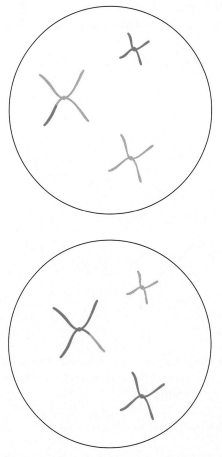

List by color the chromosomes that are in each daughter cell you made. For example, if your chromosomes are red and yellow, your daughter cells might be red, red, yellow and yellow, yellow, red. (For chromosomes that underwent crossing over, list the predominant color.)

Check with other students to see whether their daughter cells are the same as yours. If there are different combinations, can you explain why? (Hint: Review the metaphase I sketch of someone whose daughter cells turned out differently than yours.)

Possible combinations: RRR/YYY, RRY/YYR, RYY/YRR, or RYR/YRY.

The alignment of the tetrads at metaphase I, and therefore the pole that each homologue moves to, is random.

Crossing over during prophase I is one source of genetic variation that results from meiosis. The random alignment of tetrads at metaphase I is another source. With only three pairs of chromosomes, there are eight different combinations of chromosomes in the gametes (two for each possible alignment). Imagine how many possibilities exist with 23 pairs of chromosomes!

Each cell that results from meiosis I has one member of each homologous pair, which is the desired outcome for gametes. Why isn't this the end of meiosis?

Because the chromosomes are still doubled.

Prophase II
During prophase II, a spindle apparatus forms again.

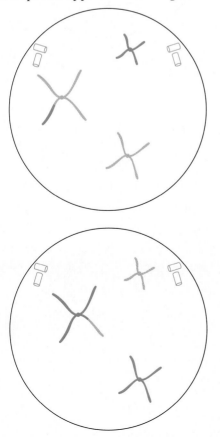

Metaphase II
With each chromatid attached to spindle fibers, the chromosomes line up in the center of the cell.

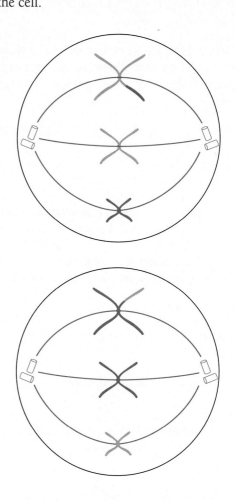

Anaphase II
During this anaphase, the sister chromatids move to opposite poles.

Telophase II and Cytokinesis
The division of chromatids is completed, and the cytoplasm divides to form new daughter cells. Move your chromosomes into the new daughter cells that will be formed when cytokinesis is completed. Make a sketch below.

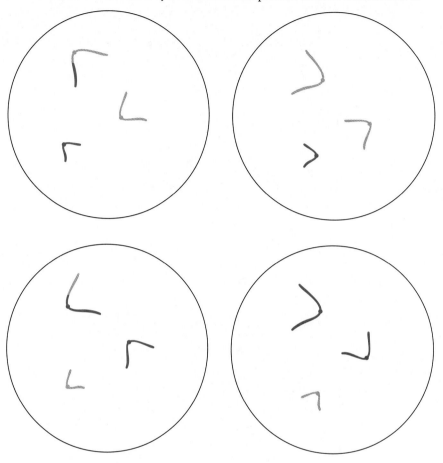

How many chromosomes does each cell have?

   3

How many homologous pairs does each cell have?

   None; each cell has only one member of each homologous pair.

Are these cells haploid or diploid?

   Haploid

List by color the chromosomes that are in each daughter cell you made.

Check with other students to see whether their gametes turned out the same as yours. Make a complete list of all possible chromosome combinations in the gametes.

   There are 8 possible combinations. Again using red and yellow pop bead chromosomes, the combinations are: RRR, YYY, RRY, YYR, RYY, YRR, RYR, YRY.

Activity B: Viewing Meiosis in Organisms

You will view professionally prepared slides to see what actual cells look like during the phases of meiosis. If you were to use animal tissue for this exercise, what parts of the animal would be used to show meiosis?

The Prep Guide recommends purchasing prepared slides either of lily anthers or ovaries or of grasshopper testes.

    Ovaries or testes

If you were to use plant material, which parts would you look at to see meiosis? (Hint: Where does sexual reproduction take place?)

    Students may not know about the angiosperm life cycle, but they should be able to guess that the flower is the general structure involved. Anthers and ovaries are the sites of meiosis in flowering plants. There is a diagram of the angiosperm life cycle in Lab Topic 12.

Procedure

1. Your instructor will supply you with prepared microscope slides to examine. Begin by using scanning power to locate the specimen. Switch to low power to identify an area that includes cells undergoing meiosis.

2. Using high power, sketch the positions of the chromosomes during different stages of meiosis in the spaces below. Label your drawing with the name of the stage, as best you can identify it.

Lily microspore slides show cross sections of immature anthers with four pollen sacs. Lily ovulary slides are cross sections of ovaries showing two megaspore mother cells (megasporocytes) in each of the three carpels. Both of these are developing structures that include cells undergoing mitosis. Be sure students know where to locate the microspores or megaspores so they do not mistake the mitotic figures for meiotic ones.

Why isn't it as easy to identify the stages of meiosis in cells as it was to identify the stages of mitosis?

    It's difficult to distinguish meiosis I from meiosis II.

Questions for Review

1. What purpose does mitosis serve in development and growth? (Hint: A zygote is the first cell formed after conception, but by adulthood a human body has many trillions of cells.)

    Mitosis results in duplicate cells, which enables the body to grow from a single cell all the way to adulthood.

2. After an adult has all his trillions of cells, does he still need the process of mitosis? Explain.

    Yes, because cells must constantly be replaced.

3. Which phase(s) of mitosis could the karyotype in Figure 8.1 have been made from? Explain your reasoning.

    The sister chromatids are attached, so it can't be anaphase or telophase. The chromosomes only appear in this configuration in prophase and metaphase.

4. Besides the root tips, where else in plants would you expect mitosis to take place?

   Mitosis must occur anywhere there is growth in the plant. Get students to think about what parts of plants grow. For example, flower buds form and produce flowers; leaf buds form and produce leaves. The roots of plants are constantly growing, and stems may grow taller. Also, the plant begins life as an embryo, which was itself produced by mitosis in the seed.

5. Meiosis is from a Greek root meaning "to diminish or lessen." Why is this an appropriate term?

   Meiosis reduces the number of chromosomes.

6. Summarize the sources of genetic variation in meiosis.

   Crossing over and how the tetrads line up in metaphase I result in variation.

7. Which phase of meiosis is responsible for the genetic variation of gametes demonstrated in the introduction to Exercise 8.2?

   Metaphase I (random alignment of tetrads)

8. Most people who have Down syndrome have an extra copy of chromosome 21 (trisomy 21), which is a result of an egg or sperm that carried an extra copy of chromosome 21. How could a gamete with an extra chromosome come about?

   Failure of the tetrad to separate in anaphase I or failure of the sister chromosomes to separate in anaphase II would result in a gamete with an extra chromosome.

9. You have homologous pairs of chromosomes because you got one member of each pair from each parent. Your parents got their chromosomes from your grandparents, and so on. But are the chromosomes that are passed along through the generations exactly the same? From example, is your maternal chromosome #1 exactly the same one that your mother received from one of her parents? Explain.

   Maybe not. During prophase I of meiosis, the maternal and paternal chromosomes may have exchanged segments of DNA (crossing over), resulting in genetic differences between the chromosome in the mother's body cells and in her gametes.

Chromosomes and Cell Division
Instructor Notes

Organization

Students should work in pairs for the DNA extraction and pop bead simulations. Each student should do his or her own microscope work.

Time requirements for this lab topic depend on how much understanding students bring from their lecture class. If students have not covered the phases of mitosis and meiosis in lecture and you wish to make that the focus of the lab, you should probably omit the DNA extraction (Exercise 8.1, Activity A). If your laboratory emphasis is on the general concept of DNA as the genetic material and how it is passed on during cell division, then you may wish to spend more laboratory time on the DNA extraction and on making and viewing slides of mitotic and meiotic figures. Exercise 8.2, Activity A and Exercise 8.3, Activity A could then be done as homework. If possible, let students take home baggies of pop beads. If that is not feasible, there are perfectly clear illustrations of mitotic and meiotic phases in the textbook that they can use as a basis for completing the exercises.

Introduction (5 minutes)

Briefly describe the composition of the genetic material and the purposes of mitosis and meiosis. Give a quick overview of the activities students will be doing.

Prep for Procedures

- Show students where the materials are for DNA extraction (Exercise 8.1, Activity A) and have them perform the procedure through step 8. During the 15-minute waiting period, go through the karyotype (Exercise 8.1, Activity B) with the class.

- If students do not come from lecture with an understanding of the phases of mitosis, take the class through Exercise 8.2, Activity A step by step, pausing to let them make their drawings. It is very useful to have a metal board of some kind (for example, a small blackboard or a cookie sheet) so that you can demonstrate the chromosome positions.

- After finishing mitosis with pop beads, explain the chromosome squashing technique and point out the locations of materials for making onion root tip slides. Demonstrate how to make root tip squashes.

- Once students have identified mitotic figures in onion root tips, go on to meiosis. Again, lead the class through the phases and end up by having them view slides of plant and/or animal meiosis.

During the Laboratory Period

Exercise 8.2, Activity A

If students are not performing the simulation as a class, you will need to circulate among them checking on progress. When a student needs help, try to give it quickly and move on to avoid getting stuck giving one or two students a private tutorial on the whole process. A good technique is to locate someone who has gotten past the stage the student is having trouble with and send him or her over to help.

Exercise 8.2, Activity B

If you are short of time for the onion root tip slides, tell each student to locate a specific phase. Hand out cards with the names of the phases printed on them. When a student has placed his or her assigned phase at the tip of the pointer, he or she should tape the appropriate card to the microscope. Students can then walk around the room and make their sketches from different slides.

Exercise 8.3, Activity A

When you are ready for the class to begin Exercise 8.3 (Meiosis), perform the brief introductory exercise demonstrating the potential for variation in the gametes.

If students are not performing the pop bead simulation as a class, you may need to stop everyone and explain the alignment of tetrads in metaphase I as a source of variation.

Exercise 8.3, Activity B

Since students are probably not familiar with the sexual organs of flowers, it is a good idea to sketch the cross sections of tissue that students will view and show them where to look for cells undergoing meiosis.

Images of meiotic stages in lilies can be viewed at http://biology.about.com/library/gallery/blmeiosisimageslilyI.htm. If this URL fails, a Google search for "lily meiosis" will turn up several course Web sites that include micrographs of lily micro- and megasporocytes.

Summary

- Review the different purposes and outcomes of mitosis and meiosis.
- Review the sources of genetic variation in meiosis.
- Have students answer the Questions for Review.

For a 2-Hour Lab

See the suggestions for organization above.

Alternatively, you could split the exercises into two lab periods, performing Exercises 8.1 and 8.2 during one lab period and Exercise 8.3 during the next period.

Investigative Extensions

The Prep Guide recommends growing the onion sets in the dark, then harvesting root tips as soon as they are brought into the light. This recommendation is based on the traditional practices at institutions where I've worked. Does mitosis follow a circadian cycle? Is there a best time of day to find mitotic figures?

Does the cell spend equal amounts of time in all of the stages, or do some stages last longer than others? Within a single root tip meristem, count the number of cells in each phase of mitosis.

Some sources report good results finding mitotic figures with root tips from hyacinth and paperwhite narcissus bulbs. Compare these to results from onions.

DNA yields can be quantified by allowing the spooled DNA to air dry on filter paper for a day or two. It can then be weighed. Wheat germ is used for the procedure in this lab manual because it has a greater concentration of DNA than other sources. But a Google search for "DNA extraction" will turn up numerous protocols for extracting DNA from a variety of sources, including onions, bananas, peas, liver, calf thymus, and dog testes. Each of these tissues contains a lot of other material in addition to DNA, so there are additional grinding and filtration steps. Protocols range from a high degree of sophistication and purity to the quick-and-dirty extraction in Exercise 8.1. So if you would like to make DNA extraction the focus of student-designed experiments, numerous investigations involving comparison of tissues or techniques are possible.

Acknowledgment

The DNA extraction protocol in Exercise 8.1 was provided by Barbara Nicodemus, biology teacher at Walhalla High School in South Carolina.

Mendelian Genetics

Introduction

The study of genetics involves learning how traits are passed from one generation to the next. Humans have been interested in genetics since the beginning of agriculture. The modern science of genetics began in the 1860s with investigations by Gregor Mendel. His experiments with pea plants provided the foundation for understanding mechanisms of heredity. In this lab topic you will learn how the pattern of inheritance of some traits follows the principles discovered by Mendel. You will also perform a cross using Wisconsin Fast Plants™, which are members of the mustard family, and interpret the results according to Mendelian principles.

 Bring a calculator for this lab topic.

Before coming to this lab, refer to your textbook to write a definition for each of the following terms:

Gene:

Unit of DNA that governs a particular trait.

Locus:

Location of a gene on a chromosome.

Allele:

Alternative form of a gene.

Dominant allele:

Allele whose expression masks the expression of an alternative allele at that locus.

Recessive allele:

Allele whose expression is masked by the expression of an alternative allele at that locus.

Instructor Overview
In this lab topic students will review the terminology of Mendelian genetics by writing out definitions before the lab, do exercises to review basic principles, and review the principles of segregation and independent assortment to predict the outcomes of monohybrid and dihybrid crosses using a Punnett square. The hands-on portion of the lab is a dihybrid cross using Wisconsin Fast Plants. Recessive mutations that produce yellow-green leaves and anthocyaninless hypocotyls are used. F_1 (heterozygote) seeds are planted and grown to maturity. Students perform cross-pollinations between the plants, collect and germinate the seeds to determine the phenotypic characteristics of the F_2 generation, and compare their results to the expected phenotypic ratio using a chi-square "goodness of fit" test.

Fast Plants materials must be purchased from Carolina Biological Supply. Plants must be started at least 7 weeks before the data are collected. See the Prep Guide for more information about the specific materials and equipment required to perform this lab.

Genotype:

The genetic makeup of an individual; a list of the alleles that an individual carries.

Phenotype:

The observable expression of genetic composition.

Homologous chromosomes:

Chromosomes that are morphologically alike and which carry genes for the same traits.

Diploid:

Having two sets of chromosomes that are homologous pairs. One member of each pair was donated to the zygote by the egg; the other member came from the sperm.

Haploid:

Having only one set of chromosomes, as in an egg or sperm.

Heterozygote:

A diploid individual having two different alleles at a locus on the two homologous chromosomes.

Homozygote:

A diploid individual having the same alleles at a locus on both homologous chromosomes.

Outline

Time Requirements
Review terms: 20–30 minutes
Exercise 9.1: 40–60 minutes
Exercise 9.2
  Introduction: 10 minutes
  Activity A: 15 minutes, 7 weeks before lab
  Activity B: 10 minutes, 6 weeks before lab
  Activity C: 5 minutes each day for 3 days, 5 weeks before lab
  Activity D: 15 minutes, 1 week before lab
  Activity E: 50–60 minutes

Exercise 9.1: Mendel's Principles
  Activity A: The Chromosomal Basis of Segregation
  Activity B: Predicting the Outcome of a Monohybrid Cross
  Activity C: The Chromosomal Basis of Independent Assortment
  Activity D: Predicting the Outcome of a Dihybrid Cross
Exercise 9.2: A Dihybrid Cross Using Wisconsin Fast Plants
  Activity A: Planting F_1 Seeds
  Activity B: First Observations
  Activity C: Pollination
  Activity D: Germinating the F_2 Seeds
  Activity E: Collecting and Interpreting Data from the F_2 Generation

EXERCISE 9.1
Mendel's Principles

Objectives

After completing this exercise, you should be able to

1. Define and explain Mendel's principle of segregation.
2. Define and explain Mendel's principle of independent assortment.
3. Follow one pair of alleles through meiosis and determine the resultant gametes from a diploid cell with two homologous pairs of chromosomes.
4. Follow two unlinked pairs of alleles through meiosis and determine the gametes (or spores) that result from a diploid cell with two homologous pairs of chromosomes.
5. Solve problems involving monohybrid and dihybrid crosses.

Activity A: The Chromosomal Basis of Segregation

Diploid cells contain two sets of **homologous chromosomes.** One set, or one member of each pair, comes from each parent. Each pair of homologous chromosomes carries **genes** that govern the same traits. For example, in pea plants, flower color is determined by a single gene *F,* which can have two different forms, *F* or *f,* called **alleles.** A gene is a segment of DNA, and alleles have different sequences of nucleotides in that segment. Every cell in the diploid plant has two copies of the gene, one on each member of a homologous pair of chromosomes. These two versions of the same gene may be alike **(homozygous)** or different **(heterozygous).** The genetic makeup of the cell is known as its **genotype.** In this example, a cell with two *F* alleles would have the genotype *FF.* A cell with two *f* alleles would have the genotype *ff.* If the cell has one *F* and one *f,* its genotype is *Ff.* The genotypes *FF* and *ff* are homozygous, and the genotype *Ff* is heterozygous. All of the cells of the plant should have the same genetic composition.

In pea plants, the *F* allele is **dominant** over *f:* When one *F* is present it masks the *f* allele. The *f* allele is called **recessive.** A capital letter is used to denote the dominant allele. A pea plant having either the *FF* or *Ff* genotype has the **phenotype,** or outward appearance, of purple flowers. The phenotype that results from the *ff* genotype is white flowers.

Genetic traits are passed from one generation to the next by reproduction. When animal cells undergo meiosis (Lab Topic 8) they produce **gametes,** which are haploid. When plant cells undergo meiosis they produce **spores,** which then become the plants that produce gametes. This difference between plant and animal reproduction is explained in detail in Lab Topic 12 (Plant Diversity).

In the next example we will follow a gene *R* that codes for the tongue-rolling ability in humans. The allele *R,* which gives a person this ability, is dominant over *r.*

What is the phenotype of an individual whose genotype is *RR?*

Tongue-roller.

Figure 9.1.
Tongue-roller phenotype. This person can roll her tongue into a U-shape.

What is the phenotype of an individual whose genotype is *Rr*?

> *Tongue-roller.*

What is the phenotype of an individual whose genotype is *rr*?

> *Can't roll tongue.*

What are your phenotype and genotype?

> *Genotype is rr for students who can't roll their tongues. For students who can roll their tongues, the genotype can be either RR or Rr. But if one parent is a nonroller, the genotype must be Rr.*

Instructor Note:
If you did the pop bead simulation in Lab Topic 8, you may wish to use the pop beads again to illustrate segregation. To add "alleles," print Rs and rs on label paper and cut them out. Students should label as directed, then remove the labels at the end of the lab period.

After meiosis, each resultant gamete will have only one member of each homologous pair of chromosomes. Therefore each gamete will have only one allele for tongue-rolling.

The distribution of alleles during the formation of gametes was one of the principles described by Gregor Mendel. It is called the **principle of segregation:** The two alleles of a gene segregate, or separate, from each other so that each one ends up in a different gamete. Mendel, however, did not know about the existence of chromosomes or meiosis. Not until decades after his death was the chromosomal basis of Mendel's law discovered.

If a person's genotype is *RR*, what are the genotypes of the resulting gametes?

> *The genotype of all of the gametes is R.*

If the person's genotype is *rr*, what are the genotypes of the resulting gametes?

> *The genotype of all of the gametes is r.*

If the person's genotype is *Rr*, what are the genotypes of the resulting gametes?

> *The genotype of half of the gametes will be R. The genotype of the other half of the gametes will be r.*

Activity B: Predicting the Outcome of a Monohybrid Cross

When the genotypes of the parents are known, we may determine what gametes the parents can make and in what proportion the gametes will occur. This information allows us to predict the genotypes and phenotypes of the offspring. The prediction is simply a matter of listing all of the possible combinations of gametes. In this section you will be doing **monohybrid** crosses: Only one trait is followed.

By convention, the parental generation is called P. The first generation of offspring is called F_1. F stands for filial, which refers to a son or daughter, so F_1 is the first filial generation. If members of the F_1 generation are crossed, their offspring are called the F_2 generation, and so on.

Predict the results of the following cross (using R to denote tongue-rolling ability):

P generation: $RR \times RR$

What genotype(s) will be found in the F_1 generation?

All RR.

What phenotype(s) will be found in the F_1 generation?

All tongue-rollers.

Explain why you made these predictions.

Both parents can only produce R gametes, and therefore all of the offspring must be RR.

Predict the results of the following cross:

P generation: $RR \times rr$

What genotype(s) will be found in the F_1 generation?

All Rr.

What phenotype(s) will be found in the F_1 generation?

All tongue-rollers.

Explain why you made these predictions.

One parent can only produce R gametes and the other parent can produce only r gametes, so all of the offspring must be Rr.

In the examples given so far, each parent has only been able to produce one type of gamete, so the outcomes of the crosses are fairly simple. The **Punnett square** was devised to keep track of all possible combinations of genotypes when more than one type of gamete can be produced. Fill out the Punnett square in Figure 9.2 for the F_2 generation by crossing off-spring of the previous cross ($Rr \times Rr$).

What are the possible genotypes in the F_2 generation?

RR, Rr, rr.

What is the phenotype of each genotype in the F_2 generation?

RR *and* Rr: *tongue-rollers;* rr: *unable to roll tongue.*

What is the phenotypic ratio for this cross?

3 rollers to 1 nonroller.

Give students additional problems if you are not sure they understand the Punnett square method. There are other methods for solving this type of genetics problem. If you use a different method or if students are taught a different method in their lecture class, introduce it here in place of the Punnett square.

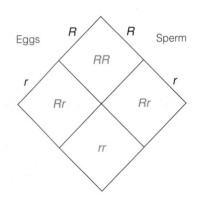

Figure 9.2.
Punnet square for monohybrid cross.

Pop bead chromosomes with allele labels are very helpful for this activity (see note for Activity A).

The cell should have two pairs of homologous chromosomes that are distinguishable from each other by size or shape. Each member of a pair should have an allele labeled R or r on one pair and A or a on the other. Both members of each allelic pair should be at the same locus.

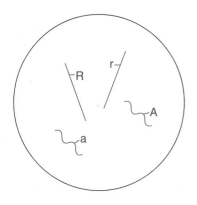

Figure 9.3.
Diploid cell carrying two unlinked genes.

Activity C: The Chromosomal Basis of Independent Assortment

Genes that are located on the same chromosome are **linked** with each other. If genes are located on separate, nonhomologous chromosomes, they are not linked, or unlinked. Unlinked genes separate independently during meiosis. For example, consider the allelic pair *R* and *r* and a second allelic pair *A* and *a*. If the *R* gene and the *A* gene are not linked, their alleles can be found in any combination in the gametes. That is, the *R* allele can be in the same gamete as either *A* or *a*. This is Mendel's **principle of independent assortment.** The word assortment in this case refers to the distribution, or sorting, of alleles into gametes.

In Figure 9.3, draw a cell that represents the following conditions: diploid; two homologous pairs of chromosomes; two unlinked genes called *R* and *A;* cell is heterozygous for both genes.

What is the genotype of this cell?

> RrAa.

R is the gene for tongue-rolling, as used in the previous examples. *A* determines arch characteristics. A person who has the dominant allele has normal arches. An individual who is homozygous recessive has flat feet.

What is the phenotype of the individual represented by the cell in Figure 9.3?

> *Tongue-roller, normal arches.*

Recall that when this cell undergoes meiosis, each gamete receives one member of each homologous pair. List the possible combinations of alleles that will be found in the gametes.

> RA, Ra, rA, ra.

In what proportion would you expect these gametes to occur?

> *1:1:1:1. (There's no reason to think that any of the gametes would occur more frequently than the others.)*

Activity D: Predicting the Outcome of a Dihybrid Cross

The outcomes of **dihybrid** crosses (following two traits) can be determined just as they are for monohybrid crosses. List all possible gametes each parent can produce and make all possible combinations of those gametes. When determining what the gametes will be, remember that each gamete must have one member of each homologous pair of chromosomes. If you are considering an *A* gene and an *R* gene, for example, each gamete must have one allele for the *A* gene (either *A* or *a*) and one allele for the *R* gene (either *R* or *r*). Try the following examples.

Genotype: *RRAA*   Gametes: <u>RA</u>

Genotype: *RrAA*   Gametes: <u>RA and rA</u>

Genotype: *rrAa*   Gametes: <u>rA and ra</u>

Genotype: *RrAa*   Gametes: <u>RA, Ra, rA, and ra</u>

To predict the outcome of a cross, construct a Punnett square to accommodate the gametes from each parent. For example, cross two heterozygous parents using the Punnett square in Figure 9.4.

Give additional examples to be sure students understand how to determine the gametes for any cross. A common student mistake is to think that there can be Rr and Aa gametes. If necessary go back to the cell drawing or pop bead chromosomes and review the homologous pairs.

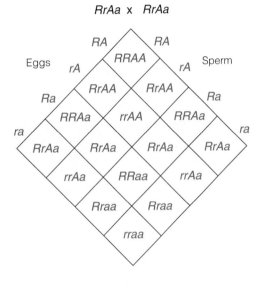

Figure 9.4.
Punnett square for dihybrid cross.

List the possible genotypes of the offspring:

RRAA, RRAa, RRaa, RrAA, RrAa, Rraa, rrAA, rrAa, rraa.

List the possible phenotypes of the offspring and the expected ratios in which they would occur.

9 normal arches, tongue-rollers; 3 normal arches, nonrollers; 3 flat feet, tongue-rollers; 1 flat feet, nonroller.

A couple with the genotypes *RrAa* and *RrAa* have 16 children. Ten of them have normal arches and can roll their tongues; the other six have normal arches and can't roll their tongues. Why doesn't this family match the expected ratio? (Hint: If the probability of having a male child is $1/2$, why can one family have seven daughters and no sons?)

The sample size is too small. The larger the number of progeny counted, the more likely the expected ratios will be fulfilled. That's why we typically use fast-breeding, prolific organisms (such as plants, mice, or fruit flies) to demonstrate Mendelian principles, rather than slow-breeding humans who produce only a small number of offspring.

Probability will be explored further in Lab Topic 10. For this lab, the point should be made that small sample sizes (such as families) will not necessarily show the expected phenotypic ratio. This may also help students understand the results of the Fast Plants experiment. Students often fail to understand that each offspring is an independent event. Thus each child has a $1/4$ chance of having flat feet, regardless of how many siblings have already been born with normal arches.

EXERCISE 9.2
A Dihybrid Cross Using
Wisconsin Fast Plants

Objectives

After completing this exercise, you should be able to

1. Identify the wild type and anthocyaninless (green stem) and yellow-green mutant forms of Fast Plants.

2. Explain how to obtain F_2 plants from F_1 seeds.

3. Perform a chi-square test on data from a genetic cross and interpret the results.

4. Explain the results of the Fast Plants experiment in terms of the genotypes and phenotypes of the F_1 and F_2 generations.

There is a particular form that is used to designate Fast Plant genotypes. For example, a plant that is heterozygous for anthocyanin is referred to as anl/ANL. For the purposes of this lab, you may allow your students to use the familiar designations of uppercase letter for dominant and lowercase letter for recessive.

In this exercise you will perform a dihybrid cross and determine the inheritance pattern by comparing your results to the expected phenotypic ratio for this type of cross. You will use a plant of the genus *Brassica* in the mustard family that completes its life cycle faster than most other plants. This plant variety, which is related to cabbage, cauliflower, and brussels sprouts, was developed at the University of Wisconsin and is known as Wisconsin Fast Plants. Fast Plants have several mutations that are clearly distinguishable from the normal (wild type) for the variety. You will use two of those mutations in your experiment.

The stems of wild-type seedlings have a red or purplish tinge that is caused by anthocyanin pigments. If you performed Lab Topic 2, pH and Buffers, you may recall using the anthocyanin pigments of red cabbage as a pH indicator. The mutant form does not produce any anthocyanin, so it has a green stem (see the left photograph in Color Plate 1).

If the mutation is a simple recessive gene, what are the possible genotypes of a plant that produces anthocyanin?

AA or Aa (or you could use P for purple stem: PP or Pp).

What is the genotype of a plant that does not produce anthocyanin?

aa.

Like most plants, the leaves of Fast Plants produce plenty of chlorophyll, which makes them appear green. A mutant form has yellowish-green leaves (see the right photograph in Color Plate 1).

If the mutation is a simple recessive gene, what are the possible genotypes of a plant that has green leaves?

GG or Gg.

What is the genotype of a plant that has yellowish-green leaves?

gg.

What is the genotype of a plant that is heterozygous for both genes?

AaGg.

What is the phenotype of a heterozygote?

Has anthocyanin (a reddish or purple stem); leaves are green rather than yellow-green.

If both mutations are unlinked, simple recessive alleles and two heterozygous parents are crossed, what ratio of phenotypes is expected in the offspring? See Figure 9.4 if you need to review the expected outcome of a dihybrid cross.

9 anthocyanin, green leaves; 3 anthocyanin, yellowish leaves; 3 no anthocyanin, green leaves; 1 no anthocyanin, yellowish leaves.

In this exercise you will perform a dihybrid cross between two heterozygotes and compare the phenotypic ratio of the offspring to the expected ratio. Fast Plants are only fast relative to other plants. In fact, the experiment takes about 7 weeks to complete. You will perform different stages of the experiment during several labs prior to doing Lab Topic 9, when you will collect data.

Read the Growing Instructions booklet that is supplied with the Fast Plants kits from Carolina Biological Supply to supplement the instructions given here. All instructions must be followed carefully for plants to grow and mature according to schedule. Light banks containing no fewer than six 40-watt fluorescent tubes must be used to provide enough light. The plants must be illuminated 24 hours a day and watered by hand for the first 3 days after planting. It may be more convenient to designate a technician, TA, or faculty member to water the plants than to arrange for students to do it, especially if planting is done during Wednesday, Thursday, or Friday labs.

You will perform the following steps:

- Plant the seeds, which represent the F_1 generation.
- Record the phenotypes of the seedlings.
- When the plants flower, cross-pollinate them.
- Wait for the seeds to mature and harvest them.
- Plant the seeds, which represent the F_2 generation.
- Record the phenotypes of the F_2 seedlings.

During this time, of course, the plants must be watered to remain healthy. You will be responsible for ensuring that your plants survive to produce data.

Activity A: Planting F_1 Seeds

7 Weeks Before Activity E

Your lab team will be responsible for one quad. A quad is a very small styrofoam pot with four compartments (Figure 9.5a). In order for Fast Plants to complete their life cycle rapidly, they require special growing conditions. Fertilizer pellets will be added to the soil, and the plants will need to receive constant light.

You should plant wild-type and homozygous mutant seeds in order to demonstrate phenotypes.

Procedure

1. Push one wick through the hole in the bottom of each compartment so that the tip sticks through about 1 cm (Figure 9.5a).
2. Add potting soil so that each compartment is about half full.
3. Place three fertilizer pellets in each compartment.

See Instructions for Planting Fast Plants at the end of the Instructor Notes for this lab topic.

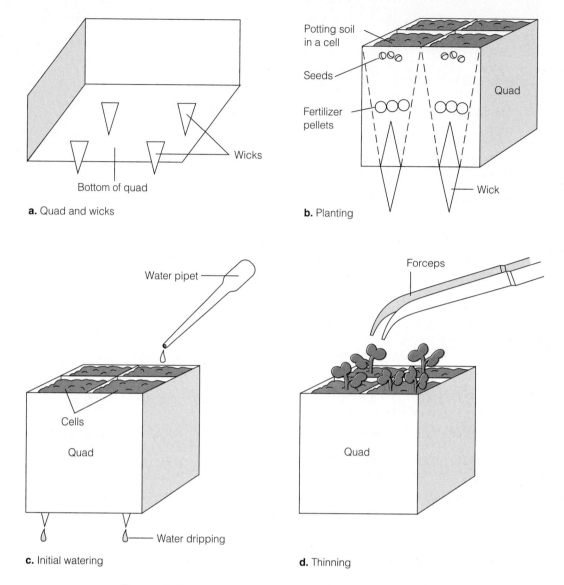

Figure 9.5.
Planting and thinning Fast Plants.

4. Add potting soil so that each compartment is loosely filled.

5. Use a fingertip to make a slight depression in the soil of each compartment.

6. Put three seeds in each depression.

7. Sprinkle a little more potting soil in each compartment so that the seeds are covered.

8. Using a waterproof pen, write the date and your team's initials on a pot label and insert it in one of the compartments of your quad.

9. Use a pipet to water each compartment gently until you see water dripping from the wicks (Figure 9.5c).

10. Place your quad under the lights on the water mat along with those of other lab teams. The water mat draws water from the reservoir beneath it. In turn, the wicks will draw water into each quad.

Caring for Plants

First 3 Days After Planting

Plants should be watered from above with a pipet for the first three days after planting (Figure 9.5c). This may be handled by your instructor. It is critical to keep the soil moist while the seeds are germinating.

If you are doing the watering for the first 3 days, use a wash bottle of water rather than a pipet to speed up the process.

Daily Thereafter

Check the soil moisture in your quad daily. Your instructor may do this for you. If it is dry, pick up the quad and add water with a pipet until water drips from the wick. If the mat beneath your quad is dry, ask your instructor to resoak it.

Even if your plants have wilted, don't give up. Water the plants from above with a pipet. Then place the quad in about 1 cm of water in a sink or dish tub and leave it until the plants recover. However, if the plants are already brown and crispy, it is too late to save them.

If the plants will not stand upright as they grow taller, you may support them with a small wooden stake.

Again, it may be more convenient for you to have a staff member check on the plants rather than having students come in to do it. If you do want students to assume this responsibility, have teams schedule their members to take turns. Fill the water reservoirs on Friday and the plants should be fine over the weekend. See the Growing Instructions booklet if you notice problems with insect damage. Maintain a distance of 5–8 cm between the tops of the plants and the bulbs.

Activity B: First Observations
6 Weeks Before Activity E

When the seedlings are approximately 1 week old, compare them to the wild-type plants grown by your instructor to be sure that they express the correct phenotype. Destroy any plants that lack anthocyanin in the stems or have yellow-green leaves. That is, destroy any plants that express the mutant phenotype. Why should these plants not be used for your experiment?

There is no need to record data for the F_1.

> *They are not heterozygotes, and we want to count offspring from a heterozygote cross. (They are probably the result of self-pollination by one of the homozygous parents.)*

Thin the plants so that there is only one per compartment in your quad (Figure 9.5d). Leave the most vigorous-looking plant to survive.

Preparation for Pollination

Each group should prepare a bee stick to be used to transfer pollen from one flower to another. As you will learn in Lab Topic 18, Flowers, Fruits, and Seeds, bees are natural pollinators, so their bodies contain small hairs that pick up pollen easily. Alternatively, a soft paintbrush can be used for pollination.

The Fast Plants kit includes dead honeybees to use for the bee sticks. The bee sticks must be made at least 2 days in advance of use because some glues contain a solvent that is toxic to the pollen. You might find it more convenient if the prep staff or you, yourself, make the bee sticks rather than taking lab time to have the students do it. Don't assemble the bee sticks near the plants, and don't leave them to dry near the plants. Have a piece of Styrofoam available for students to stick their toothpicks in. The bottom of a Styrofoam cup could be used to hold the bee sticks for each lab section. The developers of Fast Plants invented the bee-stick method to emphasize the relationship between plants and their insect pollinators. A small, soft paintbrush is just as effective and may be used in place of bee sticks.

 Do not perform this procedure near the plants. The glue may contain a chemical that will prevent pollen germination.

Procedure

1. Hold a dead bee by its wings. Remove its head, legs, and abdomen. The part remaining is the thorax. The thorax will have a small hole at each end where the other body parts were removed.
2. Put a small drop of glue on the tip of a toothpick.
3. Insert the tip of the toothpick into the hole at one end of the thorax.
4. Stick the other end of the toothpick into a piece of styrofoam and allow the glue to dry.

Activity C: Pollination

Approximately 5 Weeks Before Activity E

Flowers will begin to open about 2 weeks after planting. Pollinate three times, 2 days apart. You may exchange pollen with the plants in your quad or with plants in other quads.

> **✳ Do not cross-pollinate with the demonstration plants grown by the instructor.**

Begin pollinating when two or three flowers open on most plants. Flowers are ready to pollinate when there is a dusting of yellow pollen on the anthers, indicating that the anthers have opened. You may not be able to pollinate exactly 5 weeks before the data collection lab because flowers may mature earlier or later than expected.

Procedure

1. Rub the bee thorax gently over the anthers (male part) of a flower to obtain pollen.
2. Transfer the pollen to a new plant by brushing the bee on the stigma (female part) of a flower on another plant.
3. Repeat 2 days and 4 days later.
4. After you have pollinated three times, remove all new flower buds that appear so that only the flowers that you have pollinated will produce seeds. Check plants two or three times a week for this purpose. This may be handled by your instructor.

Harvesting Seeds

It takes approximately 3 weeks for the seed pods to mature. During that time, you or your instructor should continue to ensure that your plants have enough water and continue to remove new flower buds and shoots.

Two weeks before the data collection lab (approximately 3 weeks after pollination), remove the plants from the watering system so the seed pods will dry out.

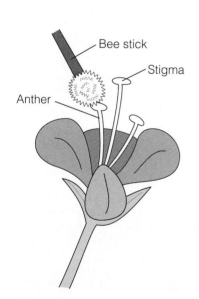

Figure 9.6.
Reproductive structures of Fast Plants.

Activity D: Germinating the F_2 Seeds

1 Week Before Activity E

The seeds that you harvest are the F_2 generation. In order to determine the phenotypes of these plants, you will germinate them on filter paper. There is no need to grow the plants in soil as you did with the F_1 generation because you will be able to determine the phenotypes from week-old seedlings.

Procedure

1. Remove the dry seed pods from each plant.
2. Over a collecting pan, roll each pod between your thumb and forefinger to remove the seeds.
3. Use a pencil to write the date and your lab team's initials on one end of a filter paper disk.
4. Place the filter paper in the top of a petri dish, which is larger than the bottom.
5. Use a wash bottle to wet the filter paper thoroughly.
6. Place approximately 25 seeds in rows across the top two-thirds of the dish.
7. Fit the bottom of the petri dish into the top and stand it upright in a reservoir (Figure 9.7). There should be about 2 cm of water at the bottom of the reservoir.
8. Empty the contents of each quad into the container provided.
9. Return the quad and the label to your instructor.

The petri dishes can be laid flat under the lights if you have enough space, avoiding the use of reservoirs. However, the seeds will need to be watered daily. If you use this method, use two or three layers of filter paper in each dish to hold moisture better. Save the extra seeds in case catastrophic neglect or some other misfortune befalls the germinating F_2 before data are collected.

There is very little soil in each quad; nevertheless you may want to save it for other uses. If so, have a collection container ready. If not, have students empty their quads into a trash container.

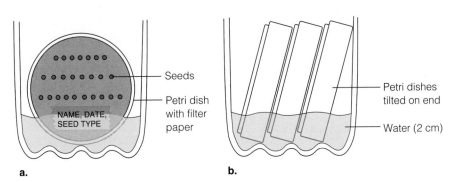

Figure 9.7.
Germination of Fast Plant seeds.

a.

b.

Activity E: Collecting and Interpreting Data from the F_2 Generation

Data Collection Lab

Procedure for Collecting Data

1. Count the number of seedlings that show each of the phenotypes listed in Table 9.1. Score a seedling as having anthocyanins if the stem has even a slight reddish or purple tinge. Ask your instructor to show you examples of each phenotype if you are in doubt. (The photographs on Color Plate 1 illustrate the phenotypes.) Using forceps, remove each seedling from the filter paper as it is counted so you won't count it again.
2. Record your data in Table 9.1 on the next page.

As discussed in an earlier section of the lab, when there is a small number of observations the data may not display the expected phenotypic ratio. For example, if the parents are both heterozygous for tongue-rolling, they may have two children who are tongue-rollers and two who are not instead of the 3:1 ratio expected of a monohybrid cross. What is the

Table 9.1

Data from Dihybrid Cross Between *AaGg* Parents

Phenotype	Number Observed (your data)
Anthocyanins (purple stem), green leaves	
Anthocyanins (purple stem), yellow-green leaves	
No anthocyanins (green stem), green leaves	
No anthocyanins (green stem), yellow-green leaves	

When you review the answers with students, be sure they understand why "/$_{16}$" is used in each part of the ratio. The Punnett square in Figure 9.4 has 16 boxes because there are 16 possible combinations of gametes. Nine of those 16 combinations result in the same phenotype (anthocyanin, green); three result in the anthocyanin, yellow-green phenotype; and so on. The total of all possible outcomes is 16 (9 + 3 + 3 + 1 = 16). This is important for students to understand because to determine the expected number for a chi-square test, students must multiply the sample size by $^9/_{16}$, $^3/_{16}$, and so on (see Table 9.2).

expected phenotypic ratio in the dihybrid cross you performed with the Fast Plants? (Hint: Review the Punnett Square in Figure 9.4.)

Anthocyanin, green:

9/16.

Anthocyanin, yellow-green:

3/16.

No anthocyanin, green:

3/16.

No anthocyanin, yellow-green:

1/16.

Chi-square is a statistical test that determines how well the actual data fit the expected ratio. The test is performed by calculating a value called chi-square. A probability table is then used to determine whether this value fits the expected ratio. Small deviations from the expected ratio can happen due to chance. The chi-square test tells us whether the difference between the observed data and the expected data can be attributed to chance. If it can, then the data *do* fit the expected ratio for the F_2 generation of a dihybrid cross; the deviation is not significant. If the deviation is not due to chance, then it must be explained some other way. For example, a large deviation could mean that the inheritance does not follow a simple dominant-recessive pattern.

Since statistical analysis is not covered in this lab manual, the chi-square test has been simplified as much as possible.

The example in Table 9.2 shows how to calculate chi-square. The data recorded from the experiment, the number of each phenotype observed, is recorded in the first column. Then the expected number of each phenotype is calculated, based on the total number of seedlings that were counted (177 in the example). The total number is multiplied by the expected ratio for each phenotype. Then the deviation of the observed number from the expected number in each category is calculated. Finally the deviations from all four categories are added. The result is the chi-square value.

Table 9.2

Example Calculation of Chi-Square

Phenotype	Number Observed	Number Expected	Deviation = $\dfrac{(\text{Observed} - \text{Expected})^2}{\text{Expected}}$
Anthocyanins, green	102	$177 \times \frac{9}{16} = 100$	0.04
Anthocyanins, yellow-green	40	$177 \times \frac{3}{16} = 33$	1.48
No anthocyanins, green	27	$177 \times \frac{3}{16} = 33$	1.09
No anthocyanins, yellow-green	8	$177 \times \frac{1}{16} = 11$	0.82
Total seedlings observed = 177.		Sum of all deviations = chi-square = 3.43.	

The chi-square value is then compared to Table 9.3 to determine whether the observed deviation from the expected ratio is likely to have occurred by chance. Table 9.3 is an abbreviated version of a chi-square table that includes only the row of values that pertain to your data. If you want to perform a chi-square test for some other purpose, you should obtain a complete table from a statistics book.

There are 3 degrees of freedom (df) in these data (number of classes – 1). The only part of the chi-square table that is reproduced in Table 9.3 is the row of probabilities for df = 3.

If the probability is greater than 0.05, we can say that the differences between the observed data and the expected ratio are due to chance. In the example, the chi-square value of 3.43 corresponds to a probability of between 0.50 and 0.30. This tells us that the observed ratio is not significantly different from the expected ratio. The data in this example would support the hypothesis that the anthocyaninless and yellow-green phenotypes are caused by a recessive allele.

Procedure for Interpreting Data

1. Record the class data in the first column of Table 9.4.

2. Add up the total number of seedlings observed for the experiment.

3. Calculate the expected number for each phenotype. To do this, multiply the total number by $\frac{9}{16}$, $\frac{3}{16}$, or $\frac{1}{16}$, as shown in Table 9.2. This calculation determines the share of the total that is expected for each phenotype. Remember, the expected ratio is derived from the Punnett square for a dihybrid cross (Figure 9.4).

Table 9.3

Chi-Square Table for Dihybrid Cross

	Probability That Deviation Occurred by Chance										
	0.95	0.90	0.80	0.70	0.50	0.30	0.20	0.10	0.05	0.01	0.001
Chi-Square =	0.35	0.58	1.01	1.42	2.37	3.66	4.64	6.25	7.82	11.34	16.27
	Deviation from expected ratio is explained by chance.								Deviation is too great to be due to chance.		

Source: Fisher, R. A., and F. Yates. *Statistical Tables for Biological, Agricultural and Medical Research*, 6th ed., Table IV. Edinburgh: Oliver and Boyd, Ltd.

Table 9.4
Chi-Square Test of Class Data

Phenotype	Number Observed (class data)	Number Expected	Deviation = $\dfrac{(\text{Observed} - \text{Expected})^2}{\text{Expected}}$
Anthocyanin, green		____ × $^9/_{16}$ = ____ (total #)　　　(# exp.)	
Anthocyanin, yellow-green		____ × $^3/_{16}$ = ____ (total #)　　　(# exp.)	
No anthocyanin, green		____ × $^3/_{16}$ = ____ (total #)　　　(# exp.)	
No anthocyanin, yellow-green		____ × $^1/_{16}$ = ____ (total #)　　　(# exp.)	
Total seedlings observed (class) = _____ .		Sum of all deviations = chi-square = _____ .	

4. Calculate the deviations of the observed values from the expected values using the formula shown in the last column of Table 9.4.

5. Find the sum of the deviations. This number is the chi-square value.

6. Compare the chi-square value to the second row of Table 9.3 to determine the probability that the observed deviations can be explained by chance.

What is the probability?

Do your results provide evidence that the anthocyaninless and yellow-green mutants are recessive alleles? Explain your answer.

Any probability greater than 0.05 indicates that the deviation of the observed ratio from the expected ratio can be attributed to chance. That is, it is not a significant difference. The results from this cross between heterozygotes should show that the data do fit a 9:3:3:1 phenotypic ratio, which is evidence that the mutant alleles are recessive to the wild type.

Questions for Review

To practice doing monohybrid and dihybrid crosses, work the following problems. Each of the human traits used in the problems is determined by a single gene locus that has two different alleles, one dominant and one recessive. Show all of your work for the problems.

Dominant traits	**Recessive traits**
Freckles (*F*)	No freckles (*f*)
Astigmatism (*A*)	Normal vision (*a*)
Ability to roll tongue (*R*)	Cannot roll tongue (*r*)
Normal arches (*A*)	Flat feet (*a*)
Widow's peak (*W*)	Straight hairline (*w*)

1. A man who has normal vision marries a woman who is heterozygous for astigmatism. What are the possible genotypes and phenotypes that their children can have?

    Given: P genotypes aa (male) × Aa (female):

    Male gametes are all a. Female gametes can either be A or a. F_1 genotypes: Aa and aa. F_1 phenotypes: astigmatism and normal vision.

2. Two people who have normal arches produced a child who has flat feet. What is the genotype of the child?

    aa.

    What is the genotype of the father?

    Aa.

    What is the genotype of the mother?

    Aa.

    Explain how you arrived at your answers.

    If two individuals with the dominant phenotype produce offspring with the recessive phenotype, the parents must both be heterozygotes (each had to contribute one recessive allele).

3. A woman has a widow's peak, but she does not know her genotype. She marries a man who has a straight hairline and they have 13 children. Nine have widow's peaks and four have straight hairlines. What are the genotypes of the parents? What are the genotypes of the children? Explain how you arrived at your answers.

    Given: P genotypes W? (female) × ww (male) and F_1 genotypes W? and ww:

    The mother's genotype must be Ww, because the ww children must have received one w allele from her. The genotype of the children who have widow peaks is also Ww, since the only allele the father could donate was w.

4. A couple who both have the ability to roll their tongues have a son who is also a tongue-roller. The son is very curious as to whether he is homozygous or heterozygous for the tongue-rolling trait. How would he go about finding out?

    He would have to have children by a woman who cannot roll her tongue (rr). (This is called a test cross.) If he had at least one child who couldn't roll her tongue, he would know his genotype was Rr. Of course, it could take many children before he was reasonably certain of the results. In other words, he'd have to be pretty darn curious.

5. A man who has no freckles and flat feet marries a woman who is homozygous dominant for both traits. What is the man's genotype?

    ffaa.

    What genotype(s) will his gametes have?

    fa.

    What is the woman's genotype?

    FFAA.

    What genotype(s) will her gametes have?

    FA.

    What genotype(s) will their children have?

    FfAa.

    What will their children's phenotype(s) be?

    Freckles, normal arches.

6. A woman who is heterozygous for astigmatism and cannot roll her tongue marries a man who has normal vision and is homozygous dominant for tongue-rolling. What are the possible genotypes and phenotypes of their children for these traits?

    Given: P genotypes Aarr (female) × aaRR (male):

    Possible F_1 genotypes: AaRr, aaRr. Possible F_1 phenotypes: astigmatic, tongue-roller; normal vision, tongue-roller.

7. A man has freckles and can't roll his tongue. His wife has no freckles and can roll her tongue. All of their many children have freckles. About half can roll their tongues, while the other half can't. What are the probable genotypes of the parents, and what gametes can each parent produce?

   Given: P genotypes ffR? (female) × F?rr (male) and F₁ genotypes F?R? and F?rr:

   Father's genotype: FFrr, father's gametes: Fr. Mother's genotype: ffRr, mother's gametes: fR and fr. Father is probably FF, since no homozygous recessive offspring have shown up. (Still, this is not certain.) Mother is heterozygous for tongue-rolling, since there are homozygous recessive offspring.

8. A woman who has normal arches and a straight hairline has children with a man who has flat feet and a widow's peak. One of their children has a straight hairline and normal arches, one of their children has a straight hairline and flat feet, one has a widow's peak and normal arches, and one has a widow's peak and flat feet. What are the genotypes of the parents?

   Given: P genotypes A?ww (female) × aaW? (male) and F₁ genotypes A?ww, aaww, A?W?, aaW?:

   In order to get children who are homozygous recessive for flat feet, the woman must be heterozygous because one of the a alleles comes from her. Same reasoning for the father's genotype for widow's peak.

9. The F₁ seeds for the Fast Plants experiment were heterozygous for both traits (anthocyanin and leaf pigment). How were these heterozygote seeds produced?

   By crossing a homozygous dominant parent with a homozygous recessive parent.

LAB TOPIC 9

Mendelian Genetics
Instructor Notes

The Wisconsin Fast Plants experiment used in this lab topic must be started at least 7 weeks before Lab Topic 9 is performed. Only a few minutes of lab time is required during each of the weeks before Lab 9. The plants can be started earlier and the F_2 seeds held until the week before the lab if you wish. Be sure to read the instruction booklet that comes with the Fast Plants kit from Carolina Biological Supply for details on these procedures. Following is a brief outline of the events in the Fast Plants experiment.

■ 7 weeks before lab: Plant seeds (~15 minutes).

■ 6 weeks before lab: Observe phenotypes of F_1 and thin plants; prepare bee sticks (~10 minutes). Team members can split up the tasks.

■ 5 weeks before lab: Cross-pollinate flowers (over a period of 6 days, ~5 minutes each time).

■ After pollination is completed, pinch off new flowers and shoots every few days.

■ 2 weeks before lab: Remove plants from watering system so seed pods will dry.

■ 1 week before lab: Sow seeds in petri dishes on filter paper (~15 minutes).

■ During lab: Record phenotypes of F_2 generation and perform chi-square test (50–60 minutes).

Planting, first observations, and the initial pollination can be done during the regular weekly lab meetings. Watering, two subsequent pollinations, and pinching off new growth must be done outside lab time. These are not time-consuming activities, and you may find it more convenient to have the prep staff or lab instructors perform these tasks rather than to provide students access to the room where the plants are kept.

Organization

Background for Fast Plants Experiment (60–90 minutes total)

Review the definitions at the beginning of the laboratory. Students should have filled out their lab manuals before class, but you should review to be sure that they understand each term. Try to involve the students rather than delivering a lecture. One technique is to have students spend several minutes with their lab teams agreeing on the best answers. Then call on a representative from each team. When that person has given an answer, check with the rest of the class for correction or concurrence before you accept the answer.

Exercise 9.1 consists of worksheet exercises. You can work through them with the class or you can have students complete them by working first with their lab teams and then discussing the answers as suggested above. Give students additional crosses to work using Punnett squares to be sure they understand the method.

Data Collection and Interpretation for Fast Plants Experiment (50–60 minutes)

- Review the introductory material in Exercise 9.2.
- Have each lab team count and record the number of each phenotype in their petri dish.
- Explain what the chi-square test is used for and work through the example in the lab.
- Pool class data in Table 9.4 and have students perform the chi-square test and interpret the results. You might want to have them perform a chi-square test on their own group's data as well to see the effect of sample size. The lecture professor can collect and present data from all of the lab sections (see next section).

Integration of Fast Plants Experiment into the Lecture Class

The Fast Plants experiment can easily be incorporated into the lecture class. When presenting Mendelian genetics, the lecturer should use the Fast Plants mutants as an example along with Mendel's peas. This will help students relate their lecture material to an example that they have observed. After the lab sections have collected data, the lecturer can assemble and present the data from all of the sections and demonstrate the chi-square test on this large sample. Students will see the effect of sample size on achieving the expected ratio from a cross. When the experiment is repeated in subsequent years, the lecturer can further enlarge the sample size by adding in data from previous classes.

Instructions for Planting Fast Plants

1. Put the potting soil in a dishtub and moisten it before students use it.
2. Prepare the water mats by soaking them in water to which you have added 2 to 3 mL liquid detergent. Squeeze out the water; then soak again. Repeat twice. Then soak the mats again and, without squeezing them out, lay one mat on the lid of each water reservoir and smooth it out (see the Growing Instructions booklet for an illustration). The mats should be wet throughout, and should have no air pockets.
3. Fill each reservoir with water.
4. Place a copper sulfate square in each reservoir to retard algal growth.
5. Lights should be 5–8 cm above the tops of the plants.
6. Plant the parental-type seeds as a demonstration. Mark your quad so that it can be easily recognized (for example, put a big blotch of color on the label). When students pollinate their plants, they must avoid cross-pollinating with the parental types.

For a 2-Hour Lab

Students can complete Exercise 9.1 as homework. The only preliminary information that is absolutely necessary for them to know is the expected phenotypic ratio for the Fast Plants cross.

Forensic Application of Molecular Genetics

 You will need a calculator for this lab!

Introduction

You already know that DNA is the genetic material and that its sequence of nucleotides holds the "recipe" for making proteins. But it turns out that only about 3% of our genes are actually the instructions to make functional proteins. The rest of the nucleotide sequences are nonsensical or irrelevant. Collectively, they are known as "junk" DNA. Like old magazines or newspapers that have accumulated over decades in someone's attic, there is no use for junk DNA in our bodies. But scientists have found a way to turn trash into treasure, based on the fact that each of us has our own unique collection of junk. Although we vary genetically in many of the functional genes, we have even more variability in certain types of these nonsense sequences. There are thousands of sites in our chromosomes where they have been inserted. Although numerous people may share the same bit of junk at a single site, we are extremely unlikely to share the same ones at each of several sites. Statistical comparisons are used to determine how likely it is that two samples of DNA came from the same source. Thus, some of the most effective techniques for DNA typing—otherwise known as DNA fingerprinting or DNA profiling—are based on our unique patterns of junk DNA.

In this laboratory, you will learn about one of the methods of DNA typing that is currently widely used. You will then learn how to compare the results of DNA analysis to statistics for a population group. To complete the laboratory, you will apply your knowledge of DNA profiling to solving a crime.

Instructor Overview
This lab topic introduces students to the use of STR (simple tandem repeat) analysis in forensic DNA profiling. First students learn what STRs are and use paper models to simulate cutting the DNA with restriction enzymes and using electrophoresis to sort the segments. They review Mendelian inheritance to apply STR analysis to paternity testing. Then they learn how allele frequencies are used to calculate the probability of a match between DNA samples, using a missing persons case as an example. In the final exercise, they use all the applications of DNA profiling that they have learned to solve a mystery.

Outline

Exercise 10.1: DNA Profiling Using STR Analysis

    Activity A: The STR Technique

    Activity B: Application of STR Analysis to Paternity Testing

    Activity C: Using Population Genetics to Match DNA

Exercise 10.2: Application of DNA Profiling

Time Requirements
Exercise 10.1:
    Activity A: (20–30 minutes)
    Activity B: (15–20 minutes)
    Activity C: (30 minutes)
Exercise 10.2: (60 minutes)

E X E R C I S E   1 0 . 1
DNA Profiling Using STR Analysis

Objectives

After completing this exercise, you should be able to

1. Explain what STRs are and why they are useful for DNA profiling.
2. Describe the role of restriction enzymes in STR analysis.
3. Describe the role of electrophoresis in STR analysis.
4. Apply the laws of Mendelian inheritance to STR alleles to determine parentage.
5. Use allele frequencies to calculate the probability of an individual having a particular genotype.

Activity A: The STR Technique

One type of junk DNA that has proven useful in DNA profiling is called STR (simple tandem repeat). A STR (pronounced "star") is a locus or place in the DNA where a short segment of nucleotides—for example, ACG—is repeated numerous times. We all share the same sequence of repeated nucleotides (ACG, in this case) but the *number* of repetitions varies. For example, in one allele there might be seven repetitions: ACGACGACG ACGACGACGACG. In another allele there might be 25 repetitions—ACG ACGACGACGACGACGACGACGACGACGACGACGACGACG ACGACGACGACGACGACGACGACG—while a third allele might have 33 repetitions. More than 8,000 STR loci have been identified in the human genome.

STRs are microsatellites with 2–5 bp repeats.

In contrast to the simple dominant/recessive loci that you are most familiar with, which have only two possible alleles (for example, *A* and *a*), STR loci have *many* different alleles. As a result, the frequency of any particular allele within a population may be quite low. For example, at one locus called D3S1358 that is used in DNA profiling, the Caucasian population has 11 possible alleles whose frequencies range from 0 to 0.258. To put it another way that will be useful later in the lab, the probability that a Caucasian has a "15" allele (15 repetitions) at D3S1358 is 0.258.

You may need to explain "allele frequency" to your class. It is the proportion of a particular allele in the gene pool. For example, suppose there are 500 individuals in a population. For any given locus there are 1,000 total alleles (no matter how many different alleles there are) because each diploid individual has two alleles. Let's call a locus STR1 for an example. If 100 of the alleles at STR1 are "7" (7 repetitions), then the frequency of "7" is 100/1,000 = 0.1. If another 500 of the alleles are "25," then the frequency of "25" is 500/1,000 = 0.5. If the remainder of the alleles are "43," then the frequency of "43" is 400/1,000 = 0.4. The usefulness of STR alleles stems from the fact that many of them have very low frequencies. A frequency of 0.005 for allele X means that only 5 of the 1,000 alleles in the hypothetical population of 500 individuals are "X."

Recall from Lab Topic 9 (Mendelian Genetics) that each person has the same locus on two homologous chromosomes. Just as an individual could be homozygous or heterozygous at a locus such as the one for tongue-rolling, an individual can be homozygous or heterozygous for STR loci. An individual who has two identical alleles—for example, both chromosomes have 7 repeats—would be homozygous. An individual who has seven repeats on one chromosome and 25 repeats on the homologous chromosome would be heterozygous.

The methods that scientists use to analyze an individual's alleles involve isolating the repeating segments and sorting them by size. Although in actual practice the techniques are complicated to perform, it is fairly easy to visualize the results with paper models. For the sake of simplicity, we will use single strands of DNA instead of the actual double-stranded molecule.

Procedure

Make simulated segments of DNA that contain STRs.

1. Your instructor will give you four nucleotide sequences. Cut them into single strips. These paper strips represent small portions of chromosome pairs from two individuals. Keep in mind that these segments come from much longer strands of DNA.

Isolate the STRs.
Restriction enzymes are molecules that chemically cut the DNA segments into smaller pieces by breaking bonds. There are many different restriction enyzmes. Each one "recognizes" a different base sequence in the DNA where it makes its cut.

2. You will "cut" the DNA strands with two different restriction enzymes. One enzyme recognizes "ACTATG" and cuts the DNA between the T and the G. Find that sequence in your DNA strands and cut each strip of paper there.

3. The other restriction enzyme recognizes the sequence CACTAT and cuts after the first C and before the A. Find that sequence in your DNA strands and cut each strip of paper there.

4. You have now isolated the STRs for these chromosomes. Keep only the STRs and discard the other DNA.

What is the STR sequence?

GCC

Sort the STRs by size.
Scientists sort the pieces of DNA with a technique called electrophoresis that uses a slab of gel-type material. DNA samples are lined up in lanes on the gel like a row of swimmers on the blocks before a race. The gel is then subjected to an electrical field, which causes the DNA samples to travel in a straight line from their starting points. The gel resists the molecules as they move through it, so shorter DNA segments travel farther than longer ones. After the samples have been "run," techniques to visualize specific bands of DNA are used. An example is shown in Figure 10.1. Notice that a "standards" lane is included on the gel. Pieces of DNA of known sizes are run alongside the unknown DNA samples so that the sizes of the unknown sample segments can be identified by comparison.

I have simplified the actual procedures of a STR analysis to get to the main point: visualizing the results in terms of comparing the number of repetitions present. I have not included any information about the initial procedures of isolating and amplifying the alleles. And whereas in actual practice numerous markers are analyzed simultaneously, students will only deal with one locus at a time throughout these lab exercises.

DNA molecules have a negative charge. The molecules migrate toward the positive pole. The larger the molecule, the more difficult it is to make its way through the gel, so the shorter segments move farther than longer ones.

Figure 10.1.
Gel showing two DNA samples plus standards.

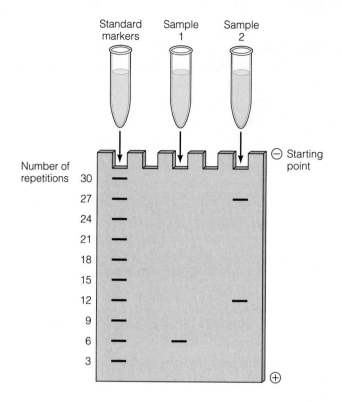

Keep in mind that an actual DNA sample contains all of the DNA, not just the tiny bit we are interested in. The visualization technique highlights the bands of interest. In a real DNA profile, several loci might be "read" on the same gel. We will stick with one locus per gel through the lab to clarify the results.

We will use numbers that correspond to the size of the DNA segment, or the number of repetitions, to designate alleles throughout the lab. The alleles in Figure 10.1 would be called 6, 12, and 27.

Each band for Sample 1 and Sample 2 in Figure 10.1 represents an allele for this STR locus. How many different alleles does Sample 1 have?

1 allele

How many repetitions are there in Sample 1's allele(s)?

6 repetitions

How many different alleles does Sample 2 have? How many repetitions are there in the allele(s)?

2 alleles. One has 12 repetitions and the other has 27.

Which sample is homozygous?

Sample 1 is homozygous for allele 6. (Review the definitions of homozygous and heterozygous from Lab Topic 9 if necessary.)

5. On a blank sheet of paper, label a lane for the DNA from Person A and another lane for the DNA from Person B. Label a third lane with standard markers ranging up to 25 repetitions. (Remember that the largest standard will be closest to the starting point.)

6. Arrange the STRs you "isolated" in the appropriate places on your paper "gel" and secure them with tape. This is the DNA profile or fingerprint.

What is Person A's genotype at this locus?

5,17

What is Person B's genotype at this locus?

15,23

Activity B: Application of STR Analysis to Paternity Testing

STR analysis has many different applications, including missing persons investigations, matching a suspect with evidence in criminal cases, wildlife crimes, determining the zygosity of twins, and identifying human remains at disaster sites such as the World Trade Center. In all of these applications, the general objective is to determine the probability that one sample matches another (see Activity C). STR analysis coupled with the laws of Mendelian inheritance is also used in paternity and pedigree testing.

You may want to review the concepts of diploidy and homologous pairs of chromosomes.

STR alleles are inherited just like the more familiar genes discussed in Lab Topic 9—for example, the genes for tongue-rolling, widow's peak, and astigmatism. Although there are only two known alleles at the "tongue-rolling" locus and there are numerous STR alleles at a given locus, each individual still only has two alleles. One allele was inherited from the mother and one was inherited from the father. As an example, let's use a hypothetical STR locus, STR1. Suppose a man's genotype at this locus is 6,6 and his wife's genotype is 12,27. Use a Punnett square to show the combinations of alleles possible in their children.

What are the possible genotypes of their children, and what is the probability of each genotype?

Offspring will be either 6,12 or 6,27, with equal probability for each.

Punnett square for cross 6,6 × 12,27

	6	*sperm*
eggs 12	6,12	
27	6,27	

Use the gel in Figure 10.2 to show the STR analysis of the parents and two children for this locus. Give each child a different genotype.

Figure 10.2.
Gel for STR analysis.

Notice that each child gets one allele from each of the parents, but none of the children has a genotype identical to the parents. This type of information is useful in determining paternity and other relationships. Figure 10.3 shows a simple example of how STR analysis can be useful in a paternity case.

Figure 10.3.
STR analysis of paternity case.

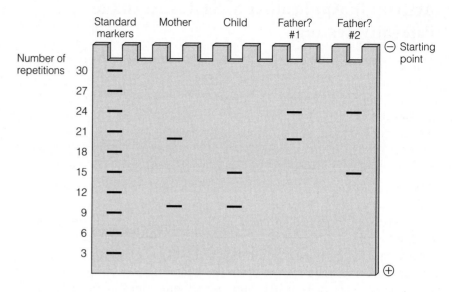

Which man is excluded as a possible father? Explain.

> *Father? 1 can't be the child's father. His genotype is 24,20. The child doesn't have either of these alleles.*

What about the other man? Is he the child's father?

> *Possibly. The child's genotype is 15,10 and the mother's is 20,10. The child got her "10" allele from her mother. Father? 2's genotype is 24,15 so he could have contributed the 15 allele. However, that is only a possibility.*

Activity C: Using Population Genetics to Match DNA

Although many STR alleles are rare (have a low frequency), the fact that two DNA samples have one allele in common is not proof that the DNA came from the same source. For example, suppose that for a locus called STR1 the frequency of the "27" allele is 0.02. That's a low frequency, just 2%, or 2 out of every 100 of the alleles in the population for STR1. But that kind of probability is not going to be enough to convince a jury. In order to get the one-in-a-billion kind of odds that are beyond coincidence, several loci are used. When the multiplication rule of probability is applied, the odds grow enormously. (The multiplication rule states that the probability for the occurrence of two independent events is the product of their separate probabilities.) So analyzing DNA samples is just the first step. It is the statistical analysis that provides convincing proof.

The first ingredient needed for the statistical analysis of DNA is the frequencies for the alleles that are to be included in the analysis. Allele frequencies vary by ethnicity and geographic area, so scientists have established databases for a variety of populations. Forensic DNA samples can then be compared to a relatively narrow group of people. By using several different loci, conclusive identifications can be obtained. For example, there are DNA databases that include allele frequencies for U.S. Caucasians, U.S. African-Americans, U.S. Mexican-Americans, U.S. Asians, German Caucasians, U.K. Caucasians, and a few dozen other groups. Some categories are further subdivided by region, such as U.S. Southeast Hispanic and U.S. Southwest Hispanic. Statistical analysis is then based on these known allele frequencies. There are also databases for animals such as cats, dogs, horses, and primates other than humans.

For example, suppose an unidentified body has been found. The general description of the body is similar to those of two missing persons. The first step is to compare the genetic makeup of the missing persons with a DNA sample from the body. DNA for the missing persons may be obtained from the roots of hair left in their hairbrushes or even cells clinging to their toothbrushes. (Lacking a direct sample, DNA from close relatives could be used, but that's more complicated so we'll assume good samples can be taken from the possessions of the missing persons.) The second step is to apply statistics to determine whether the DNA from the body matches the DNA from each of the missing persons. Both missing persons are U.S. Caucasians, so that is the database used for the allele frequencies. Table 10.1 compares the DNA profile of the two missing persons to that of the body that was found.

Table 10.1
Results for 4 STR Loci on DNA Samples from Body and Two
Missing Persons

Locus	Allele	Frequency in U.S. Caucasians	Present in Body	Present in Missing Person 1	Present in Missing Person 2
STR1	9	0.15	X	X	X
STR2	10	0.13	X	X	
STR2	6	0.28			X
STR3	12	0.06	X	X	X
STR3	9	0.02	X	X	
STR3	4	0.13			X
STR4	17	0.14			X
STR4	6	0.04	X	X	

Keep in mind that each individual has two chromosomes containing the same locus. If only one allele is detected, then the individual is homozygous at that locus.

STR locus 1 is represented here by just one allele, "9," and the same allele is present in all three samples. What is the genotype of

The body—*9,9*

Missing Person 1—*9,9*

Missing Person 2—*9,9*

We are now converting allele frequencies to genotype frequencies. A genotype frequency is the proportion of the population that has a particular genotype. Again, this could also be expressed as the probability that a member of the population has a particular genotype.

For homozygotes, genotype frequency or probability can be determined by multiplying the allele frequencies. (This follows the rule of multiplication. The probability of receiving a "9" allele from one parent is not affected by the allele contributed by the other parent.)

The frequency for the "9" allele in the U.S. Caucasian population is 0.15. What is the probability that an individual is homozygous for this allele?

$0.15 \times 0.15 = 0.0225$

Is STR1 helpful in determining the body's identity?

> *No. Although the homozygous 9,9 genotype has a low frequency, all three DNA samples have the same genotype.*

These DNA samples have two alleles at STR locus 2, 10, and 6. What is the genotype of

The body—*10,10*

Missing Person 1—*10,10*

Missing Person 2—*6,6*

Is STR2 helpful in determining the body's identity?

> *Yes, it proves that the body is not Missing Person 2. However, it does not prove that the body is Missing Person 1.*

The probability that a U.S. Caucasian has the "10" allele for STR2 is 0.13. What is the frequency of individuals that are homozygous for this allele?

$0.13 \times 0.13 = 0.0169$

STR1 and STR2 are inherited independently, so the probability of having a combination of genotypes is also determined by the multiplication rule. That is, the probability of a combination of genotypes equals the product of their separate probabilities.

What is the probability that an individual is homozygous both for the "10" allele at STR2 and the "9" allele at STR1?

$0.0169 \times 0.0225 = 0.00038025$

State the probability in a different way: In a population of 10,000 people, how many would be expected to be homozygous for both of these alleles?

3.8 out of 10,000

There are 3 alleles at the STR3 locus: 12, 9, and 4. What is the genotype of

The body—*12,9*

Missing Person 1—*12,9*

Missing Person 2—*12,4*

Is STR3 helpful in determining the body's identity?

> *Yes. Missing Person 2 is again excluded because he has an allele that the body does not have.*

The probability calculation for heterozygotes is 2 (frequency of allele *A* × frequency of allele *B*). Referring to a Punnett square for a heterozygote cross may help you understand why. There are two ways to produce a heterozygous offspring. Let's say the STR3 locus is on chromosome 1. The "12" allele could be on the maternal chromosome 1 and the "9" allele on the paternal chromosome 1, or vice versa. In other words, the genotype could be 12,9 or 9,12. The heterozygote appears twice in the Punnett square.

This calculation follows from the Hardy-Weinberg equation. The population databases used for forensic analysis are also shown by statistical tests to be in Hardy-Weinberg equilibrium.

What is the probability of having the alleles 12 and 9 at STR3? (That is, what is the genotypic frequency?)

> *2 (0.06 × 0.02) = 0.0024*

These DNA samples have two alleles at STR locus 4, 17, and 6. What is the genotype of

The body—*6,6*

Missing Person 1—*6,6*

Missing Person 2—*17,17*

What is the probability that an individual is homozygous for the "6" allele?

> *0.04 × 0.04 = 0.0016*

It appears from the comparison of alleles in Table 10.1 that the unidentified body is Missing Person 1. How sure can we be that the body is in fact Missing Person 1 and not some other, unknown, missing person? Calculate the probability of an individual having this genotype (including all 4 loci).

> *0.0225 × 0.0169 × 0.0024 × 0.0016 = 0.00000000146016*
> *(approximately 1.46×10^{-9})*

State the probability in a different way: In a population of 1,000,000,000 (one billion) how many people would be expected to match the body's genotype at these four loci? (In other words, what is the probability of getting a match by chance?)

> *1.46 in one billion*

The FBI uses 13 core STR loci, which enables investigators to show that the odds of two DNA samples sharing the same genotype by chance alone are vanishingly small.

EXERCISE 10.2

Application of DNA Profiling

Objective

After completing this exercise, you should be able to

1. Obtain "DNA profiles" and interpret and apply the results.

See the Instructor Notes at the end of the lab about formulating teams.

Now that you have learned the fundamentals of interpreting STRs and calculating probabilities, you will use these techniques to solve a mystery. Your instructor will divide you into investigative teams. Use the background information from the newspaper articles to develop hypotheses about the situation. Your instructor will give you DNA samples to analyze to test your hypotheses. When everyone has finished, each team will present evidence in support of its conclusions.

Procedure

Who fears the revelation of dark secrets from the past? Who sent the anonymous letters? Who is Mr. X and how did he end up in a coma? From the information given in the news articles, formulate hypotheses about the people and events described.

Students should look for clues in the articles. If they are stuck as to what they should hypothesize about, the questions below are more specific. Hypotheses will vary.

How will DNA profiles help you test these hypotheses?

Answers depend on the hypotheses made. For example, students may say that the DNA on the knife and cigarette ends can test a hypothesis about whom Mr. X fought with. DNA samples from the anonymous letters can test hypotheses about their sender, and so on.

The DNA samples are printed in the Prep Guide for you to photocopy. At this point, you may decide that students don't need to actually cut out the alleles. They can count the number of repetitions without cutting.

Your instructor will give you "DNA samples" from a variety of sources. Analyze them and record your results in Tables 10.2 and 10.3. Then answer the questions below, using the allele frequencies given in Tables 10.4 and 10.5 to calculate probabilities to support your conclusions.

On the basis of structural characteristics, detectives decided that the cat hair in the letters and on Mr. X came from one of three breeds: Persian, Turkish Angora, and Birman. To calculate probability, you can assume that police are correct in identifying these three as the only possible breeds.

Grandview Gazette

The Unofficial Word from Grandview Arts and Technical College

Assault Victim Remains in Coma

A man who was found unconscious in the cemetery at the edge of campus remains unidentified and in a coma at this time. Police continue to seek witnesses who may have seen or spoken to this man before the assault. Mr. X, as the hospital staff call him, is a Caucasian possibly in his late 20s or early 30s, 5'10" tall and weighing about 170 pounds. He has brown hair and eyes and a full beard. Because of facial injuries he apparently suffered in a fight, police don't believe that releasing a current photo

would produce any leads. In their investigation to date, the only local connection police have found was a newspaper clipping in the man's pocket describing the drug patent obtained by GATC Professor Jack Watkins a few years ago. Dr. Watkins was unable to identify him. No identification was found in Mr. X's wallet, which contained the stub of a bus ticket from Saskatoon, Canada, a faded photograph of a young woman, a receipt from a private detective agency, and a small amount of money. The

Royal Canadian Mounted Police in Saskatchewan have been contacted and inquiries are under way.

The circumstances of Mr. X's injuries are as mysterious as his identity. Since several cigarette ends were found at the scene, police believe that Mr. X and his assailant had a conversation before the fight. They have concluded that a knife found beneath Mr. X belonged to him, since only his fingerprints were on it. There was fresh blood on the knife blade, but no cuts or

stab wounds were found on Mr. X's body. Police further theorize that Mr. X's severe head injury, which resulted in coma, happened when he fell and hit his head on a rock during the scuffle.

One promising lead is the cat hair that was found on his clothing. So far, police have determined that the hair is from a long-haired breed and have identified some possibilities to test. If the hair proves to be from an unusual breed, police may be able to trace the man's identity through the breed registry.

GATC Faculty Targeted by Letters

A STORM OF MYSTERIOUS anonymous letters has blown into the sleepy town of Grandview. For the past several months, rumors have circulated that two well-known campus figures have been the targets of sinister letter-writing campaigns. For the first time, police have publicly acknowledged that the recipients are popular GATC biology professor Jack Watkins and Diane Creek, GATC's Dean of Students.

Dean Creek's administrative assistant alerted police to the first suspicious letter, which arrived while the Dean was out of town. She has received two subsequent letters, one at her campus office and one at her home address. Police have refused to comment on the content of the letters. When the *Gazette* attempted to ask her about them, she would only say "The matter is in police hands. I have no comment." Anonymous sources in the GATC administrative offices, however, have revealed to the *Gazette* that Dean Creek is extremely upset that the police are involved. The administrative assistant who gave the first letter to authorities was fired two days later, allegedly for chronic tardiness. Dean Creek also refused to comment about the firing, citing confidentiality of personnel issues and her former assistant's pending grievance.

Dr. Watkins was willing to talk about the letters he received, but his description of the contents sheds little light on the subject. "Cryptic," he declared, "totally mystifying to me. I wouldn't have bothered the police with them if I hadn't heard rumors about other anonymous letters." However, the police are not yet sure whether the same person is responsible for all of the letters.

The letters themselves have offered few clues to their origin. According to police sources, they were all printed by computer on plain paper and mailed in ordinary, inexpensive envelopes. They all had different postmarks, leading police to hypothesize that they were sent through a mail-forwarding service. None of the letters bears fingerprints other than those of people who handled the letters locally. However, there is some forensic evidence that will be useful when a suspect is identified. The state crime lab may be able to obtain DNA from saliva on the envelope seals and stamps. In addition, one of the letters to Dr. Creek had a few strands of human hair stuck to the envelope where its glue had gotten damp. Since this letter was received at her home, testing is under way to exclude the recipient and her family as sources of the hair. Some of the letters also contained cat hairs.

GATC People in the News

More Honors for Professor Watkins

Dr. Jack Watkins, Professor of Biology, has been named a Grandview Distinguished Professor, the highest accolade that Grandview Arts and Technical College offers. The award recognizes faculty for outstanding scholarly contributions to the College.

Dr. Watkins's career at GATC has indeed been illustrious. He has received three teaching awards. In 1999 he earned a patent—and the thanks of a grateful nation!—for a drug that

cures the common cold. With characteristic generosity, he has donated some of his patent income to the scholarship fund here at GATC. His alma mater, Great Plains University in Colorado, where he also spent several years as a faculty member, has also benefited from his philanthropy. A lifelong bachelor dedicated to his work, Dr. Watkins carried out his groundbreaking research in a home laboratory

(continued on p. 2).

Creek for President?

As GATC searches for a new President, it may find the best candidate in its own backyard. Dr. Diane Creek, who has served GATC as Dean of Students for the last six years, is a finalist for the position. A graduate of Granite College and Midwestern University, Dean Creek held faculty and administrative positions at Great Plains University, Saskatchewan Provincial College, and Southwestern State University before coming to GATC.

To find out more about our Dean of Students and possible President-to-be, the *Gazette* interviewed Dean Creek recently. We met in

her campus office, which she has decorated with unique homey touches. Pictures of her husband Jay, who owns a chain of pet stores, son David, a member of the Grandview police department, daughter Mary, a veterinarian, and son Alex, who works with his father, adorn her desk. Given her family's occupations, perhaps it is not surprising that there is also a photo gallery of cats. Dr. Creek explained that she breeds and shows Persian cats as a hobby. She pointed out Cookie, a silky silver cat with startlingly luminous copper eyes whose official name is Best Cook's Star Rainbow Flower

(continued on p. 3).

Table 10.2

Analysis of Cat DNA. For each STR locus, enter the genotype of each DNA sample.

STR Locus	Cat Hair on Mr. X	Cat Hair in Letter
1	7,7	7,7
2	5,5	5,5
3	23,23	23,23
4	2,2	2,2
5	15,8	15,8

Table 10.3

Analysis of Human DNA Samples. For each STR locus, enter the genotype of each DNA sample.

STR Locus	Mr. X	Blood on Knife	Blood on Rock	Cig. End 1	Cig. End 2	Saliva on Creek Letter	Saliva on Watkins Letter	Hair on Creek Letter	Dean Creek	Mr. Creek	David Creek	Alex Creek	Mary Creek	Prof. Watkins
1	16,16	16,16	16,16	16,16	16,16	16,16	16,16	20,16	20,16	20,16	16,16	20,16	20,20	16,16
2	15,6	15,6	15,6	15,6	15,6	15,6	15,6	15,15	15,15	15,10	15,6	15,15	15,10	10,6
3	12,12	12,12	12,12	12,12	12,12	12,12	12,12	12,12	12,4	12,12	12,12	12,12	12,12	12,9
4	17,7	17,7	17,7	17,7	17,7	17,7	17,7	17,11	17,11	24,17	17,7	17,11	24,17	24,7

Table 10.4
Allele Frequencies for Cat STR Loci

STR Locus	Allele	Persian	Turkish Angora	Birman
1	7	0.02	0.07	0.60
	3	0.38	0.15	0.04
2	10	0.06	0.21	0.22
	5	0.13	0.09	0.00
3	23	0.00	0.11	0.35
	13	0.17	0.13	0.03
	3	0.24	0.01	0.07
4	19	0.29	0.04	0.10
	13	0.11	0.03	0.21
	9	0.03	0.00	0.19
	2	0.13	0.27	0.08
5	15	0.04	0.17	0.16
	11	0.19	0.05	0.09
	8	0.06	0.20	0.16

Table 10.5
Allele Frequencies for Human STR Loci

STR Locus	Allele	Frequency in Caucasians*
1	20	0.24
1	16	0.07
2	15	0.03
2	10	0.17
2	6	0.10
3	12	0.12
3	9	0.02
3	4	0.21
4	24	0.13
4	17	0.11
4	11	0.09
4	7	0.04

* Assume that all characters are Caucasian.

Answer each of the following questions and **explain all of the evidence that supports your conclusion.** Where two DNA samples match, calculate the probability that the match occurred by chance.

What kind of cat did the cat hairs in the two samples come from?

> *They came from a Turkish Angora cat. Allele "5" for the STR2 locus is not found in Birman cats (that is, allele frequency is 0), so the cat could not be a Birman because both samples were homozygous for the "5" allele. Allele "23" for the STR3 locus eliminated the Persian breed in the same way.*

What is the probability of the combination of genotypes in the two cat hair samples occurring? That is, out of one billion cats of this breed, how many would you expect to share this genotype?

> *0.00000000238, or 2.38 per billion cats*
>
> *STR1 $0.07 \times 0.07 = 0.0049$*
> *STR2 $0.09 \times 0.09 = 0.0081$*
> *STR3 $0.11 \times 0.11 = 0.0121$*
> *STR4 $0.27 \times 0.27 = 0.0729$*
> *STR5 $2(0.17 \times 0.20) = 0.068$*

Whose (human) hair was on the letter sent to the Creek house?

> *The hair DNA matches that of Alex Creek. The probability of this combination of genotypes is 0.0000000117, or 1.17 in 100,000,000.*
>
> *Frequency of STR1 20,16 is $2(0.24 \times 0.19) = 0.0456$*
> *Frequency of STR2 15,15 is $0.03 \times 0.03 = 0.0009$*
> *Frequency of STR3 12,12 is $0.12 \times 0.12 = 0.0144$*
> *Frequency of STR4 17,11 is $2(0.11 \times 0.09) = 0.0198$*

Who wrote the anonymous letters?

> *The DNA samples from the saliva on the letters matches both Mr. X and David Creek. The probability of this combination of genotypes is 0.00000000373, or 3.73 in 1,000,000,000.*
>
> *Frequency of STR1 16,16 is $0.07 \times 0.07 = 0.0049$*
> *Frequency of STR2 15,6 is $2(0.03 \times 0.10) = 0.006$*
> *Frequency of STR3 12,12 is $0.12 \times 0.12 = 0.0144$*
> *Frequency of STR4 17,7 is $2(0.11 \times 0.04) = 0.0088$*
>
> *In addition, the cat hair found on Mr. X matches the cat hair found inside one of the letters. We can surmise that Mr. X owns a Turkish Angora cat that is the source of both samples. This evidence points to Mr. X as the author of the anonymous letters.*

Who assaulted Mr. X?

> *The blood on the rock and the saliva on the cigarette ends match Mr. X (only a 0.0000000373 frequency for this genotypic combination), so that evidence is not helpful. The best evidence comes from the blood on the knife. Since no cuts were found on Mr. X, the blood must belong to his assailant (whom Mr. X wounded during the struggle). Strangely, the DNA from that blood matches Mr. X (again, a frequency of 0.0000000373 for this genotypic combination). It also matches David Creek, despite the fact that only 3.73 out of 1,000,000,000 people would be expected to share this genotype. Either the genotypic similarity is an astounding coincidence, or Mr. X and David Creek are actually genetically identical—identical twins.*

Who is Mr. X?

> *David Creek's identical twin, which also makes him Dean Creek's son.*

What does Dean Creek have to hide?

> *Mr. Creek is not David's father. For STR2, David has a "6" allele that he could not have gotten from either Mr. Creek or Dr. Creek. He also has a "7" allele at STR4 that did not come from either Mr. Creek or Dr. Creek. Since David Creek and Mr. X are identical twins, Dr. Creek is Mr. X's mother.*

What does Professor Watkins have to do with the case?

> *He could be David (and Mr. X's) biological father. He has the "6" allele for STR2 and the "7" allele for STR4. The information in the news stories raises the possibility that Watkins was a student at Great Plains University when Dr. Creek was on the faculty there.*

Questions for Review

1. The frequencies of many known human alleles are very high compared to those of the STR alleles. For example, the frequency of the *i* allele in the ABO blood-typing system is 0.67 and the frequency of the *A* allele is 0.26. Why are the allele frequencies so low for the STR alleles?

   > *These sites are highly variable—there are numerous alleles in the population. That's what makes them so useful.*

2. Explain how the following are used in STR analysis:

Restriction enzymes—*Restriction enzymes are used to "cut" the STR (the repeating segment of DNA) out of the long DNA strand that makes up the chromosome.*

Electrophoresis—*Electrophoresis is a technique that separates DNA fragments on the basis of size.*

3. In matching DNA found at a crime scene to a suspect's DNA, why isn't it enough to show that the DNA samples have the same alleles? Why must statistical analysis be used?

When two DNA samples have the same alleles at a locus, it could just be coincidence. Even when two DNA samples have the same alleles at 10 loci, it could just be a coincidence. Statistics are used to show just how big a coincidence that would have to be: Is this genotype shared by 1 in 100 people, or 1 in a million?

LAB TOPIC 10

Forensic Application of Molecular Genetics
Instructor Notes

Organization

Introduction

Before beginning this lab topic, be sure that students understand the following terms: nucleotide, locus, allele, genotype, homozygote, and heterozygote.

It is best to lead the class through Exercise 10.1 together. Begin with a brief introduction to STRs and why they are useful in forensics. If you watch the newspapers for a couple of weeks prior to teaching the lab, you are likely to find a current (possibly local) case that involves DNA profiling that you can use in your introduction.

Prep for Procedures

Exercise 10.1, Activity A

This activity uses paper models of DNA. Be sure to relate the procedures that students use to the actual techniques of restriction digests and electrophoresis.

The nucleotide sequences are printed at the end of this lab. Photocopy them and give each pair of students a set of all four sequences.

Exercise 10.1, Activity B

Review the use of Punnett squares to predict the outcome of a genetic cross. Emphasize that each allele an offspring has must have come from one of its parents. That's the basis of paternity testing.

Exercise 10.1, Activity C

Explain the role of statistics in DNA profiling. Students should be able to follow through the rest of the activity on their own, but if they are struggling you may want to lead the class through the calculations to expedite matters.

Exercise 10.2

Divide the students into teams of investigators. The number of teams depends on your students and the time available. If students have struggled with Exercise 10.1 or if you don't have much time remaining, make the teams large so that the work will get done quickly. If students will be able to handle the exercise without much input from you, make the teams smaller. Smaller teams make the activity more fun, since each student will have a larger piece of the puzzle. Smaller teams also facilitate better communication among team members.

Each team of investigators will need a complete set of DNA samples (16 total). Masters for photocopying are in the Prep Guide.

Summary (20 minutes)

As a summary, have teams of investigators present their conclusions. They should also explain the evidence that supports their conclusions. If everyone has reached the same conclusion, avoid repetition by asking each team to address a different question.

For a Two-Hour Lab

You can shorten the lab by not having the students do probability calculations. This would eliminate Exercise 10.1, Activity C and some of the work involved in Exercise 10.2. However, the probability calculations are very important to understanding how DNA evidence can prove a case. You should demonstrate a few calculations for the students so they can see how the chances of two DNA samples being an exact match become smaller and smaller as more and more loci are considered.

Alternatively, after doing the activities in Exercise 10.1 during the lab period, you could have students perform Exercise 10.2 as a homework assignment. Since there are more than a dozen DNA samples to process and some of them are repetitive, you should assign teams of students to work together on the exercise. If available, e-mail or chat would be a good way for students to share their results and work out the solution outside of class time.

Nucleotide sequences for 10.1, Activity A

Photocopy for each student

Person A

......CGACTATGCCGCGCCGCGCCACTATCG............

......CGACTATGCCGCGCCGCGCCGCGCCGCGCCGCGCCGCGCCACTATCG............

Person B

......CGACTATGCCGCGCCGCGCCGCGCCACTATCG............

......CGACTATGCCGCGCCGCGCCGCGCCGCGCCGCGCCGCGCCGCGCCACTATCG............

Bacteria and Disease

Introduction

Members of the two prokaryotic domains, Bacteria and Archea, are not visible to the unaided eye. The organization of their cells is simpler than that of the much larger eukaryotic cells. However, prokaryotes display a wide variety of metabolic activities. Some are autotrophic. That is, they can synthesize their own organic compounds using energy from light (photosynthetic) or inorganic chemicals (chemosynthetic). Heterotrophic prokaryotes acquire their energy by feeding on organic molecules. Cellular respiration is also diverse—there are at least a dozen types of fermentation pathways. Because of this diversity, prokaryotes are found in many different habitats. For example, some are important members of soil and aquatic communities and some inhabit the skin and bodies of animals. Despite the fact that the vast majority of prokaryotes cause no harm to humans and some even perform essential functions, most of our experience is with those that cause disease. About half of all human diseases are caused by members of the domain Bacteria, including strep throat, food poisoning, leprosy, tuberculosis, gonorrhea, and even ulcers, to name just a few. In this lab topic you will learn about some aspects of bacteria, especially pathogens (agents of disease).

Outline

Exercise 11.1: General Features of Bacteria

    Activity A: Bacterial Shapes

    Activity B: Colonies

Exercise 11.2: Investigating Microorganisms and Disease

    Activity A: Gram Stain

    Activity B: Antibiotic Sensitivity Test

    Activity C: Antibiotic Resistance

    Activity D: Transmission

Instructor Overview
In this lab topic, students get a general introduction to bacteria through a review of structural differences between prokaryotes and eukaryotes. Students view slides that show the coccus, bacillus, and spirillum forms of bacteria and a bacterial capsule. Next, students observe the diversity of bacterial colonies obtained by exposing nutrient agar to various sources of inoculum. (If the class is small enough, students can expose the plates themselves the week before this lab is performed.) In Exercise 11.2, students perform a Gram stain on bacteria from the plaque on their own teeth. They observe the results of an antibiotic sensitivity test and relate the results to the Gram-staining characteristics of the bacteria. Then they look at the growing problem of antibiotic resistance. In the final procedure, the whole class participates in a simulation of disease transmission. Class members exchange "bodily fluids" with each other, determine who has been "infected," and trace the transmission route of the "infection." Students discuss how the results relate to the spread of actual diseases such as AIDS.

Time Requirements
Exercise 11.1:
    Activity A: 20 minutes
    Activity B: 20–30 minutes
Exercise 11.2:
    Activity A: 30 minutes
    Activity B: 15–20 minutes
    Activity C: 20 minutes
    Activity D: 30–40 minutes

E X E R C I S E   1 1 . 1
General Features of Bacteria

Objectives

After completing this exercise, you should be able to

1. Describe how prokaryotic cells differ from eukaryotic cells.
2. Draw and name the three major morphological types of bacteria.
3. Explain how bacterial colonies arise and how they may be useful in classification.
4. Explain what the exposure test tells us about the presence of bacteria in the environment.

If available, use electron micrographs of bacteria and eukaryotic cells to contrast their size and complexity.

Members of the Domain Bacteria are prokaryotic. All other organisms are eukaryotic, including the protistans and fungi that are also considered microorganisms. **Prokaryotes,** as you may recall from studying cells, are much smaller than eukaryotes and lack their complex organization. Chiefly, they do not have the specialized membrane-bound organelles such as mitochondria or chloroplasts that are found in eukaryotes. Prokaryotes do have ribosomes, which function in protein synthesis. Although they have the same function as eukaryotic ribosomes, the ribosomes of prokaryotes are somewhat different in composition. These differences are useful for developing antibacterial drugs: Prokaryotic ribosomes can be targeted while eukaryotic ribosomes are unharmed.

Most bacteria have a cell wall. Many are also surrounded by some sort of capsule of gelatinous material, which may play a role in pathogenicity. For example, in the pneumonia-causing bacterium, forms that have a capsule have the ability to cause the disease, while those without a capsule do not.

In using drugs to combat pathogens, scientists want to select chemicals that interfere with some process or part of the bacterial cell without harming the host (human) cells. For example, if a drug that inhibits glycolysis were administered to a patient, it would damage his or her own cell function as well as killing the bacteria. Considering the differences between bacteria and animal cells discussed above, what aspects of bacterial structure might be targeted?

Several possibilities are mentioned: ribosomes (because bacterial ribosomes are different from eukaryotic ribosomes), the cell wall (because animals do not have a cell wall), or the capsule (important to virulence of some bacteria).

Activity A: Bacterial Shapes

Most bacteria have one of the three shapes illustrated in Figure 11.1: coccus (spherical), bacillus (rod), or spirillum (spiral). The characteristic shapes help microbiologists classify and identify bacteria.

a. ×6,300 **b.** ×4,200 **c.** ×8,750

Figure 11.1.
Bacterial shapes. In addition to the coccus (a), bacillus (b), or spirilla (c) shape of an individual bacterium, aggregations of bacteria in clusters or strings may provide an additional characteristic for classification.

In the following procedure you will look at prepared slides in order to distinguish the three bacterial shapes. You will also view a slide that is stained to show the capsule of a bacterium.

Procedure

1. Locate the bacteria first at low power and focus.
2. Change the objective to high power and center several bacteria in the field of view. You may need to adjust the fine focus slightly, but don't adjust the coarse-focus knob.
3. If your microscope is equipped with an oil immersion lens, turn the nosepiece so that no objective is directly over the slide. Place a small drop of immersion oil on the area of the slide that you were viewing. Then turn the oil objective into place and focus. Again, you may need to adjust the fine focus slightly.

Students generally need assistance to find cocci and bacilli at high power on typical student-model microscopes (spirilla are larger and easier to locate). If student scopes don't have oil immersion lenses, try to obtain one or two scopes that do have them for your demonstrations.

4. Check with your instructor to make sure you are looking at the right thing, and then draw the bacteria in the space below.

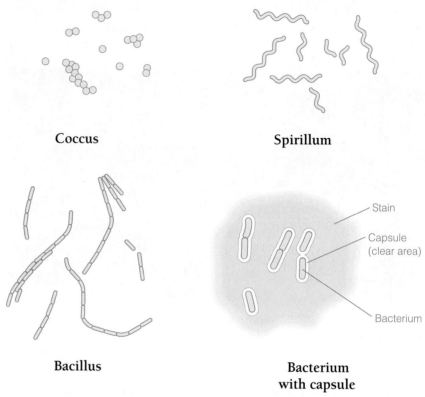

Coccus   **Spirillum**

Bacillus   **Bacterium
                                with capsule**

Can you see any internal structures in the bacteria? Explain.

> *No, that's the point: bacteria are very small, so shape is about all you can see with a light microscope. For example, Escherichia coli is ~1μ in diameter, while a rat liver cell may be 20μ. The volume of most eukaryotic cells is 1,000–10,000 times larger than that of prokaryotes.*

If you were looking at a eukaryotic cell at high power on your microscope, what structures might you see?

> *In viewing slides in previous labs, students probably have seen the nucleus in animal cells and chloroplasts, vacuoles, and cell walls in plant cells.*

 If you used an oil immersion lens, be sure to clean the oil off the lens and the slide with a clean piece of lens paper.

Activity B: Colonies

Bacterial form clearly does not provide much information for classification. Much higher magnification would have to be used to see any characteristic other than shape, though sometimes aggregations (the ways that cocci cluster together) can be informative. One way to "magnify" bacteria is to look at many millions of bacteria together rather than looking at a single cell.

Bacteria may divide as often as once every 10 minutes, so a single bacterium that finds itself in a hospitable environment can give rise to a population of millions in 24 hours. While the individual bacterium is not easily seen, the **colony**, or the population of bacteria derived from one or a few cells, is visible to the unaided eye. The appearance of the colony, such as its color and shape, may be useful in describing the species of bacteria.

Bacteria are found everywhere. Last week, petri dishes of agar, a gel nutrient medium, were exposed to various sources of microorganisms such as the sole of a shoe and a dog's paw. After exposure, the dishes were kept in a favorable environment, with the result that each dish now contains an array of colonies that arose from the bacteria present in the various sources. Fungal spores, single cells that are capable of germinating and growing into colonies of fungus, are also found everywhere, so the petri dishes also contain fungal colonies. In general, the colonies that have a fuzzy appearance are fungi, while the rest are bacteria.

In this exercise, you will compare the diversity of microorganisms from various sources.

If bacteria and fungal spores are found everywhere, why don't we see colonies of bacteria and fungi everywhere?

> *Most sites do not offer the nutrition and environmental conditions that would be conducive to microbial growth. However, microorganisms are present and will quickly exploit any environmental change that favors growth.*

For example, healthy bronchii are a poor site for bacterial growth and reproduction, but excess mucus production resulting from an irritation such as cigarette smoke provides an excellent medium for bacterial growth.

Procedure

1. Get a petri dish from your instructor and count the number of different types of colonies present. Use a dissecting microscope to examine the colonies more closely. Figure 11.2 should help you distinguish how many different types there are.

If your class is small enough that the prep is feasible, have students take plates home the week before this lab and make their own exposure tests. Explain how to inoculate the plates. Airborne bacteria and spores can be sampled by leaving the plate open on a counter, for example. An object, for example a toothbrush or a dog's nose, can be touched directly to the agar. Or a moistened Q-tip can be used to swab a surface, for example a drinking fountain, then streaked on the agar.

 Do not open the petri dishes. They could be harboring allergenic or pathogenic microorganisms.

Figure 11.2.
Characteristics used to differentiate bacterial colonies.

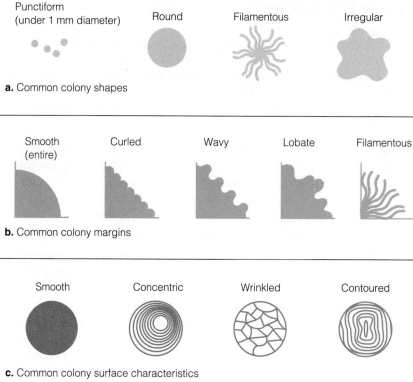

a. Common colony shapes

b. Common colony margins

c. Common colony surface characteristics

2. Record your data in Table 11.1.

3. When you have finished with one dish, trade with another lab team so you can examine dishes from a variety of sources. Record your data in Table 11.1.

4. Compare the diversity of microorganisms found in different sources.

Which source shows the greatest diversity? Which has the least diversity?

The answers to these questions depend on the sources used. In general, sources that are exposed to many different places (such as the sole of a shoe) will develop a greater diversity on the plates than sources whose exposure is more restricted (such as a toothbrush).

Table 11.1
Diversity of Bacterial Colonies from Different Sources

Source	Number of Different Types of Colonies

Which sources have the most similar types of bacteria? Can you explain why?

> *Sources that are related in some way will share some of the same microorganisms. For example, the keys and money carried by an individual would be expected to have some common microorganisms. The paws of different dogs might be expected to yield common soil microorganisms.*

Can you identify a type that appears in all the petri dishes?

EXERCISE 11.2
Investigating Microorganisms and Disease

Objectives

After completing this exercise, you should be able to

1. Explain what the Gram stain is and why it is useful.
2. Define antibiotic and give examples of how antibiotics work.
3. Explain how the antibiotic sensitivity test works.
4. Explain how the antibiotic sensitivity test could be used to determine the effectiveness of a drug.
5. Explain what antibiotic resistance is, how it arises and spreads, and why it causes serious concern.
6. Explain how diseases can be transmitted, and how to trace an epidemic to its source.

Activity A: Gram Stain

There are two major types of bacterial cell walls that can be differentiated by a procedure known as the Gram stain (named for the person who invented it). This is one of the most useful procedures in bacterial identification. Gram-positive bacteria tend to be more sensitive to most antibiotics than gram-negative bacteria, though there are major exceptions. Some antibiotics, however, work better on gram-negative bacteria. Broad-spectrum antibiotics work on both gram-positive and gram-negative bacteria.

In Activity A, you will perform a Gram stain. In Activity B, you will look at the effect of various antibiotics on cultures of gram-positive and gram-negative bacteria.

 Wear safety glasses while you are performing the following procedure.

Procedure

1. Take a clean slide and make a small circle in the center of it with a grease pencil. Then turn the slide over so the unmarked side is up.

2. Place a very small drop of water in the circle. With the flat end of the toothpick, scrape some tartar from your teeth (your back teeth will probably have the most bacteria). Stir it into the drop of water and spread it over the circled area.

   Dispose of the toothpick in the container provided.

3. Let the slide air dry.

4. Pick up the slide with a clothespin and heat fix it by passing it rapidly through the flame of an alcohol lamp several times. If you don't do this, your bacteria will be washed away at the next step, but you don't want to fry them, either!

   Be careful to keep everything but your slide away from the flame.

5. Place a small petri dish upside down in a large petri dish, which will hold the excess stain. Set the slide across the small petri dish and flood it with crystal violet stain. (See Figure 11.3a). Leave the stain on the slide for one minute.

6. Drain off the crystal violet stain and rinse the slide gently with water from the wash bottle (see Figure 11.3b).

Figure 11.3.
(a) Staining dish. The small petri dish is a platform for the slide; the large dish catches excess stain.
(b) Rinsing the slide. Use a gentle stream of water to rinse stain from the slide.

Microscope slide

Use wash bottle of water to rinse slide

Large petri dish   Small petri dish

a.   b.

7. Place several drops of Gram's iodine on the slide and let it sit for 1 minute.

8. Wash the iodine off the slide as in step 6.

9. Hold the slide at an angle and apply 95% alcohol drop by drop until the alcohol running off is clear rather than blue. This should take 10–30 seconds—don't overdo it!

10. Rinse the slide briefly with water. At this point the gram-positive bacteria are stained violet or blue and the gram-negative bacteria are colorless. The next step, called a counterstain, allows you to see the gram-negative bacteria.

11. Flood the slide with safranin, the counterstain, for 30 seconds.

12. Rinse the slide gently with water. Hold the slide upright to let as much water as possible drain off; then let it air dry before you look at it. No coverslip is used on these slides. Gram-positive bacteria will be violet or blue; gram-negative will be red.

13. Draw the bacteria you see in Table 11.2. Record the Gram reaction and structural type for each. A species of *Streptococcus* (gram-positive cocci in chains) is the cause of tooth decay. Did you find any in your plaque?

Students are likely to find squamous epithelial cells on their slides; these provide an excellent contrast to the bacteria, once students understand what they are. If the microscopes' optics are poor, students may have trouble distinguishing between red- and purple-stained bacteria.

Table 11.2
Bacteria in Plaque

Bacteria (sketch)	Gram Reaction (positive or negative)	Type
° ° ° ° ° °	+	*Cocci*
••• •••	–	*Cocci (chains)*
` ` ` ´ ` `	–	*Bacillus*

 When you have completed the exercise, dispose of the excess stain and your slide as directed by your instructor.

Activity B: Antibiotic Sensitivity Test

The human immune system is very effective at checking the growth of pathogens that invade the body. However, it has limitations. A pathogen that does manage to reproduce rapidly in the body may overwhelm the immune system's ability to respond, resulting in serious illness or death. Antibiotics, drugs that either kill or inhibit the growth of bacteria, assist the immune system. Once antibiotics bring the population of bacteria under control, the immune system can regain the upper hand and finish off the rest.

When you think about how many times you and your family have taken antibiotics, it is hard to conceive of a world without them. But before the development of penicillin in the mid-1940s, deadly diseases such as bubonic plague, cholera, and syphilis had no cure. Even strep throat was potentially lethal. For example, consider the bacterium that causes tuberculosis (TB). A disease known at least since 1000 BC, TB infects the lungs. Symptoms include a bloody cough, fever, and a gradual wasting away of the body that ends in death. Before antibiotics, TB claimed countless victims, including many famous ones such as the poet John Keats, authors Emily Brontë, Franz Kafka, and George Orwell, and Doc Holliday, legendary figure of America's Old West. The antibiotic streptomycin, introduced in the 1940s, proved to be the miracle cure for tuberculosis. Other effective antibiotics were introduced in the next decade.

As discussed in Exercise 11.1, antibiotics should interfere with pathogens without affecting the cells of the body. Thus, their action usually relies on targeting some aspect of cell structure or function that differs between prokaryote and eukaryotes.

Why would antibacterial agents usually not be effective against fungi?

> *Fungi are eukaryotic.*

The antibiotics most familiar to us are antibacterial agents, probably because we are more likely to have had bacterial than fungal diseases. Some of these antibiotics, including erythromycin, tetracyclines, and compounds related to streptomycin (gentamycin, kanamycin, and neomycin) work by binding to bacterial ribosomes. How does this affect bacteria?

> *Binding to ribosomes inhibits protein synthesis.*

Why would eukaryotic organisms not be affected by antibiotics?

> *The drugs are specific for prokaryotic ribosomes, which are different from eukaryotic ribosomes.*

Why would antibiotics not be effective against viruses?

> *Viruses are not cells and do not contain any of the components that are targeted by antibiotics.*

Many antibiotics in use today were originally derived from certain types of fungi and bacteria that live in the soil, such as *Penicillium* and *Streptomyces*.

Suggest a reason why *Streptomyces* and *Penicillium* might have evolved the ability to produce antibacterial chemicals.

> *To compete against fungi or bacteria in the soil; to deter fungal or bacterial pathogens themselves.*

In this exercise you will determine the effectiveness of different antibiotics in inhibiting the growth of bacteria. Two petri dishes have been prepared with cultures of bacteria, one with a gram-positive species (*Streptococcus aureus*) and one with a gram-negative species (*E. coli*). Small disks soaked in various antibiotics were placed on the dishes. Each disk is coded with the name of the antibiotic; note the key to the antibiotics next to the dishes.

Be sure the prep staff provides a key to the antibiotic code. The disks are too small to accommodate the entire name.

If the bacteria growing on the plate are sensitive to a given antibiotic, their growth will be inhibited and no bacteria will be observed growing near that disk. That is, the antibiotic will cause a **zone of inhibition** as shown in Figure 11.4.

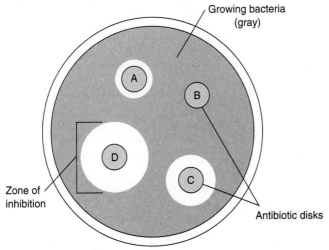

Figure 11.4.
Antibiotic sensitivity test. In this example, antibiotic D is most effective and antibiotic B is least effective against these bacteria.

Procedure

1. Measure in millimeters the diameter of the zone of inhibition for each antibiotic.

 Do not open the petri dishes because they may have been contaminated by pathogenic or allergenic microorganisms.

2. Record the results in Table 11.3.

Table 11.3
Results of Antibiotic Sensitivity Test

Antibiotic	Zone of Inhibition (mm)	
	Gram-Negative Bacteria	**Gram-Positive Bacteria**

Do gram-positive and gram-negative bacteria react similarly to the same antibiotic? Explain the results you have recorded above.

Results depend on the antibiotics used. Penicillin is mainly effective against gram-positive species, but there are such major exceptions as Treponema pallidum (which causes syphilis) and Neisseria gonorrhoeae (which causes gonorrhea). Both those organisms, plus species of gram-negative cocci, are sensitive to penicillin. Ampicillin is a chemically altered form of penicillin which is effective against a broader range of species; it will inhibit E. coli. The cephalosporins, which are structurally similar to penicillin and are also obtained from fungi (and which also act by inhibiting cell wall synthesis), are broad-spectrum antibiotics. Tetracyclines are effective against a broad spectrum of gram-positive and gram-negative bacteria. Kanamycin, gentamycin, erythromycin, and streptomycin are antibiotics derived from related species of filamentous bacteria. Streptomycin is primarily used against gram-negative bacteria, but erythromycin, kanamycin, and gentamycin are considered broad-spectrum drugs.

How might a medical lab use this test?

A bacterial sample (for example, from a throat swab) can be cultured on a plate with antibiotic disks. The results will reveal which antibiotic would be most effective against the infection.

Activity C: Antibiotic Resistance

As treatments became available for formerly incurable diseases, the development of antibiotics seemed almost miraculous. It even seemed possible that some pathogens could be wiped out entirely. For example, experts confidently predicted the total eradication of tuberculosis by 2025. Indeed, in industrialized countries where treatment was available, cases of TB became very rare. But in the 1980s, TB cases in the United States as well as around the world showed a marked increase. In 1993, the World Health Organization declared the disease, which kills between two and three million people each year, a global emergency. Why did the health care community swing so rapidly from optimism to alarm over antibiotic use?

Bacteria are organisms that are subject to natural selection by their environment—in this case, the human body. They can evolve. Changing their environment by introducing antibiotics into it creates an opportunity for evolutionary change. Even amid the euphoria of the new era of antibiotics, some microbiologists noticed warning signs of trouble ahead. They discovered that some bacteria harbored a natural resistance to penicillin. That is, the bacteria had some genetically determined mechanism that enabled them to escape the drug's effect.

Suggest a reason why some bacteria might naturally carry a gene that defeats an antibiotic. (Hint: Refer to Activity B. Why does the *Penicillium* fungus produce the antibiotic?)

> *If we hypothesize that the antibiotic was produced in the first place as a competitive weapon, then it is plausible that some bacteria have evolved a mechanism to counter it.*

It takes a large population of bacteria to produce disease. For example, the lungs of a person who has an active case of tuberculosis contain hundreds of millions of bacteria. Those bacteria are not genetically identical to each other, and as a result they also vary phenotypically. The phenotypic differences may include an ability to circumvent the effects of antibiotics. Some bacteria may produce an enzyme that degrades or deactivates an antibiotic. Other bacteria may have the ability to prevent a drug from entering their cells, or they may be able to disguise the drug's molecular target, or even to expel the antibiotic before it has a chance to work.

The different writing implements provided for this activity simulate the genetic and phenotypic differences in the bacterial population.

If you don't have time to include this activity, you could do a brief demonstration of the principle by using water-soluble (e.g., Vis-à-vis) and permanent (e.g., Sharpie) pens on an overhead transparency.

Procedure

1. Use the pens and pencils provided to fill the diagram of lungs below with dots or circles representing bacteria.

The first "antibiotic" is the kneaded eraser.

2. The erasers represent antibiotics. Your instructor will tell you which one to "administer" first. Rub very lightly at first over the entire area, then use increasing pressure. But don't bear down so hard that you tear the paper!

Describe the results, including which bacterial types have been completely erased, which are fainter but still present, and which are as bold as ever.

> *Results will depend on how hard students press when they make their bacteria, and how vigorously they rub. In general, the lightest pencil dots will be erased, other pencil dots will be erased to some degree, and the dots made with the pen will be unchanged.*

When you take an antibiotic, it begins killing some bacteria immediately. But some are harder to kill than others—it takes a higher concentration of the drug to do the job. When you don't finish your antibiotic prescription, you leave a population of the tougher bacteria unscathed. This gives the bacterial population a chance to resume rapid reproduction. As a result, your immune system may be overwhelmed again, causing a recurrence of your illness.

The pencil eraser is the second antibiotic.

3. Repeat step 2 using the second "antibiotic" designated by your instructor.

Describe the results, including which bacterial types have been completely erased, which are fainter but still present, and which are as bold as ever.

4. Repeat step 2 using the third "antibiotic" designated by your instructor.

Describe the results, including which bacterial types have been completely erased, which are fainter but still present, and which are as bold as ever.

> *This antibiotic is an ink eraser, which will clear away everything but the marker pen. You might compare this to vancomycin, which is currently the antibiotic of last resort. There are now vancomycin-resistant strains of some bacteria.*

5. Fill the lung diagram with "bacteria" that have survived all of the antibiotic treatments.

When this individual passes the disease on to someone else, which bacteria will the newly infected person acquire?

> *The type that is most resistant to antibiotics.*

In the process of natural selection, members of the population that are better suited to their environment are more likely to survive and reproduce. When the environment—in this case, the human body—changes, the phenotype that is best adapted to the environment changes. The resistant bacteria reproduce rapidly, so within a short period of time, the person is once again suffering from a full-blown infection.

Strains of the tuberculosis bacterium that are resistant to multiple antibiotics have emerged since the 1990s. These strains are more difficult and costly to combat (almost a billion dollars were spent controlling an outbreak in New York City in the early 1990s). But that is just one example of a pathogen that has developed antibiotic resistance. Others include bacteria that can cause pneumonia, toxic shock, gonorrhea, food poisoning, internal infections, and (parents, take note!) childhood ear infections. To make the situation even more alarming, the resistance genes are carried on plasmids, small circles of DNA that can be readily passed from one *species* of bacteria to another.

Activity D: Transmission

Exercise 11.1 showed that microorganisms are found in many different places, but that some are more likely to be found in one place than another. Pathogens are most likely to be found on or in the vicinity of their human hosts, where they can easily be transmitted to new hosts. For example, tuberculosis is highly contagious. From its breeding grounds in the lungs, it spreads by coughing, sneezing, spitting, or even talking. It only takes a small number of inhaled bacteria to start an infection. Other pathogens may be transmitted by direct contact (including sexual contact), by contaminated objects, or by excretory products and body fluids. Animal carriers are responsible for some disease transmission. In addition, certain diseases can be transmitted from mother to fetus in the uterus or during birth.

Many times when we are ill we have no idea how we contracted the infection. If there is a local outbreak of a disease, though, we are aware that something is "going around" and can often identify "whose" cold we got. For example, small epidemics on college campuses frequently get started at the beginning of the semester or after a vacation. A student returns to school with a cold he got from his little sister and passes it on to his roommate, who transmits it to six people in his biology lab, who take it back to their roommates, and so on.

Epidemics of some diseases are cause for serious concern, and health departments work to identify the transmission routes. This may involve determining common factors among the victims (for example, in an outbreak of food poisoning) or tracing all contacts of the victims (for example, in sexually transmitted diseases). In this exercise, you will simulate transmission of an infection and trace its course through the class.

If there is an odd number of students in the class, you should participate also.

The "infected" bottle contains 0.1N NaOH; all the others contain dilute HCl. The NaOH is strong enough that the solution will turn red with phenol red even if only one pipetful is added to ~5 mL of HCl. Dilute HCl is used to make sure that uninfected solutions turn yellow (tap water often tests red with phenol red). Students remove an aliquot of solution from the stock bottle to use for the exercise, so stock bottles should not become contaminated unless students do it deliberately. The test tubes can be rinsed and reused throughout the week of labs with no problem as long as they are not washed with detergent, which is alkaline and clings to the glass if the test tubes are not rinsed thoroughly. Pasteur pipets can also be reused. You should gather them all together in a handful and run water through them until they are thoroughly rinsed, and then leave them large-end-down to drain.

Procedure

1. Each student has a stock bottle of solution. One student's solution is different from all the rest, but the difference is not visible.

2. Transfer 3 pipetsful of solution from your stock bottle to a clean test tube, using a pasteur pipet. You can think of this as the aerosol droplets from a sneeze or as the exchange of bodily fluids in a sexual contact.

3. Find a person at random in the class and exchange a dropperful of solution with him or her. That is, you put a dropperful of your solution into that person's test tube, who then puts a dropperful of his/her solution into your test tube.

4. Record the name of your contact below.

5. When your instructor says to, repeat steps 3 and 4 twice more, so you have contacted three *different* people.

   List your contacts.

   First contact: _____

   Second contact: _____

   Third contact: _____

6. Put a dropperful of phenol red into your test tube and record the color: _____

   Phenol red is an acid/base indicator: Acidic solutions turn yellow, and basic solutions turn red when phenol red is added. The "infected" solution was basic, and all of the others started out acidic. The solutions in the test tubes of "infected" individuals will now be basic, or red.

There is a transparency master of Table 11.4 in the Prep Guide.

7. Record in Table 11.4 the names and contacts of infected individuals in the class.

8. Determine the transmission routes and answer the following questions.

What was the *maximum* number of people in your lab class who could be infected after each round of contacts?

First round: ___2___

Second round: ___4___

Third round: ___8___

Table 11.4
Record of Contacts of "Infected" Persons

Infected Person	Contacts		
	First	Second	Third
1	16	13	2
2	3	11	1
3	2	9	14
7	18	5	9
9	5	3	7
11	12	2	13
13	19	1	11
14	4	15	3

It is a good idea to work through an example like the following one with your students before asking them to figure out the transmission route from the in-class exercise: Only infected persons figure in the transmission route, so you can strike out all contacts who are not infected (16, 18, 5, 12, 19, 4, and 15). This shows that in round 1, only 2 and 3 were infected. One of them infected the other, and in the next round infected two more people: 2 infected 11 and 3 infected 9. In the third round, each of the infected individuals contacted a noninfected individual, giving us the maximum number of infected people at the end. We can determine who infected whom, with the exception of knowing the original carrier. That can only be determined by testing a sample from each of their stock bottles. How can there be six infected people in round 2 when the maximum is four? The table shows who contacted whom in each round. A second-round contact person was not necessarily infected at that time. In this example, 2, 3, 9, and 11 were infected at the end of the second round. In the third round, these four infected 1, 14, 7, and 13. Five was 7's second-round contact, but didn't become infected because 7 himself wasn't infected until the third round.

Explain your answers.

If you start with one infected person, after one round of contacts that person will have infected one other person. After a second round, each of those two people could have infected another, making a total of four infected. After a third round there could be eight. In other words, the maximum infection rate is that each round doubles the number of infected individuals.

After the three rounds of contacts, what would be your chances of finding an uninfected partner? How would your chances change after a fourth round?

The answers depend on the size of your lab section. The point to make is that the more extensive the "experience" of your potential partners, the more likely you are to encounter one who was infected by an earlier contact.

Why might you find that fewer than the maximum number were infected?

Individuals who are already infected can contact each other in the second or third round, so no new victim is added to the pool of carriers. This may happen if students stay in small groups rather than circulate freely around the room, and it illustrates what can happen in real situations.

How might the results have been different if you had been given 5 minutes and were told to make as many contacts as you wanted?

If the original carrier were "monogamous" and contacted only one partner, only two people would be infected at the end. If the infected person was a member of a small group of students (such as one lab table) who only contacted each other, the infection would be limited to that group. If the carrier were the friendly type who contacted people all over the room, the infection would be spread to a greater extent. While contacting more people is riskier than contacting fewer people or being monogamous, note that monogamy is no protection if your partner is already infected—and you can't tell whether he or she is or not.

In this lab exercise you could not avoid contacting an infected individual because you couldn't tell who was infected until it was too late. What are some actual diseases where this might be the case?

There are many, of course, but the most important point to make is that HIV carriers may give no indication of their infection. This is true of some other sexually transmitted diseases as well; women especially may be completely asymptomatic.

Questions for Review

1. Name several different structures that eukaryotes (for example, animal cells) have that prokaryotes (for example, bacteria) do not have.

   True nucleus and any membrane-enclosed organelles or components of the cytomembrane system (for example, mitochondria, lysosomes, Golgi bodies, endoplasmic reticulum).

2. How and under what conditions do bacterial colonies arise?

   Colonies arise from one or a few bacteria that find a habitat favorable for growth and reproduction.

3. Smoking irritates the cells lining the respiratory tract, causing them to produce more mucus than usual. How is this related to the fact that smokers have more respiratory infections than do nonsmokers?

   Mucus provides moisture and nutrients for bacterial growth.

4. What is an antibiotic?

   An antibiotic is a drug that kills bacteria and other microorganisms.

5. Give the most common examples of how antibiotics work.

   Antibiotics inhibit cell wall synthesis and bind to ribosomes.

6. What part of the bacterial cell is stained by the Gram stain? Why is this procedure useful?

   Cell wall. It can help determine what antibiotic to use (also used for classification).

7. When an antibiotic binds to bacterial ribosomes, how (specifically) does that affect the bacterium?

   It inhibits protein synthesis.

8. Explain how the antibiotic sensitivity test works.

   Filter paper disks impregnated with an antibiotic are placed on a dish of nutrient agar that has been inoculated with a single type of bacteria. In the exercise, students used a gram-positive species and a gram-negative species. The plates are then placed in a favorable environment so that the bacteria can grow. If an antibiotic is effective, there will be a clear area (no bacterial growth) around the disk containing that antibiotic.

9. How could the antibiotic sensitivity test be used to determine whether a particular drug is effective in treating a particular bacterial disease?

   A culture of the bacteria can be obtained. The bacteria can then be grown in contact with the drug to be tested; results will show whether the drug is effective in inhibiting bacterial growth.

10. Explain how the process of natural selection results in populations of bacteria that are not harmed by antibiotics.

    If some members of the bacterial population have a genetically determined mechanism to survive an antibiotic, then treating the patient with that drug kills the susceptible bacteria and leaves the resistant ones alive. The resistant ones can then reproduce, resulting in a large population of antibiotic-resistant bacteria.

11. List several ways that disease can be transmitted.

    Direct contact with infected person; contact with contaminated objects, excretory products, and body fluids; animal vectors; transmission from mother to fetus.

12. Using the exercise you did in lab as an example, explain how an epidemic can be traced to its source.

    In lab, students determined whether or not each individual was infected. Students then examined the contacts made by each infected person and were thus able to trace the course of the epidemic.

13. What is another strategy epidemiologists might use to determine the source of a disease?

Epidemiologists usually do not have the luxury of having so much information at once. By having individuals who have sexually transmitted diseases list their sexual contacts, then contacting those persons and having them list their contacts, and so on, investigators can at least identify and treat affected individuals.

Epidemiologists may also work by determining common factors among victims, including where they live, where or what food they have eaten, and where they obtain their drinking water. The job is easier if the cause of the disease is known.

LAB TOPIC 11

Bacteria and Disease
Instructor Notes

Organization

Introduction (10–15 minutes)

Review the differences between prokaryotic and eukaryotic organisms. If possible, use some visual display such as a PowerPoint presentation or 2 × 2 slides to illustrate the differences in size and complexity. Include slides that show students what they will see when they view bacteria on the compound microscope.

Explain how colonies of microorganisms are formed and what students will see on the plates.

Prep for Procedures (5 minutes)

- Students should work in pairs throughout the lab.
- Safety glasses should be worn when students are performing the Gram stain. Students should also be instructed on proper disposal of stains and used slides and should be cautioned to be careful when using the alcohol lamps.
- Caution students not to open the petri dishes used in Exercise 11.1, Activity B (Colonies). It is possible that a pathogenic bacterium was part of the inoculum. In addition, students may be allergic to fungal spores. The petri dishes for Exercise 11.2, Activity B (Antibiotic Sensitivity Test) should also remain sealed.
- Students should read the introduction to Exercise 11.2, Activity D (Transmission) before it is performed in class.

During the Lab Period

When students have finished the other procedures, have everyone participate in the transmission exercise (Exercise 11.2, Activity D).

The results should be discussed immediately after completing the procedure.

Summary (15 minutes)

- Review the morphological types of bacteria.
- Discuss students' answers to the questions on Exercise 11.1, Activity B (Colonies).
- Review results of the antibiotic sensitivity test (Exercise 11.2, Activity B).
- Review antibiotic resistance (Exercise 11.2, Activity C).

For a 2-Hour Lab

Eliminate having students perform the Gram staining procedure (Exercise 11.2, Activity A). Gram-negative and gram-positive bacteria can be used to illustrate the morphological types in Exercise 11.1, Activity A. Exercise 11.2, Activity C can be done as a demonstration.

Investigative Extensions

Students can easily extend the exposure test in Exercise 11.1 by selecting their own sources of inoculum. Have petri dishes of nutrient agar available for students to take home (see directions in the Prep Guide). Besides comparing the results from different sources, students can compare inoculations made before and after some cleaning procedure (for example, swab the drinking fountain before and after it is cleaned with Lysol; inoculate a plate with saliva before and after a mouthwash is used; inoculate using coins before and after they have been soaked in Clorox).

Students can also investigate the effects of environmental conditions on bacterial growth. Petri dishes can be subjected to different temperatures, light intensity, or photoperiod, for example. The agar can be made up with buffers to vary the pH. For these types of investigation, it is best to inoculate the plates from single-species cultures of bacteria (which can be obtained from Carolina Biological Supply).

Reference

Dickey, Jean. 1989. A quick and easy simulation of disease transmission. Amer. Biol. Teacher 51(6):364–365.

Plant Diversity

Introduction

The plant kingdom arose from aquatic algal ancestors more than 400 million years ago. A survey of the kingdom is a study in the evolutionary history of plants as they successfully made the transition from water to land. Figure 12.1, on the next page, shows the phylogenetic relationships among the four major plant groups you will study in this lab topic.

As we describe the divisions we will note a series of anatomical adaptations for living on land, including:

- Vascular tissue, which is specialized for transporting water and nutrients inside the plant.

- A root system, which anchors the plant and absorbs minerals and water from the soil.

- A cuticle and/or cork layer on above-ground parts, which retards water loss.

- The presence of pores in the above-ground parts that allow for gas exchange.

Most of this lab topic, however, will focus on the other major evolutionary trend: the ability of plants to reproduce on land. In the earliest plants, sperm were required to swim to the egg in order to achieve fertilization and thus needed to remain in or near water. For plant reproduction to be completely successful on land, some other means of bringing the gametes together was necessary. Pollen, which brings the sperm-producing cell to the egg-bearing female structure via air, solved this problem.

The evolution of the seed, which allowed plants to provide their offspring with a good start in the hostile terrestrial environment, was another critical development.

In this lab topic you will study four representative life cycles—moss, fern, pine, and flowering plant—and see how each group's reproductive adaptations contributed to its success. You will also have an opportunity to see living representatives of several other groups of plants.

Outline

Exercise 12.1: The Alternation of Generations Life Cycle

Exercise 12.2: Seedless Nonvascular Plants

Exercise 12.3: Seedless Vascular Plants

    Activity A: Club Mosses

    Activity B: Horsetails

    Activity C: Ferns

Instructor Overview
This lab topic illustrates one of the fundamental themes of plant evolution: the reproductive modifications that enabled plants to colonize terrestrial habitats. Students are first taught the alternation of generations life cycle pattern. Each life cycle that is studied (moss, fern, pine, and angiosperm) is illustrated by a display of living and preserved specimens that represents the life cycle components. The parts of the life cycles are consistently described with the same terms used in the textbook: sporophyte, sporangium, spores, gametophyte, gametangium, gametes, zygote, and embryo. Students identify the points at which mitosis, meiosis, and fertilization take place in each life cycle; they color the diagrams to show which structures are haploid and which are diploid. Emphasis is thus placed on the evolution of the alternation of generations pattern rather than on the structural details of specific life cycles.

   Improvements in anatomical structures that represent adaptations to the terrestrial habitat are noted but not described in detail. In addition to the plant divisions that are represented by life cycles, several other groups are illustrated by specimens in the lab, including club mosses and horsetails (seedless vascular plants whose ancestors were major components of the Carboniferous forests) and cycads and ginkgos (early seed plants).

Time Requirements
Exercise 12.1: 20 minutes
Exercise 12.2: 20 minutes
Exercise 12.3: 30 minutes
Exercise 12.4: 30 minutes
Exercise 12.5: 30 minutes

Exercise 12.4: Vascular Seed Plants—Gymnosperms
    Activity A: Cycads
    Activity B: Ginkgo
    Activity C: Conifers
Exercise 12.5: Vascular Seed Plants—Angiosperms

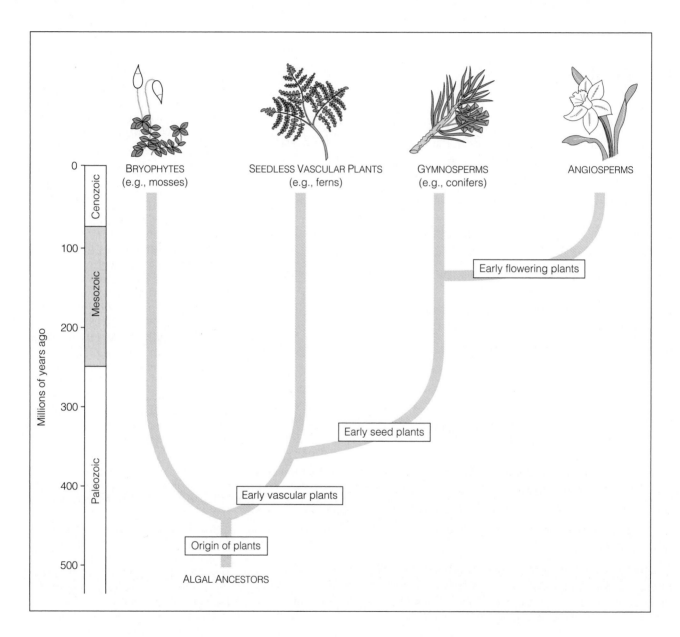

Figure 12.1.
Phylogenetic tree of major plant groups. The bryophytes were the earliest land plants and are represented in modern times by the mosses. The first vascular plants were seedless and gave rise to seed plants during the Paleozoic era, over 300 million years ago. The most recent addition to the plant kingdom is the angiosperms, or flowering plants.

EXERCISE 12.1
The Alternation of Generations Life Cycle

Objective

After completing this exercise, you should be able to

1. Explain the general pattern for the alternation of generations life cycle.

The process of sexual reproduction involves **meiosis**, production of haploid cells that contain only one set of chromosomes (**1n**). Meiosis is followed by **fertilization**, restoration of diploid cells, which have two sets of chromosomes (**2n**), by fusion of haploid cells. In animals such as ourselves the process is familiar and straightforward. Our diploid bodies have specialized cells that undergo meiosis to produce eggs or sperm, which are called gametes. The gametes join during fertilization to form the diploid zygote from which a new individual develops.

Plants, however, go through an **alternation of generations**, as shown in Figure 12.2. When a diploid plant undergoes meiosis, it produces haploid **spores**, not gametes. Spores are made in a specialized structure called a **sporangium.** (The suffix -angium is from a Greek word that means "container.") The spores undergo mitotic divisions to produce a plant that is entirely composed of haploid cells.

Cells in the **gametangium,** a specialized structure in the haploid plant, undergo mitosis to produce gametes. Male and female gametes (sperm and egg) must then be brought together during fertilization. Animal sperm, as you probably know, are motile: They swim through fluid to reach the egg. Some plant sperm are also motile, but if this is the case then the plant requires a film of water for reproduction. More advanced plants have evolved sperm that do not require water for transport, as you will see in this lab. In any case, when the sperm fertilizes the egg, a zygote is formed. The zygote grows by mitotic cell divisions into a diploid plant.

Because the diploid plant produces spores, it is called a **sporophyte.** The haploid plant, which produces gametes, is known as a **gametophyte.** In some groups of plants the sporophyte and gametophyte are both autotrophic, or capable of photosynthesis, and thus are able to live independently of each other. In other plants, one generation is dependent on the other for nutrition—that is, one generation is heterotrophic.

What characteristic of a plant indicates that it is autotrophic? (Hint: What is necessary for photosynthesis?)

A green color (chlorophyll).

The alternation of generations is shown in Figure 12.2 as a linear progression through time because each sporophyte gives rise to a gametophyte, which then gives rise to a *new* sporophyte. However, we could just as easily trace a life cycle from gametophyte to gametophyte, and in fact this is done when the gametophyte phase of the life cycle is dominant.

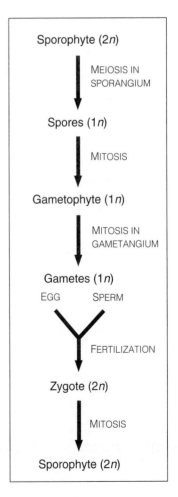

Figure 12.2.
Alternation of generations life cycle.

Diagrams of life cycles are often drawn as circles to emphasize that you can start at any point in the life cycle and follow it through. You should learn the alternation of generations pattern thoroughly so that you can understand a life cycle regardless of the format used to present it.

On Figure 12.3, fill in the blanks to indicate where meiosis, mitosis, and fertilization (events) occur and where the spores and zygote (structures) are found in the life cycle.

Figure 12.3.

Review of the structures and events in the alternation of generations life cycle.

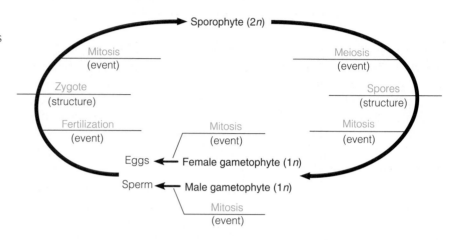

The displays are set up to illustrate as much of each life cycle as possible with living and preserved material. They should be organized to correspond with the diagrams; for example, the display of the moss life cycle begins with moss gametophytes. Following that are microscopes with slides of male and female gametangia (structures where gametes are produced). All the displays should be tagged with labels corresponding to the structures students are supposed to identify. Finally, arrows between the displays indicate where mitosis, meiosis, and fertilization occur. As far as possible, the diagrams themselves will be drawn from the actual specimens that students will see. The Prep Guide gives detailed instructions for setting up the life cycle displays.

In this laboratory you will view specimens representing the different stages of each life cycle. A diagram of these structures in the lab manual will help you fit the pieces together and see the entire life cycle. For each life cycle, you will be asked to identify events (meiosis, mitosis, and fertilization), whether the tissue is haploid or diploid, and certain structures, including the gametophyte, gametes, zygote, sporophyte, and spores.

You should approach each life cycle like a puzzle. First, locate something you recognize—for example, fertilization. Once you do that, you know what happens next in the general life cycle, and you can deduce the rest from that.

Keep in mind that:

- Only diploid cells can undergo meiosis.
- Meiosis occurs only once in the life cycle.
- Gametes are produced by mitosis.
- Fertilization is necessary to restore diploidy.

The diagrams in the lab manual show the life cycles as circular, indicating that the same events are repeated in each generation. It is more convenient in the laboratory, however, to map out each life cycle display in a linear manner. The beginning point is either a gametophyte or a sporophyte (whichever is dominant) and the end point is a *new* gametophyte or sporophyte.

EXERCISE 12.2
Seedless Nonvascular Plants

Objective

After completing this exercise, you should be able to

1. Identify in living materials or on a diagram all of the structures and events in the moss life cycle.

The **bryophytes** (Greek *bryon,* moss) are relatively small plants that live in habitats where water is available at least during part of the year, since water is necessary for reproduction. They have been called the amphibians of the plant kingdom because amphibians (frogs, toads, and salamanders) have a similar requirement for water. There are 24,000 species of bryophytes. They are most common in tropical and subtropical regions but are found in a wide variety of habitats, including deserts. Since they are extremely sensitive to air pollution, their absence from otherwise suitable habitats is an indicator of pollution.

The descriptions of plants given here refer to the gametophyte phase of the life cycle, since that is predominant. Most bryophytes lack **vascular tissue,** the specialized food- and water-conducting tissue found in the higher plants. They do not have true stems, leaves, or roots, which places severe limitations on their size. Nevertheless, botanists use the terms leaf and stem to refer to their leaflike and stemlike structures. The rootlike organs are called rhizoids (Greek *rhiza,* root; *oid,* like).

You are probably already familiar with mosses, which are leafy and upright. The bryophytes also include a group of plants called liverworts. (*Wort* is part of numerous plant names; it is derived from *wyrt,* an Old English word meaning "plant.") Some liverworts are leafy, but many are flat (sort of liver-shaped) and undifferentiated into leaves and stems, like the specimen in the lab. Another liverwort is shown on Color Plate 2.

Procedure

1. Examine the liverwort under the dissecting microscope and sketch it in the margin of your lab manual.

    Notice the pores in the liverwort's surface. What do you suppose is the function of the pores?

    Gas exchange.

    Having open pores is a problem for land plants, since water the plant needs is readily lost through the pores. Most plants have **stomata** instead. Stomata are special pores that can be closed off by the surrounding cells. Some bryophytes do have stomata, but they generally remain open all day so they are not really effective at preventing water loss. Higher plants open and close their stomata to balance carbon dioxide intake with water loss.

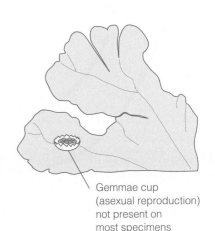

Gemmae cup
(asexual reproduction)
not present on
most specimens

2. Look at the moss life cycle in Figure 12.4, which represents the bryophytes. Label the arrows that indicate where mitosis, meiosis, and fertilization occur (mitosis occurs more than once).

3. Observe the following structures in the living material and label them on Figure 12.4:

Female gametophyte	Zygote
Male gametophyte	Embryo
Egg (female gamete)	Sporophyte
Sperm (male gamete)	Spores

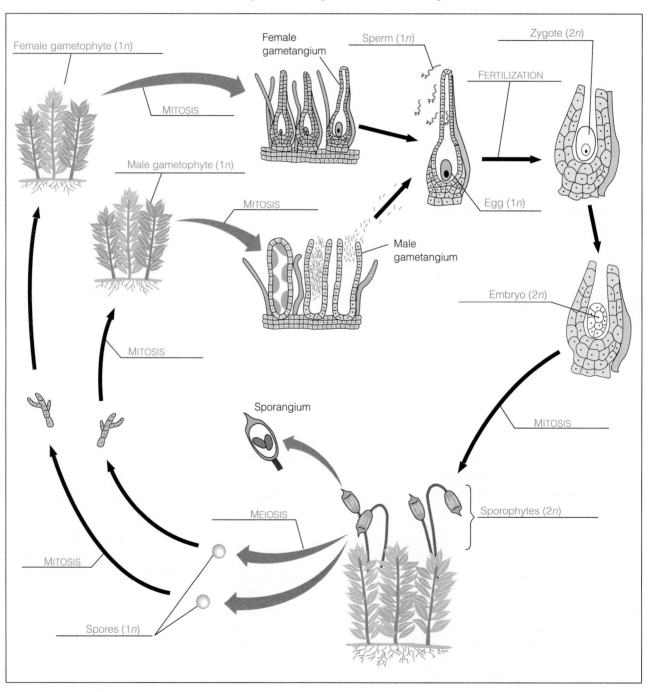

Figure 12.4.
Moss life cycle.

4. With a colored pencil, color all of the **haploid** structures in Figure 12.4.

   How does the sperm get to the egg?

   It swims.

   How does the gametophyte obtain its food (is it autotrophic or heterotrophic)?

   The gametophyte is green, so it is autotrophic.

   How does the sporophyte obtain its food (is it autotrophic or heterotrophic)?

   The sporophyte is brown, so it is heterotrophic. Since it is growing out of the gametophyte, we can assume that it obtains its nutrition from the gametophyte.

   Which phase of the life cycle is larger and lives longer?

   The gametophyte.

EXERCISE 12.3
Seedless Vascular Plants

Objectives

After completing this exercise, you should be able to

1. Identify representative members of the three divisions of seedless vascular plants.
2. Identify in living material or on a diagram all of the structures and events in the fern life cycle.

In vascular plants we see increasing refinements of the adaptations for life on land. As the name implies, members of this group have specialized vascular tissue consisting of xylem and phloem for the transport of water and food. Cuticles, which are fatty or waxy layers, are a more sophisticated means of protecting leaves and stems from disease-causing organisms and desiccation. The spores of these plants are also more protected from drying out than are the spores of bryophytes.

Seedless vascular plants make up only a small part of today's plant kingdom, but during the Carboniferous period, 360–290 million years ago, seedless forests dominated the landscape. Some of the groups that have living representatives are listed below, but most members of the seedless vascular plants are extinct.

Use the margin of your lab manual to sketch the specimens that are on display in the lab.

Advise students to observe and sketch the plants themselves; the goal is to be able to identify any member of this group, not just the specimen on display. Thus students who memorize that the horsetail was the "big plant in a green pot" may be confounded when they are shown a different specimen or a slide for a quiz.

Activity A: Club Mosses

Lycopodium may be used instead of (or along with) Selaginella.

The five living groups of club mosses, which include 1,000 species, have very small leaves and roots arising from an underground stem. The largest genus is *Selaginella;* one of the 700 species of *Selaginella* is on display in the lab. An example of another genus of club moss, *Lycopodium,* is shown on Color Plate 2. The living club mosses are all herbaceous; that is, they never develop wood. Of the extinct forms, some were woody and treelike and were dominant plants of the great Carboniferous forests, which formed the extensive coal deposits that are mined today.

Activity B: Horsetails

Horsetails were also dominant plants of the coal-forming age. Today, however, only one genus, *Equisetum,* survives. One of the 15 species is illustrated on Color Plate 2. Because of siliceous deposits in the epidermal cells, horsetails have a sandy, abrasive texture and were used in colonial and frontier times to scour pots and pans. Feel the roughness of the stem of the specimen in the lab. Horsetails are quite common in moist areas. Look for them near streams and ditches.

Activity C: Ferns

Although we are familiar with most of the 12,000 species of ferns as woodland plants, there are also aquatic and even desert forms. In size they range from a water fern with leaves 2 cm long to the giant tree fern, which may have leaves up to 5 m in length. The roots of ferns originate from underground stems. The leaves, called fronds, also arise from the underground stems, and they have more vascularization than the other seedless vascular plants. In most species, spores are produced in rusty patches called sori on the underside of a leaf or in large clusters on a leaf that is specially modified for reproduction. Because sori give the frond a spotted appearance, many people mistake these reproductive structures for some type of disease.

Procedure

It is interesting to observe the fern's spore dispersal mechanism. When most of the spores are mature the sporangium dehydrates and tears open, forcefully ejecting the spores. You can observe this by placing sporangia under a dissecting scope. The warmth of the lamp is sufficiently dehydrating to initiate the process.

1. On Figure 12.5, label the arrows that indicate where mitosis, meiosis, and fertilization occur (mitosis occurs more than once).
2. Observe the following structures in the living material on display and label them on Figure 12.5:

Sporophyte	Egg (female gamete)
Spores	Sperm (male gamete)
Gametophyte	Zygote

3. With a colored pencil, color all of the haploid structures in Figure 12.5.

How does the sperm get to the egg?

It swims.

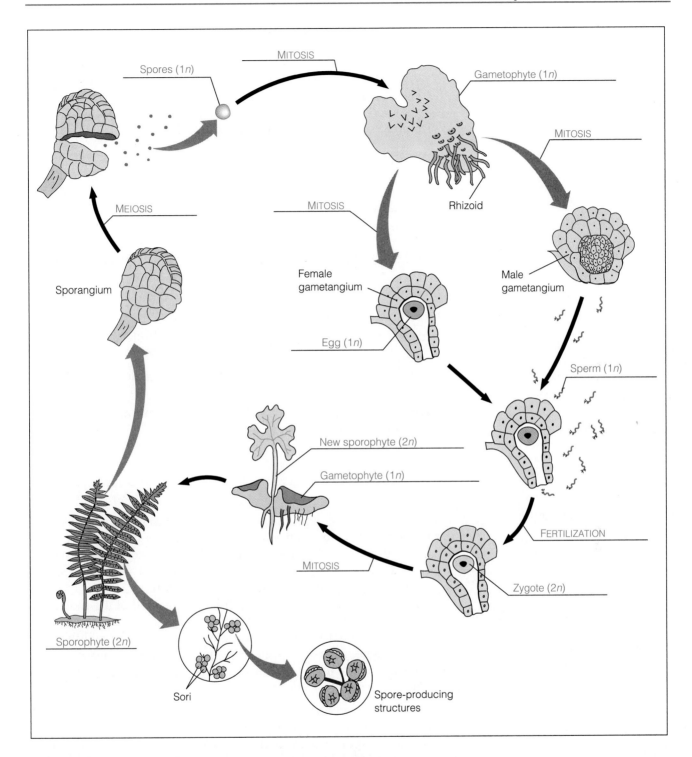

Figure 12.5.
Fern life cycle.

How does the gametophyte obtain its food (is it autotrophic or heterotrophic)?

The gametophyte is green, so it is autotrophic.

How does the sporophyte obtain its food (is it autotrophic or heterotrophic)?

The sporophyte is also photosynthetic. The sporophyte and gametophyte live independently of each other.

Which phase of the life cycle is larger and lives longer?

The sporophyte.

EXERCISE 12.4

Vascular Seed Plants—Gymnosperms

Objectives

After completing this exercise, you should be able to

1. Identify representative members of the four plant divisions of vascular seed plants.

2. Define homospory and heterospory and give examples of plants showing each condition.

3. Identify in living material or on a diagram all of the structures and events in the pine life cycle.

4. Identify and discuss the three important developments in plant evolution seen in the pine.

The remaining plant groups, the gymnosperms and angiosperms, share the evolutionary innovation of having seeds. In seedless plants, the minute, single-celled spores must leave the parent plant and find a suitable habitat to germinate and develop into gametophytes. In seed plants, the spores remain within the parent and the gametophytes that develop from the spores are produced within enclosed structures. Instead of the spore, it is the seed, equipped with a protective coat and a supply of food for the developing embryo, that ventures forth from the parent to establish a new plant. In gymnosperms the seeds are "naked" (Greek *gymnos,* naked; *sperma,* seed), while in angiosperms (Exercise 12.5) they are protected in containers (Greek *angion,* container).

When all of the spores produced by a plant are alike, the plant is said to be **homosporous** ("same spores"). If two kinds of spores are produced, the plant is **heterosporous** ("different spores"). Heterospory first evolved in seedless vascular plants, but most members of that group, including the representative life cycle you observed in lab, are homosporous. All of the seed plants are heterosporous. This makes their life cycles more complex, since there are "male" and "female" spores as well as male and female gametes.

The vascular seed plants also mark the appearance of **pollen,** which is an immature male gametophyte. Gymnosperms are free from the requirement of water for fertilization because the entire male gametophyte travels by air to the female gametophyte before the sperm are produced. Thus the sperm do not have to swim to the egg.

Use the margin of your lab manual to sketch the specimens on display in the lab.

Activity A: Cycads

These palmlike plants occur in subtropical and tropical regions; one species is common in Florida. An example is shown on Color Plate 2. Although now limited to 100 species in 10 genera, they were quite numerous in the Mesozoic era, 245–65 million years ago, when they served as dinosaur food. More recently, their underground stems and root-stalks were eaten by the Seminole Indians.

Activity B: Ginkgo

Ginkgo biloba (the maidenhair tree, so called because the leaves resemble maidenhair fern) is the only remaining species of this once common group. Unlike most gymnosperms, it is deciduous; the leaves, illustrated on Color Plate 2, are all shed in autumn. This tree has been cultivated for centuries by the Chinese and Japanese and is now widely planted in the United States, where its general hardiness and tolerance of air pollution make it a suitable city dweller. Like many trees, a ginkgo has either male *or* female reproductive parts. Male trees are preferred (and easily obtained by vegetative propagation) because the fleshy seed coat produced by the females has an odor similar to rancid butter.

Activity C: Conifers

The largest group of gymnosperms is the conifers or cone-bearers. Included in this group of 550 species are such economically important plants as pines, firs, and spruces. Conifer leaves have reduced surface area, sunken stomates, and thick cuticles, features that help resist drought.

You will look at a typical pine life cycle. This is the first heterosporous life cycle you have encountered, and you will immediately notice that it is more complex than the homosporous moss and fern life cycles. In addition to the male and female gametes that are produced by the male and female gametophytes, heterosporous plants also produce male and female spores. The key to understanding heterosporous life cycles is to follow the male path from beginning to end and then go back and start over with the female path.

All of the events of the conifer alternation of generations life cycle take place in reproductive structures called cones. The sporophyte produces two kinds of cones, male and female. The spores that are produced by meiosis in sporangia in male cones undergo two mitotic divisions to become the male gametophytes, which are also known as pollen grains. In certain areas of the country it is easy to appreciate just how prodigiously pines produce pollen: In the spring a thick yellowish coating is apparent on cars and sidewalks. The pollen is carried by the wind to the female cones of other trees; the female cones exude a sticky resin to trap it.

In the fern life cycle in the previous exercise (though not in all ferns), there is only one type of spore, which gives rise to gametophytes that have both male and female gametangia. In the pine life cycle, "female" spores (properly called megaspores) give rise to gametophytes that produce female gametes, and "male" spores (properly called microspores) give rise to gametophytes that produce male gametes. In pines, the female cones are usually found near the top of the tree, while the male cones are closer to the bottom on the same tree. This arrangement reduces the chances of self-pollination. In the lab displays of pine and flowering plants, there is a branch point where the male and female parts of the sporophyte are shown and the male and female pathways are followed separately.

The scales of the female cones bear ovules, which enclose the female sporangia. Cells in the female sporangia also undergo meiosis to produce spores. One of the spores from each meiotic division develops into a female gametophyte by mitosis. As the pollen germinates and produces sperm nuclei by mitosis, the female gametophyte produces an egg. One sperm nucleus reaches the egg through the pollen tube and fertilizes it. Notice in Figure 12.6 that the male gametangium has been eliminated, and the sperm is now merely a nucleus rather than a motile cell.

Through mitotic divisions, the zygote becomes an embryo. The female gametophyte remains as a food supply for the developing embryo to use when the seed germinates, and a protective seed coat is formed by the surrounding sporophyte tissues in the pine cone.

Procedure

1. Look at Figure 12.6 and label the arrows that indicate where mitosis, meiosis, and fertilization occur.

2. Identify the following structures in the living material and label them on Figure 12.6:

Sporophyte	Male gametophyte (pollen)
Female spores	Sperm nucleus
Female gametophyte	Zygote
Egg	Embryo
Male spores	

Pine reproduction is exceedingly slow. Meiosis to produce the female spores doesn't happen until a month after pollination. Then the female gametophyte does not begin to develop for a few more months, and altogether it might be a year after pollination when its development is complete. Meanwhile, the male gametophyte slowly undergoes a couple of mitotic divisions and two sperm nuclei are produced (one degenerates). It may be as long as 15 months after pollination before fertilization actually takes place.

3. With a colored pencil, color all of the haploid structures in Figure 12.6.

   Describe the difference between pollination and fertilization.

   Pollination occurs when the immature male gametophyte is carried to the female cone. Only then is the sperm (actually the sperm nucleus) produced. Fertilization occurs when the sperm nucleus fuses with the egg.

   How does the sperm get to the egg?

   Once the male gametophyte has arrived, a tube carrying the sperm nucleus grows into the female structure to meet the egg.

   How does the gametophyte obtain its food (is it autotrophic or heterotrophic)?

   The gametophyte is contained within the sporophyte and is dependent on it for nutrition (heterotrophic).

   How does the sporophyte obtain its food (is it autotrophic or heterotrophic)?

   The sporophyte is autotrophic.

   Which phase of the life cycle is larger and lives longer?

   The sporophyte.

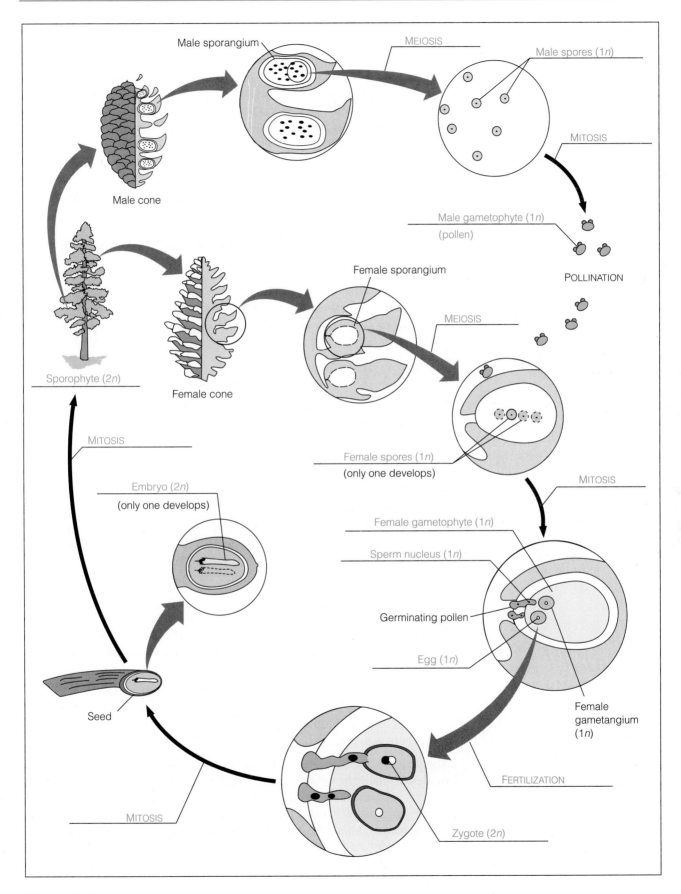

Figure 12.6.
Pine life cycle.

What advantages do seeds have over spores?

Seeds have additional protection from dehydration and the elements in the form of the seed coat, and they carry a supply of food to nourish the developing embryo. Spores are single-celled and have no special adaptations for dispersal or development of the gametophyte, though some have a fairly rugged coat and can survive for a long time.

E X E R C I S E   1 2 . 5
Vascular Seed Plants—Angiosperms

Objectives

After completing this exercise, you should be able to

1. Identify in living material or on a diagram all of the structures in the flowering plant life cycle.
2. Explain double fertilization.

The key characteristic of angiosperms is the flower. Any plant that produces flowers is a member of this group, which is also known as the flowering plants. The name angiosperm, which means "seed container," refers to the fact that the female gametophytes and then the seeds are enclosed in an ovary.

The diversity seen in all of the previously described groups is nowhere near that of the angiosperms. They range in size from the aquatic water meal at 1 mm to the giant eucalyptus trees. Flowering plants are found in a diverse array of habitats. Of the 280,000 species of plants, approximately 250,000 are angiosperms, making them by far the most successful group of plants.

Just as all the events of the conifer life cycle take place in the cone, the flower houses the entire alternation of generations for flowering plants. The flower, which is produced by the sporophyte, may have either male or female reproductive parts or both.

Male sporangia are located within the **anther,** the male part of the flower. After the male spores are produced by meiosis, each spore undergoes one mitotic division to produce a pollen grain, or male gametophyte. As in the pine, there is no gametangium.

The female reproductive structure is the **ovary.** Inside the ovary are **ovules.** Initially the ovules are the female sporangia. One cell in each ovule undergoes meiosis to produce spores. One of the spores enlarges and divides mitotically several times, and the ovule then becomes the female gametophyte. There is no gametangium. The gametophyte has eight nuclei. At one end of the gametophyte is the egg nucleus, flanked by two smaller nuclei, and at the other end are three more nuclei. The two in the center are called polar nuclei.

Pollen, which is the male gametophyte, is brought by wind, water, or animal to the female part of another flower. The pollen then germinates, and a long pollen tube containing two sperm nuclei grows down to the ovule. One of the sperm nuclei fuses with the egg nucleus to form a diploid zygote. The other sperm nucleus fuses with the two polar nuclei to form an endosperm nucleus, which is triploid because it is the fusion of three haploid nuclei. Because of these two separate fertilization events, flowering plants are said to have **double fertilization.**

The endosperm nucleus undergoes many mitotic divisions to produce a multicellular **endosperm,** which serves as a food source for the developing plant. The zygote divides and differentiates to form an embryo, and the layers of cells that surround the ovule harden into a protective seed coat. The ovary itself later matures into a fruit, which helps protect and disperse the seeds.

Procedure

1. Observe the specimens on display that represent the angiosperm life cycle and match them with Figure 12.7 on the next page.
2. On the diagram, label the arrows that indicate where meiosis, mitosis, and fertilization occur.
3. Identify the following structures:

Sporophyte	Egg
Male spore	Sperm nuclei
Male gametophyte	Embryo
Female spore	Zygote
Female gametophyte	

4. With a colored pencil, color all of the haploid structures in Figure 12.7.

How does the sperm get to the egg?

> *As in the pine, the immature male gametophyte is brought in contact with the female reproductive structure before the sperm nuclei are produced.*

How does the gametophyte obtain its food (is it autotrophic or heterotrophic)?

> *The gametophyte is contained within the sporophyte and is dependent upon it for nutrition (heterotrophic).*

How does the sporophyte obtain its food (is it autotrophic or heterotrophic)?

> *The sporophyte is autotrophic.*

Which phase of the life cycle is larger and lives longer?

> *The sporophyte.*

Pollination in conifers and the early angiosperms was effected by wind. Then insects began exploiting pollen and the pine sap found in pine cones as a food source and thus transported pollen from one plant to another. This made pollination more efficient, but also exposed the ovules to predation. The ovary was an adaptation in response to this problem. Lab Topic 18 examines flowers and seeds in more detail.

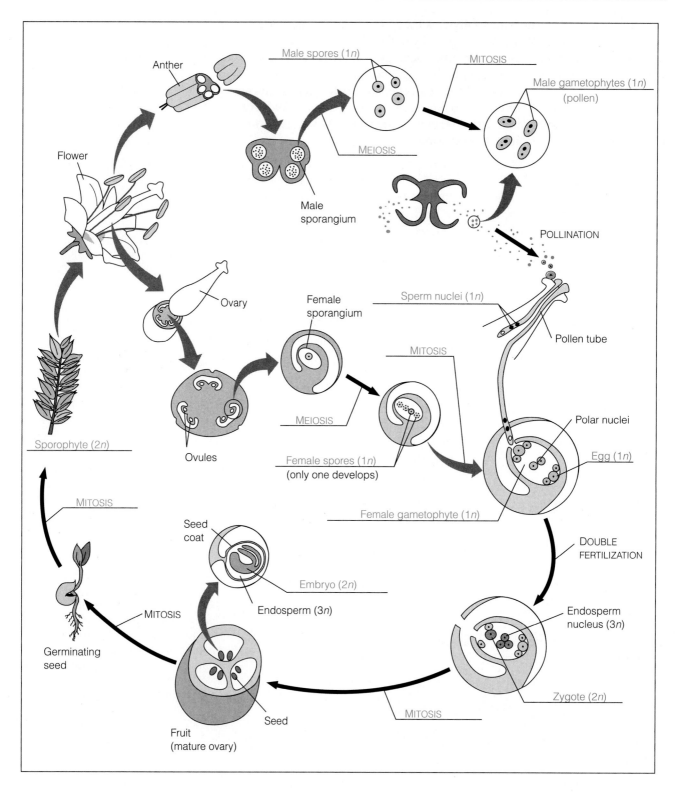

Figure 12.7.
Angiosperm life cycle.

What is the advantage of having seeds produced in an ovary?

The ovary provides protection for the ovules before fertilization and for the seeds as they mature.

Review the diagrams of the moss, fern, pine, and flowering plant life cycles on which you have colored the haploid structures. What can you say about changes in the predominance of the haploid and diploid forms in the course of evolution?

The moss life cycle is predominantly haploid. The fern life cycle also has a significant haploid phase. In the pine and flowering plant, the haploid phase is greatly reduced and is contained within the sporophyte's reproductive structures (cones or flowers).

Questions for Review

1. Why can only diploid cells undergo meiosis?

   The result of meiosis is haploid cells, each of which has one member of each homologous pair of chromosomes. Each haploid cell thus has one of each type of the organism's chromosomes. If the chromosomes in a haploid cell were divided up by meiosis, the resulting cells would lack half of the necessary chromosomes. (A haploid cell can, however, undergo mitosis with no problem because there is no "reduction division.")

2. Why are the bryophytes dependent on water for reproduction?

   The sperm must swim to the egg.

3. If a moss gametophyte has a sporophyte attached to it, is the gametophyte male or female? How do you know?

   It must be female, because the sporophyte develops at the site of egg production.

4. What characteristics would you use to distinguish a fern from a horsetail?

   The fern has leaflike fronds, while the leaves of the horsetail are very small. The fern may have sori on the underside of its fronds; the reproductive structure of the horsetail is at the tip of the stem. The horsetail stem feels grainy because of the silica in its epidermal cells.

5. How are club mosses different from true mosses?

   Club mosses have vascular tissue; true mosses do not. Both are seedless.

6. Are ferns as dependent on water as mosses are? Explain.

   Ferns still require water for fertilization, but the presence of vascular tissue to distribute water from the ground to the aboveground parts make ferns more independent of water.

7. Seedless vascular plants related to the club mosses and horsetails were once dominant forest components. When the world's climate became cooler and drier, the seedless plants were eclipsed by seed plants. Why would the seed plants have been more successful in the new climatic conditions?

   The seed gives the developing sporophyte more protection than is afforded by mosses and ferns, whose embryos grow in the vulnerable tissues of the gametophyte. Seeds are especially adaptive for a drier environment because the embryo is protected from dehydration. In addition, the seed launches the new plant away from the parent and provides a nutritional packet for the initial stages of development.

8. How did gymnosperms achieve independence from the need for water for fertilization?

   The sperm does not need to swim to the egg because the entire male gametophyte comes in contact with the female gametophyte. The male gametophyte produces the sperm after contact is made and actually grows into the female gametophyte to effect fertilization.

9. What is the advantage of an ovary?

   It protects the ovules and the developing seeds. (Lab Topic 18 describes the further advantage of the ovary maturing into a fruit that is used for seed dispersal.)

10. Compare the way seeds are made in gymnosperms with the way seeds are made in angiosperms.

    In both gymnosperms and angiosperms, the seed contains the embryo which develops from the zygote. This is the new sporophyte generation. In pines, the food source for the developing embryo is the remaining tissue of the female gametophyte (and thus haploid), while in angiosperms the endosperm is triploid tissue resulting from fusion of a sperm nucleus with the two polar bodies.

Plant Diversity
Instructor Notes

Organization

Introduction (10 minutes)

Go through the generalized alternation of generations life cycle (Exercise 12.1). Make sure students understand the terms haploid and diploid. Emphasize that gametophytes are gamete-producing plants and sporophytes are spore-producing plants. It is essential for students to master the generalized life cycle before they begin the laboratory procedures. Each life cycle they will study fits the same pattern.

You may want to illustrate how to apply the general pattern to the various life cycles by using moss as an example. On the moss life cycle, fertilization is labeled; the result is a zygote. The sporophyte develops from the zygote by mitosis (and differentiation). The sporophyte produces spores by meiosis. The spores germinate and undergo mitosis to produce gametophytes. Gametophytes produce gametes by mitosis; gametes fuse to form a zygote at fertilization. Or you can work backwards from the point you recognize: Gamete production must have immediately preceded fertilization, so the plant that produced the gametes must be the gametophyte, and so on.

Prep for Procedures (2 minutes)

Point out where the specimens are located in the laboratory and show students where to begin each life cycle. Divide up the class so that some teams begin with moss, some with fern, and some with the diversity specimens. No one should start with the pine or flowering plant life cycles.

During the Laboratory Period

When students get stuck in the life cycles, make them find a point they recognize and then work from the generalized life cycle (as described in the Introduction). Their natural tendency in an exercise like this is to want to get all the labeling done and then memorize the whole thing. You need to convince them that they only have to memorize the generalized pattern; they can then think through any of the specific life cycles.

Summary (20 minutes)

You should review the diagrams to be sure students have theirs correctly labeled. An effective technique for involving students is to give different groups transparencies of the life cycles to label at the end of the lab period and let them present the answers rather than doing it yourself. (Don't assign life cycles too early, though, or students will only complete the one they must present.)

For a 2-Hour Lab

Omit the specimens representing the various plant divisions plus the pine life cycle (Exercise 12.4). Have students read the introduction to Exercise 12.4, though, so they will appreciate the advances seen in the angiosperms.

Animal Diversity

Introduction

Our study of the animal kingdom in this lab topic will include nine of the major phyla, shown in Figure 13.1 on the next page. There are 26 other phyla (depending on which taxonomist you ask) that have been omitted. When you recall the classification hierarchy by which similar organisms are grouped together (domain, kingdom, phylum, class, order, family, genus, species), you will realize that we are painting the animal kingdom in very broad strokes. We are also able to sample only a bit of the diversity—this is a kingdom that reaches all the way from the barely organized sponges through that most complex of animals, the human.

The characteristics that distinguish animals from other organisms include multicellularity and the absence of cell walls. Most animals have the ability to move and some type of nervous system. Animals are also distinguished by their means of sexual reproduction and their life cycles. The animal characteristic that will be explored in more depth for the phyla studied in this lab topic is the heterotrophic mode of nutrition. That is, animals derive their nutrients from feeding on others (*hetero* = other; *trophic* = feeding).

Almost all animals take food into their bodies and then digest it internally. The remaining few are parasites that obtain nutrition from food that has already been digested by their hosts. You probably already know that animals can be carnivores (meat-eaters), herbivores (plant-eaters) or omnivores (eat both plants and animals). But there are also some animals that feed on decaying organic matter. And a large number of aquatic animals are suspension feeders (sometimes called filter feeders) that pass currents of water through their bodies and trap food particles from it.

Outline

Exercise 13.1: Phylum Porifera—Sponges

Exercise 13.2: Phylum Cnidaria—Jellyfish, Sea Anemones, Corals

Exercise 13.3: Phylum Platyhelminthes—Flatworms

Exercise 13.4: Phylum Nematoda—Roundworms

Exercise 13.5: Phylum Annelida—Segmented Worms

Exercise 13.6: Phylum Mollusca—Mollusks

Exercise 13.7: Phylum Arthropoda

Exercise 13.8: Phylum Echinodermata

Exercise 13.9: Phylum Chordata

Instructor Overview
The activities in this lab topic offer a small sample of these nine major animal phyla. The main characteristics and evolutionary advances are described for each phylum. In-text and review questions emphasize adaptive traits. In addition to the drawings and photos in the text, representatives of each phylum are shown in the color plates. Heterotrophy and ingestion of food are among the distinguishing characteristics of animals, so the theme of the hands-on activities is adaptations for feeding.

Time Requirements
Exercise 13.1: Phylum Porifera (15 minutes)
Exercise 13.2: Phylum Cnidaria (15 minutes)
Exercise 13.3: Phylum Platyhelminthes (20 minutes)
Exercise 13.4: Phylum Nematoda (10 minutes)
Exercise 13.5: Phylum Annelida (15 minutes)
Exercise 13.6: Phylum Mollusca (20 minutes)
Exercise 13.7: Phylum Arthropoda (30 minutes)
Exercise 13.8: Phylum Echinodermata (10 minutes)
Exercise 13.9: Phylum Chordata (30 minutes)

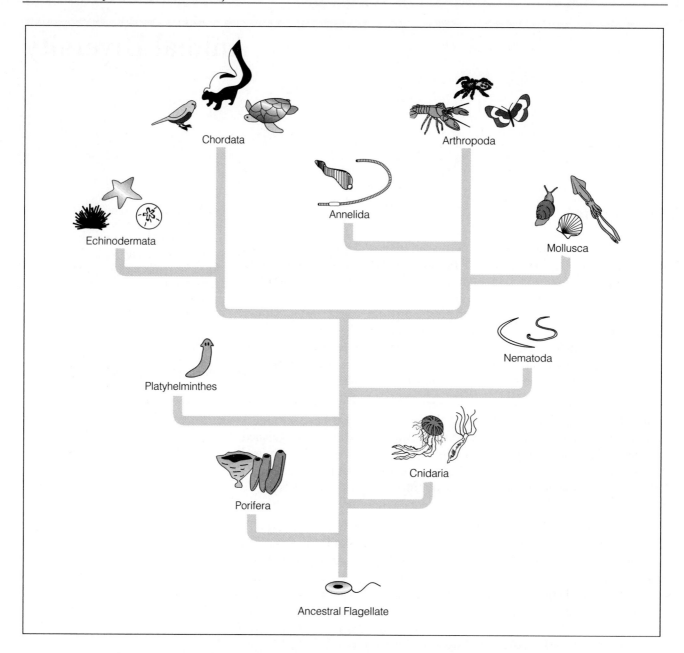

Figure 13.1.
Phylogenetic tree showing 9 of the 35 animal phyla.

EXERCISE 13.1
Phylum Porifera—Sponges

Objective

After completing this exercise, you should be able to

1. Describe how sponges obtain food.

The phylum Porifera (Latin *porus,* pore; Greek *fera,* bearing) comprises the sponges. As you can see in Figure 13.1, the sponges diverged early from the main line of animal evolution and have given rise to no other animal groups. There are approximately 9,000 species of sponges. All sponges are aquatic and most of them are marine. A vividly colored sponge is shown on Color Plate 3; its pigmentation is due to symbiotic algae. The adult forms are sessile, which means that they are stationary, but the larvae (immature animals that do not resemble the adults) are motile.

Figure 13.2 shows a diagram of a typical sponge. The body is composed of two cell layers: an outer epidermal layer and an inner layer of flagellated "collar cells." Between the epidermal layer and the layer of collar cells are amoebalike cells (amoebocytes) with various functions such as food storage, digestion, and formation of reproductive cells. Some of the amoebocytes secrete materials that form an endoskeleton, which supports the sponge. One skeletal material is a protein that forms a network of tough fibers. The other type of skeletal material is mineral crystals called spicules. Some sponges have both protein fibers and spicules and some have one or the other. For example, bath sponges are fibrous endoskeletons that have no spicules. (Luffa sponges are not animals at all, but a plant related to squash.)

Figure 13.2.
Anatomy of typical sponge.

Sponges are suspension feeders. Their bodies are perforated with the pores for which the phylum is named, and water flows into the sponge through them (Color Plate 3). There can be a remarkable number of pores. It is estimated that *Leuconia*, a sponge measuring about 10 cm tall and 1 cm in diameter, has about 81,000 of them! The water travels through a network of canals lined by collar cells (Figure 13.2) whose beating flagellae bring the water in. The "collar" of collar cells is a molecular mesh that sieves out food particles such as microscopic algae, bacteria, and organic debris. The filtered water then proceeds into the central chamber where it is expelled through a large opening called an osculum.

In the following procedure, you will observe the structure of a small marine sponge called *Grantia* (also known as *Scypha*).

Procedure

1. Get a preserved specimen of *Grantia* from your instructor and place it in a small petri dish.
2. Observe the specimen under the dissecting microscope. First use reflected light, then transmitted light.
3. Sketch the sponge in the margin of your lab manual and label the osculum.
4. Use a razor blade or scalpel to slice the sponge longitudinally so that you can see the central canal.
5. Place the specimen with the central canal facing up and look at it under the dissecting scope, again trying both light sources.
6. Sketch the specimen in the margin of your lab manual.
7. Use a compound light microscope to examine a prepared slide of *Grantia*. This thin cross section has been stained so you can see the collar cells protruding into the canal. You may also be able to see spicules.
8. Sketch the specimen in the margin of your lab manual.

Given that all sponges are filter feeders, why does it follow that all sponges are aquatic?

> *Filter feeding is only possible in an aquatic habitat since it involves capturing food borne on water currents.*

Would mobility improve the ability of sponges to capture food? Explain.

> *No. As filter feeders, they are not seeking any particular food item; food is sufficiently abundant in a single spot. The combination of currents and the pumping action of the flagella carries tremendous amounts of water through the sponge's body.*

Pores in the sponge's body can be seen using the transmitted light source.

If you can obtain a living sponge specimen, you can demonstrate the water currents by adding a bit of carmine powder or India ink to the water.

EXERCISE 13.2
Phylum Cnidaria—Jellyfish, Sea Anemones, Corals

Objectives

After completing this exercise, you should be able to

1. Name representative members of the phylum Cnidaria.
2. Sketch the polyp and medusa body forms.
3. Describe the feeding behavior of cnidarians in general and hydra in particular.

There are approximately 10,000 species in the phylum Cnidaria (Greek *knide,* nettle; Latin *aria,* like). They are all aquatic (most of them are marine), and they are all radially symmetrical. An object or an animal that is radially symmetrical can be equally divided in more than one way because its parts are arranged around an imaginary central axis. The hydra, a cnidarian shown in Figure 13.3, is an example. It can be divided into equal halves starting at any point on the perimeter, as long as the dividing line passes through its center.

Figure 13.3.
Radial symmetry.

Cnidarians have two distinct body forms: the polyp (for example, sea anemones), which is stationary, and the free-floating medusa (for example, jellyfish). Figure 13.4 shows a comparison of these forms. There are two layers of cells separated by a gelatinous material that helps support the body. Note the increased amount of this material in the medusa form—this is the "jelly" of jellyfish.

Some cnidarians have both body forms in their life cycle. If an animal is adapted to a stationary way of life, what could be the purpose of having a motile phase at some point in the life cycle?

For dispersal. The offspring can find a new territory and do not have to compete with the parents.

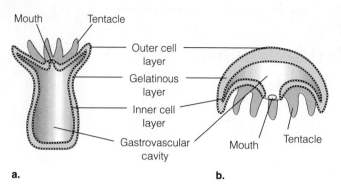

Figure 13.4.
Cnidarian body forms: (a) polyp form; (b) medusa form.

Sea anemones and corals are polyp forms of Cnidaria. Color Plate 3 shows a sea anemone waving its tentacles. The anemone's body is filled with water to provide support by hydrostatic pressure. The actual mass of tissue comprising the body is quite small. When the animal is endangered, much of the water is expelled and the anemone shrinks remarkably, drawing in its tentacles. The coral polyp itself looks very similar to an anemone. By precipitating minerals from seawater, some types of coral encase themselves in a hardened exoskeleton (see Color Plate 3). Colonial corals have left behind vast stretches of coral reefs in warm, shallow waters throughout the world. The Great Barrier Reef extends for 1,800 km along the coast of northeastern Australia and is the largest structure ever built by living organisms.

Most species of jellyfish spend most of their lives in the medusa form and the polyp stage is reduced or absent. Although most jellyfish range from 2 to 40 cm in diameter, the largest is over 2 m across and has tentacles 60 to 70 m long. Movement of jellyfish is by a beautiful rhythmic undulation of the bell, or umbrella, of the medusa, with the tentacles trailing languidly beneath. A photograph of a jellyfish appears on Color Plate 3.

Cnidarians are carnivorous, consuming prey ranging from minute zooplankton to fish. They are armed with stinging cells called **cnidocytes,** which are used in prey capture and in defense. These stinging cells are made up of coiled tubular threads, many with a barb or spine. When it receives certain chemical or tactile stimuli, the thread springs out, harpooning or entangling the prey. Some can inject poison as well. It is the stinging cells of the Portuguese man-of-war and certain jellyfish that make contact with these animals painful and sometimes dangerous for humans. The phylum name Cnidaria refers to a plant (nettle) with stinging cells that the plant uses for defense.

The hydra is a freshwater polyp form that is a popular laboratory organism. Hydra feed on insect larvae, small crustaceans, and annelid worms.

Procedure

In our labs we have had more success hatching our own brine shrimp to use as prey items than purchasing either brine shrimp or Daphnia, which are often too large. However, even large prey can be used to demonstrate prey capture behavior, since hydra will often snare a prey that is so large that it escapes before it is ingested.

1. Use a dropper to get a hydra from the container marked "hungry." Put it into a small watch glass with several drops of water from the hydra's container. (You may have to flush the hydra out of the dropper by sucking water into the dropper and then rapidly squirting it out.)

2. Examine the hydra under the dissecting microscope and sketch it in the margin of your lab manual. Reflected light works best.

3. Dip a piece of cotton thread into the dish and touch the hydra's tentacles with it. Describe what happens.

   The hydra should either not respond or retract its tentacles.

4. Soak the thread in liver juice and dip it into the water near the hydra. Describe what happens.

   In place of liver, you can use a solution of glutathione for this procedure.

   Liver contains the chemical glutathione (a tripeptide), which elicits feeding behavior. The hydra should stretch its tentacles in the direction of the thread as it senses "prey."

What type of sensory information enables the hydra to locate its prey?

   Chemical

5. Get a brine shrimp or *Daphnia* (water flea) and put it in the dish with the hydra. Describe the hydra's feeding response.

   The hydra will stretch and wave its tentacles when it senses the prey. When the prey is near enough, the hydra will fire its cnidocytes to immobilize it. (You won't be able to see the threads themselves; the prey simply appears to be stuck to the tentacles.) The hydra will then use its tentacles to stuff the prey into its mouth. The body wall distends considerably when the prey is inside the gastrovascular cavity.

✳ When you have finished with the hydra, put it in the container marked "fed," whether it consumed the prey item or not.

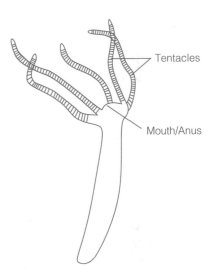

Tentacles

Mouth/Anus

Student sketch of hydra.

The digestive system of cnidarians has only one opening, which leads into the gastrovascular cavity. The undigested remains of the hydra's meal will later be ejected from the same opening that took the food into the body.

Why is this type of digestive system inefficient?

   Undigested food is mixed with digested food, which can interfere with absorption of nutrients. Food may not get completely digested before it is ejected.

Why is radial symmetry adaptive for a sessile animal in an aquatic habitat?

   The animal is surrounded by the medium that carries its food and any chemical stimuli. The animal is not going anywhere, so it is advantageous to "face" all directions at once.

E X E R C I S E 1 3 . 3
Phylum Platyhelminthes—Flatworms

Objectives

After completing this exercise, you should be able to

1. Name representative members of the phylum Platyhelminthes.
2. Explain why bilateral symmetry is more advantageous for a motile animal than radial symmetry.
3. Describe the feeding behavior of the planarian.
4. Explain how a tapeworm obtains food.

The phylum Platyhelminthes (Greek *platys,* flat; Greek *helmins,* worm) includes about 20,000 species. One unifying characteristic is the shape of these animals: Their bodies are flattened. Although they have a primitive one-opening digestive system like that of the cnidarians, this phylum has some features that are important in more advanced animals. For example, there are organs (groups of tissues that work together to perform a function) and division of labor among the organs.

Free-Living Flatworms

For this lab activity, you will observe the planarian, a freshwater flatworm. Although it is small and drably colored, its close marine relatives can be much larger (up to 30 cm), with vivid coloration. An example of a marine flatworm is shown in Color Plate 4.

Procedure

1. Get a planarian from the "hungry" container and observe it under the dissecting microscope. Make a sketch of it in the margin.
2. With only the reflected light source on, cover the side of the dish opposite the planarian with a piece of paper. What does the planarian do?

   It moves to the covered side, away from the light.

3. Put a tiny bit (just as much as you can pick up with the tips of the forceps) of egg yolk or liver in the dish with the planarian. Turn off the light while you wait for the animal to begin feeding. Describe its behavior.

   The planarian extends a muscular tube called the pharynx from midway down the ventral side of its body to feed. If the particles it is feeding on are small enough, you can see them being sucked into the tube.

Be prepared to describe for students how planaria ingest food, because even the most starved planaria may not be willing to feed during the lab period. Planarians will sometimes eagerly consume human blood when they won't take egg yolk. The color of the blood darkens the intestine so students can see where the food goes (the egg yolk is dyed red for the same purpose). If you want to prick your finger to get blood, you should arrange with the prep staff to get an alcohol swab and sterile lancet. The blood should be allowed to drop into the container from above the water line (don't immerse your finger in the planarian's water). Dispose of the lancet properly.

Due to the potential presence of blood-borne pathogens, students should not be allowed to prick their fingers.

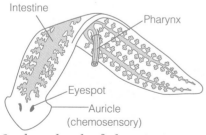

Student sketch of planarian. (Students will not label any structures.)

When you have finished with the planarian, put it in the "fed" container whether it has eaten or not.

What can you conclude about the sensory capabilities of the planarian?

It is apparently able to sense light. It might sense its prey by detecting chemicals in the water.

Bilateral symmetry is seen in the adult form for the first time in the flatworms. In a bilaterally symmetrical animal there is only one way in which to divide the animal so that both halves are the same size and shape. Humans and crayfish (Figure 13.5) are other examples of bilaterally symmetrical animals.

Figure 13.5.
Bilateral symmetry.

After observing the planarian's behavior, why is bilateral symmetry more adaptive than radial symmetry for a motile animal? (Hint: How is the body form of the planarian different from that of the hydra?)

A bilaterally symmetrical animal has a head end so it can move in a particular direction. Further, the sense organs are located in the head, enabling the animal to sense what it is approaching in the environment.

Parasitic Flatworms

Free-living flatworms such as planaria are actually a minority, comprising less than a quarter of all platyhelminth species. The rest are parasitic. Liver, lung, and blood flukes are examples of parasitic flatworms. A type of blood fluke causes schistosomiasis, a disease that is a serious threat to public health in parts of much of Africa as well as elsewhere (Figure 13.6). Swimmer's itch is caused by a related species that lives in northern North American lakes.

Figure 13.6.
Scanning electron micrograph of *Schistosoma mansoni.*

3,120x

The longest tapeworm that affects humans is the fish tapeworm, which may be 20 m long. Humans are infected by eating raw or poorly cooked fish, and in the U.S. it is most common in the Great Lakes region.

The parasitic way of life, in which nutrients are derived from a living host, requires specific adaptations for living inside the body of another organism. For example, in Figure 13.6 you can see the sucker with which the blood fluke attaches itself to the wall of a blood vessel. Tapeworms, which are found in the intestinal tracts of humans and many other vertebrates, are another group of parasitic flatworms. Figure 13.7 shows the grappling device used by one species of tapeworm to avoid being swept away by the host's normal digestive processes.

Figure 13.7.
Scanning electron micrograph of the head end of a tapeworm showing hooks and suckers for attachment.

270x

Procedure

Observe the preserved tapeworm specimen on display. Sketch it in the margin of your lab manual and describe it below.

The tapeworm is noticeably flat bodied, quite long, and the body is made up of distinct units (students may call them segments). These sections are reproductive units that will be shed from the host to continue the life cycle in another animal.

Reproduction can be an extremely complex matter for internal parasites. Why is this so? (Hint: What's the first requirement for sexual reproduction?)

The difficulty of locating a mate and the problem of dispersing offspring to new hosts complicate reproduction for endoparasites.

On the other hand, methods of obtaining and digesting food may be extremely simplified. In fact, internal parasites may have no digestive system at all. How can a tapeworm, for example, survive without a digestive system? (Hint: What is the purpose of digestion?)

The purpose of a digestive system is to break large food molecules into subunits that can be used by the cells. Parasites that live in the blood or digestive systems of their hosts have access to nutrients that have already been digested.

E X E R C I S E   1 3 . 4
Phylum Nematoda—Roundworms

Objectives

After completing this exercise, you should be able to

1. Name representative members of the phylum Nematoda.
2. Explain the advantage of having a complete digestive tract.

Although only 90,000 species of nematodes (Greek *nema*, thread; Greek *eidos*, form) have been identified, it has been estimated that there are close to half a million, living in all habitats and following various lifestyles. Where they occur (and they occur almost everywhere), their numbers are staggering. The top six inches of soil in one acre of a plowed field is estimated to contain billions of nematodes.

Nematodes are bilaterally symmetrical. They are generally small (less than 1 cm long). Free-living nematodes eat other nematodes or microorganisms, or they may feed on decaying organic material. Although most species are free-living, some are parasitic on animals (including humans) and plants. The parasitic species have been more thoroughly studied than the free-living nematodes. Examples of parasitic nematodes are hookworms, pinworms, *Ascaris* (intestinal roundworms), and filarial worms (including those known as heartworms in dogs). Some parasitic nematodes may exceed 8 cm in length.

Procedure

Observe the nematodes on display and note the characteristics they share.

There is not much distinctive about them. Large or small, nematodes are roundworms that have no apparent external features.

If you want to use a living nematode either for display or a hands-on activity, vinegar eels are a good choice. Make a wet mount from a culture (ordering information is given in the Prep Guide) and view them under the compound microscope.

The nematodes are the first animals we have seen in this laboratory that have a complete digestive tract—that is, one with two openings, a mouth and an anus (Figure 13.8). The hydra and planaria have an incomplete digestive tract.

Figure 13.8.
Complete digestive tract with little specialization.

Mouth   Pharynx   Intestine   Anus

What is the advantage of a two-opening digestive tract?

Food can be more completely processed before the remains are excreted. In a one-opening system, digested food is mixed with undigested food, so the process is much less efficient.

EXERCISE 13.5
Phylum Annelida—Segmented Worms

Objectives

After completing this exercise, you should be able to

1. Name representative members of the phylum Annelida.
2. Describe the characteristics that distinguish annelids from the flatworms and roundworms.
3. Describe the earthworm's digestive system.

There are about 15,000 species of annelids (Latin *annellus*, little ring), including marine, freshwater, and terrestrial members. Both free-living and parasitic lifestyles are represented. The earthworm is the most familiar example of an annelid, and it illustrates why this group is called the segmented worms. The annelid body consists of a head, a segmented body, and a terminal portion. The body segments are partitioned off from each other and are filled with a fluid that serves as a hydrostatic skeleton.

The largest group of annelids is the **polychaetes.** Most of them are marine and they are an important source of food for fish and crustaceans. Polychaetes have a pair of fleshy appendages on each body segment (Figure 13.9). Note that we did not see any type of appendages (external organs) in any of the more primitive phyla. In more highly organized animals, for example the Arthropoda, we will see greater specialization of these structures. The polychaete shown in Figure 13.9 is a clam worm. These animals hide out in burrows during the day and venture forth at night to search for

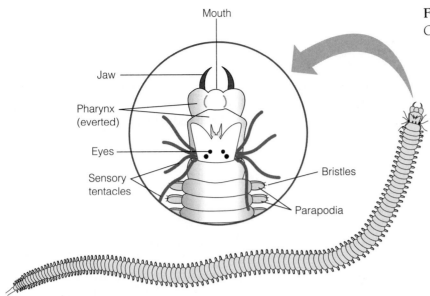

Figure 13.9.
Clam worm, a polychaete.

small animals (for example, other types of worms) to eat. Notice the predatory jaws. Other polychaetes, including the feather duster worms, build tubes to live in. They use tentacles covered with cilia to trap food such as tiny animals and decaying organic matter and transport it to the mouth.

Leeches comprise another group of annelids. Mostly found in fresh water, leeches include carnivorous species as well as the famous ecto- (external) parasites that feed on blood.

A third group of annelids includes the **earthworm.** These familiar animals feed on decaying organic matter in the soil and play an important ecological role as decomposers. Earthworms prefer a moist, humus-rich soil that they suck in through the pharynx as they burrow. Nutrients are extracted from it by the digestive tract. In marked contrast to the relatively unspecialized nematode digestive system, the earthworm digestive tract is divided into several different organs (Figure 13.10).

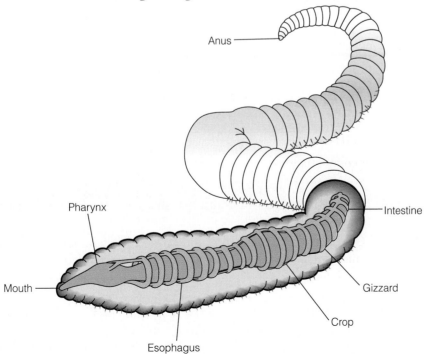

Figure 13.10.
Earthworm.

Procedure

Examine the dissected earthworm and note its specialized digestive organs. At the front end of the earthworm are the mouth and pharynx. The pharynx is a muscular organ used to draw food into the mouth. The next structure is the relatively long esophagus, a passageway between the pharynx and the crop, where food is stored temporarily. From the crop, food passes on to the gizzard where it is ground into smaller pieces.

Would you expect the gizzard to be muscular or nonmuscular? Why?

Muscular, to grind the food.

What is the advantage of having food broken up into smaller pieces?

This process exposes surface area for digestive enzymes.

The next part of the earthworm's digestive system is the intestine, which runs the remainder of the length of the worm. Enzymatic digestion takes place here, followed by absorption of the nutrient molecules.

Why is the intestine the longest organ in the digestive system?

To provide a large surface area for absorption of nutrients.

Finally, undigested material is expelled through the anus.

How does specialization of organs in the digestive system increase the efficiency of digestion?

Specialization of organs is a further improvement on the two-opening digestive system. Food stays in the digestive tract longer, newly ingested food is not mixed with partially digested food as it is subjected to enzymatic and mechanical digestive processes, and each organ has its own function to perform in digesting and absorbing food.

EXERCISE 13.6
Phylum Mollusca—Mollusks

Objectives

After completing this exercise, you should be able to

1. Name representative members of the phylum Mollusca.
2. Explain how you would distinguish among gastropods, bivalves, and cephalopods.
3. On a dissected specimen or a diagram of a clam, identify the mantle, gills, foot, and visceral mass and trace the pathway of food.

With 150,000 species identified, the phylum Mollusca (Latin *molluscus*, soft) is second only to the Arthropoda in number of species. Among the mollusks we find animals ranging in size from microscopic to gigantic, and we find fast-swimming predators, passive suspension feeders, peaceful herbivores, and parasites. Although most mollusks are marine, the phylum does include some freshwater species, and one group (snails and slugs) has been successful on land. This phylum also includes a number of species that humans enjoy as food: clams, mussels, oysters, scallops, squid, and snails.

The basic mollusk body plan includes a foot, which is mainly muscular tissue, and a visceral mass where the digestive, circulatory, respiratory, and reproductive organs are located. The mantle consists of two folds of skin that sandwich the visceral mass. The cavity created in between serves important functions in each molluscan group. In many species, the outside of the mantle secretes a protective exoskeleton, the shell. Gills extract oxygen from the water. In suspension feeders, they also collect food particles. Figure 13.11 shows the variations of body plan seen in the three major classes of mollusks.

Phylum Mollusca is categorized into classes based on the shape and use of the foot. One class is the **gastropods.** "Gastropod" means "belly foot." This group, the most abundant and diverse group of mollusks, includes snails, periwinkles, abalones, slugs, limpets, and others. Most gastropods have a single shell, for example the marine snail shown in Color Plate 4. Some gastropods, however, such as the slugs and sea hares, lack shells altogether (see Color Plate 4). The majority of gastropods are herbivores and possess a radula, which is a set of rasping teeth arranged in a tonguelike organ (see Figure 13.12). The radula is unique to mollusks. Watch a snail as it grazes on the side of an aquarium. It is scraping algae off the glass with its radula.

If possible, get a snail that students can observe in lab. The radula will be most evident if you put the snail in a tank whose sides are covered with algae.

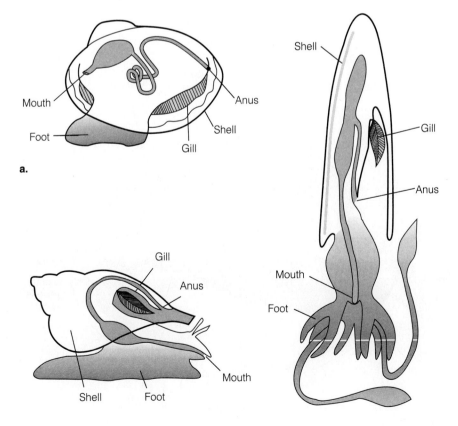

Figure 13.11.
Molluscan body plans: (a) clam; (b) snail; (c) squid.

145x

Figure 13.12.
Scanning electron micrograph of a snail radula.

Carnivorous gastropods include the oyster borer, which uses its radula to drill through the oyster's shell, and snails of the genus *Conus* (Color Plate 4), which have a radula that is specialized for stinging prey. The venom from some *Conus* species is even fatal to humans.

The squids, octopuses, and nautiluses are **cephalopods**, which means "head foot." All of these animals have well-developed heads with very advanced eyes. The foot is modified into tentacles with suction cups for grasping prey. Color Plate 4 shows a squid. Cephalopods are all carnivores and are found exclusively in marine habitats. Except for the nautilus, the shells of cephalopods are much reduced. In squids and cuttlefish the shell is small and enclosed in the mantle (the internal shell of the cuttlefish is the cuttlebone sold as a beak-sharpening device to be hung in bird cages), and the octopus has no shell at all.

Members of the **bivalve** class are "hatchet-foot" mollusks that use the foot for burrowing. Bivalves have two shells. Some familiar examples are clams, scallops, and oysters. Bivalves lack a head, eyes (except in scallops), and radula. Most of them are suspension feeders. Water enters through an opening called a siphon, and passes over the gills, which collect food particles from the water. The water is then expelled through a second siphon. Color Plate 4 shows a clam with its foot and both siphons extended outside the shell.

So that the students don't spend time struggling to open the clams, have the lab tech peg the clams ahead of time, open them before they are placed in the labs, or open the clams yourself at the beginning of the period. Open clams over a sink or dissecting pan; a lot of liquid will come out. Directions are given in the Instructor Notes.

Procedure

Examine the dissected bivalve specimen on display and identify the mantle, visceral mass, gills, and foot (see Figure 13.13).

Next, trace the path of food through the clam. Locate the structures on the dissected clam and sketch the pathway on Figure 13.13.

Figure 13.13.
Dissected clam. Top half of mantle has been cut away to reveal foot and gills.

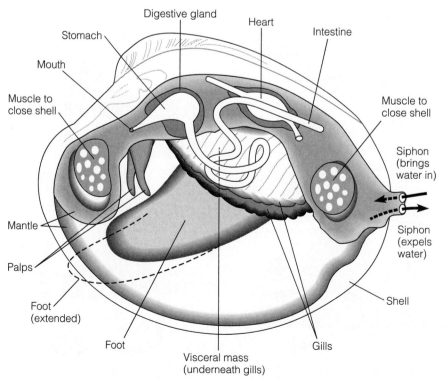

Water enters through the lower siphon and is drawn to the gills by the beating of cilia. Food particles are sieved out by the gills and conveyed by mucus threads to the palps, which direct the food to the mouth. Once in the mouth, food passes through the digestive system, including the esophagus, stomach, intestine, and anus. Meanwhile, the filtered water passes out through the other siphon.

Compare the clam's digestive system to that of the earthworm. Why do earthworms have two organs that clams don't have?

> *They are both two-opening systems and they have some of the same organs. Earthworms extract their food particles from a solid source (soil), while clams get theirs from water. Thus earthworms have a gizzard for the extra processing required, and a crop for food that is "waiting its turn" in the relatively slow process.*

E X E R C I S E   1 3 . 7
Phylum Arthropoda

Objectives

After completing this exercise, you should be able to

1. Identify representative members of the five groups of arthropods.
2. Describe the structural features that are characteristic of each of the five groups of arthropods.

With 1.1 million species, arthropods (Greek *arthron,* joint; Greek *pous, podos,* foot) are the most numerous of all animals. In fact, 85% of all animals are arthropods, 75% of all animals are insects (a class of arthropods), and 30% of all animals are beetles (an order of insects). Arthropods are also the most diverse group of animals. They are widely distributed and fill an array of ecological niches.

The name of the phylum Arthropoda refers to the jointed appendages that are characteristic of members of this group. Some of the appendages are used for locomotion, but in many groups appendages have been modified for food handling, defense, taste, touch, and reproduction. All of the arthropods are bilaterally symmetrical. Like the annelids, their bodies consist of a series of segments.

Arthropods have a chitinous exoskeleton, which is one factor that allowed arthropods to be successful in terrestrial habitats. Chitin is a tough but flexible polysaccharide that protects the soft body parts and prevents dehydration. It may be thick and armorlike (as in beetles) or soft (as in spiders). Even a hardened skeleton is soft at joints and between body segments so the body is flexible. Since the exoskeleton cannot grow, arthropods periodically shed it and grow a larger one, a process called molting.

The major groups of this diverse phylum are briefly described in this lab exercise.

The **crustaceans** include the lobster, crayfish, shrimp, *Daphnia* (water fleas), barnacles, crabs, and pill bugs (see Color Plate 5). Most crustaceans are aquatic; only the pill bug is terrestrial.

The exoskeleton of most crustaceans is supplemented by calcium salts. This is seen to an extreme degree in lobsters and crabs. Consequently, their shells are heavier and less flexible than those of other arthropods.

Thinking about various crustaceans you have encountered, how do you suppose this class got the name Crustacea?

It means crusty, and they are.

Crustaceans provide an excellent illustration of the jointed appendages that gave arthropods their name.

Procedure

1. Use a dissecting microscope to examine the crayfish.
2. Sketch the crayfish in the space below.

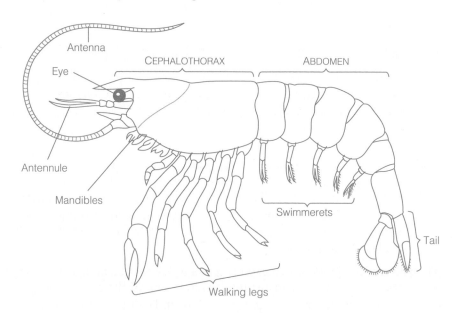

CRAYFISH

3. Describe the exoskeleton of the crayfish.

Every part of the crayfish body is covered by exoskeleton, with an especially heavy piece protecting the back (which contains the internal organs). However, the armor plating is jointed to allow movement. Students can wiggle the parts of the body and the appendages to observe their flexibility.

4. Describe the jointed appendages of the crayfish.

> *There is a pair of appendages on each segment of the body. Beginning at the head end, there are two pairs of antennae, mandibles (crushing jaws), small appendages for food handling, chelipeds (pincer claws), four pairs of walking legs, swimmerets (mostly modified for reproductive functions), and the tail (for swimming). Students may fail to note the mandibles, and of course they will not use the technical terms to describe the appendages.*

The **arachnids** include spiders, scorpions, mites, and ticks (see Color Plate 5). Most arachnids are carnivorous and have claws, fangs, poison glands, or stingers. Since spiders have no jaws and are unable to chew, they inject their prey with digestive juices. Their mouthparts are adapted for drawing in the digested fluid.

Not all spiders capture their prey with webs. For example, there are the trap-door spiders, which dig a web-lined hole in the ground with a hinged door. The door serves both as protection from enemies and as a means of locating prey, which produce vibrations as they walk over the door. Other spiders, such as jumping spiders, wolf spiders, tarantulas, and fishing spiders, hunt for prey.

Insects are the most diverse and abundant class of organisms and can be found in many habitats. They exhibit all types of feeding behaviors. Though most are herbivorous, some are parasitic, some are carnivorous, and some feed on decaying organic matter. The mouthparts of insects are specially adapted for each insect's feeding habits, so they vary greatly from one species to another.

Humans have many unpleasant interactions with insects, including those that sting, bite, suck our blood, and transmit disease. Then there are the pests that infest our homes and gardens. You are about to examine an insect that causes a lot of damage, so to balance our coverage, can you name some insects that are beneficial to humans?

> *Examples: Many insects pollinate flowers, including crop species. We eat the honey made by bees. Ladybird beetles and other insects (such as tiger beetles, praying mantises, and ant lions) eat insects that are harmful to garden plants. Insect larvae such as maggots are useful for decomposing animal waste.*

Termites are the damaging insects mentioned above, and they have an unusual digestive system. Termites feed on plant materials, including wood. But they lack the digestive enzyme needed to dismantle cellulose molecules, which are a major constituent of wood. The cellulose is digested by protists, single-celled eukaryotes that live in the termite's digestive tract. These protists belong to the group called protozoans, or animal-like protists. They are further classified as flagellates because their means of locomotion is long, whiplike structures called flagella. These inhabitants of the termite gut do have the necessary enzyme for digesting cellulose, and termites are able to absorb some of the products for their own nutrition. This type of relationship, where two organisms live in close association with each other, is called symbiosis. Since both partners in this case benefit, the relationship is further classified as mutualism.

Procedure

1. Get a termite and place it on a microscope slide.
2. Hold the termite by using forceps to grasp its body in the middle.
3. Use a second pair of forceps to grasp the very tip of the termite's abdomen (its "tail end").
4. Pull gently. You should see the termite's gut separate from its body.
5. Remove everything from the slide except the gut.
6. Put a drop of 0.9% NaCl on the gut, then a drop of Proto-Slo. Use a tip of the forceps to mix it around.

Why should you use a NaCl solution instead of water?

> *The solution of NaCl is used to preserve the osmotic balance of the protozoans. If you used water, the protozoan cells would take up water and burst.*

7. Add a cover slip and tap it lightly with the eraser end of a pencil to spread out the tissue.
8. View your slide under a compound microscope.
9. Sketch some of the protozoans you see.

Trichonympha agilis

Holomastigotes elongatum

Streblomastix

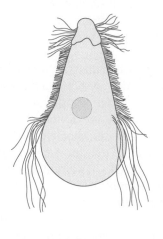

Trichonympha

EXERCISE 13.8
Phylum Echinodermata

Objective

After completing this exercise, you should be able to

1. Identify representative members of the phylum Echinodermata and describe their unique adaptations.

The 6,000 species of echinoderms (Greek *echinos*, sea urchin or hedgehog; Greek *derma*, skin) are all marine and they are all radially symmetrical as adults. Examples include sea stars, brittle stars, sand dollars, sea urchins, sea cucumbers, and sea apples. Echinoderms possess many unique adaptations. One is locomotion by means of a water vascular system (vascular refers to any vessels that convey fluid). Water pressure in the system is varied, causing the "tube feet" to extend or retract (see Figure 13.14). Echinoderms have an endoskeleton that may have spines protruding through the skin. When you pick up the shell of the sand dollar on the beach, for example, sometimes spines are still attached to it. The holes where the tube feet extended through the shell can be seen in a delicate pattern on the shell.

If at all possible, have a saltwater tank containing sea urchins or some other echinoderm so students can see the tube feet.

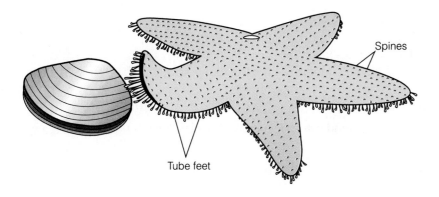

Figure 13.14.
Sea star.

Many sea stars are carnivorous, feeding on various invertebrates that either are sessile or move more slowly than they do. They are especially important predators of clams and oysters and can be an expensive nuisance to aquaculturists who farm these animals. Sea stars have an unusual way of feeding on bivalves (Color Plate 6). By wrapping its arms around the shell and pulling constantly with the suckers on its tube feet, the sea star is able to fatigue the muscles that are holding the shell closed just enough to create a small opening (less than 1 mm is sufficient). The sea star then turns its stomach inside out, inserts it into the shell, and digests the mollusk's soft body parts. When finished, it retracts its stomach into its own body.

The sea urchins, sand dollars, and sea biscuits have no arms; they are all spherical or disk-shaped. Their relationship with the star forms, however, can be seen if you picture the arms of a sea star bent backwards to meet at the tips (Figure 13.15). As in the other echinoderms, the body is enclosed in a shell that is covered by an epidermis. All of the members of this class are covered with movable spines (Color Plate 6). Although some species produce poison to paralyze prey, most sea urchins feed peacefully on algae or dead animals.

Figure 13.15.
Relationship of sea urchin (a) to sea star (b). Both sea stars and sea urchins have fivefold radial symmetry. Although sea urchins lack arms, there are five grooves where the tube feet are located.

a.   b.

The sea cucumbers (Color Plate 6) and sea apples lack both spines and arms. The skeletal elements are small and are buried in the leathery body wall. These animals lie on the ocean floor and filter feed using branched tentacles that are really modified tube feet. Some sea cucumbers are burrowers.

Procedure

Observe the echinoderm specimens on display. Sketch them in the margin of your lab manual.

Why is radial symmetry adaptive for the echinoderms?

> *As sedentary animals, they would not benefit greatly by having the sensory organs localized in one end of the body.*

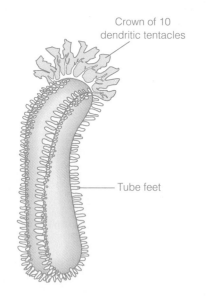

Crown of 10 dendritic tentacles

Tube feet

EXERCISE 13.9
Phylum Chordata

Objectives

After completing this exercise, you should be able to

1. Name representatives of the invertebrate members of the phylum Chordata.

2. Identify members of each of the five groups of vertebrates described in this exercise.

3. Describe the functions of the organs in the frog's digestive system.

4. Explain the significance of the amniotic egg.

5. Compare the digestive system of a typical bird to that of the earthworm and the frog.

6. Describe how the digestive tract of herbivorous mammals differs from that of carnivorous mammals.

The phylum Chordata includes a few invertebrates (animals without backbones) as well as the vertebrates. All share the common chordate characteristics: a hollow nerve cord that runs along the animal's back, a notochord, gill structures, and a post-anal tail.

One group of invertebrate chordates is the tunicates. Tunicates are marine animals commonly called sea squirts. Most of the adults are stationary suspension feeders. They are commonly found on docks, boats, and rocks, though some live in the open ocean. A group of tunicates is shown in Color Plate 6.

The largest group of chordates by far is the vertebrates, which are distinguished from invertebrates by the presence of a backbone. Although vertebrates make up only about 3% of all animals, they are probably the most familiar animals: fishes, amphibians, reptiles, birds, and mammals.

Fish are the most diverse group of vertebrates, including more than 30,000 species. The earliest fish had no jaws and survived as scavengers or parasites. The evolution of jaws brought unprecedented opportunities for exploiting food sources. The majority of fish are carnivorous. They seize and hold their prey with their teeth and then swallow it whole. Some have teeth in their throats to help grind the food. Herbivorous fish feed on aquatic plants such as grasses and other flowering plants and algae. Some fish eat both plants and animals, some are parasites, and some feed on organic debris. A large number of fish are suspension feeders. Many of them live in schools in the open sea, straining small floating plants and animals from the water.

Amphibians include approximately 4,800 species of newts, salamanders, toads, and frogs. As the name (Greek *amphi*, double; Greek *bios*, life) implies, amphibians lead a double life in terrestrial and aquatic habitats. Although important adaptations for terrestrial life have appeared, such as lungs and limbs, amphibians still require water—or at least moisture—for reproduction, and their larvae have an aquatic tadpole stage.

Most adult amphibians are carnivorous, feeding on invertebrates such as snails, worms, insects, and any other moving animal that is small enough to swallow. Frogs and toads have a tongue attached in the front of the mouth. When prey is spotted, the tongue is quickly launched and the prey is caught on the sticky end. The mouth has small teeth to hold the prey, which is then swallowed whole.

We will use the frog to illustrate the further specialization of the digestive tract in vertebrates. All vertebrates share similarities in their body plans, including in their internal organs, so the frog's digestive tract is very similar to our own.

Figure 13.16.
Digestive system of the frog.

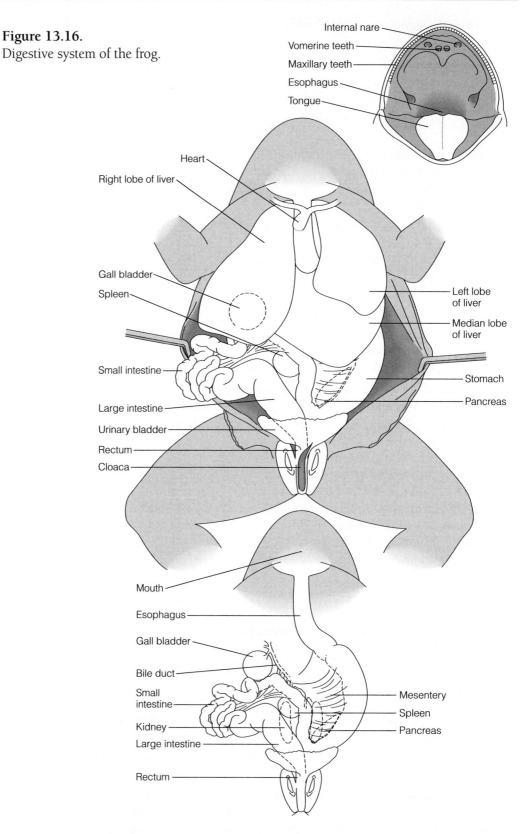

Internal nare
Vomerine teeth
Maxillary teeth
Esophagus
Tongue

Heart
Right lobe of liver
Gall bladder
Spleen
Left lobe of liver
Median lobe of liver
Small intestine
Stomach
Large intestine
Pancreas
Urinary bladder
Rectum
Cloaca

Mouth
Esophagus
Gall bladder
Bile duct
Small intestine
Mesentery
Spleen
Pancreas
Kidney
Large intestine
Rectum

Procedure

Examine the mouth of the frog. Note the teeth along the upper jaw and in the roof of the mouth. Does the frog use these teeth for chewing?

No, they are only used for holding prey.

Examine the tongue of the frog and note where it is attached. The back of the mouth cavity—the throat—is the pharynx, which connects with the esophagus. Follow the path of the esophagus from the pharynx to the stomach. (If it has not already been done for you, remove most of the liver so you can see the esophagus.) Mixing and some enzymatic digestion of the food occur in the stomach.

Next locate the small intestine. Further enzymatic digestion occurs here, followed by absorption of nutrients into the blood capillaries for distribution to the body. The inner surface of the small intestine is very wrinkled, resulting in a large amount of surface area. Why is this important?

The large surface area is needed for absorption of nutrients.

Find where the small intestine joins the large intestine. One of the main functions of the large intestine is to remove the water remaining in the material coming from the small intestine. From the large intestine, undigested remains go to the rectum to be expelled from the body.

Various glands are associated with the digestive system. The largest gland in the body is the liver. The liver produces bile, which emulsifies fats (Lab Topic 14). The bile is carried by a small duct to the gall bladder, where it is stored and released to the small intestine as needed. The gall bladder is beneath the liver and resembles a small green pea.

Also look for the pancreas, a thin, inconspicuous organ located in the membranous tissue between the stomach and small intestine. The pancreas provides the digestive enzymes needed by the small intestine.

Reptiles include approximately 6,500 species of snakes, lizards, turtles, and crocodiles. Reptiles were the first vertebrate group to achieve complete independence from the aquatic habitat. This was in part due to significant changes in reproduction. Amphibian eggs are encased in jelly and must be protected from drying out. Reptiles have a protective shelled egg that contains food and moisture for the developing embryo (Color Plate 6). This amniotic egg represents a major breakthrough for terrestrial life.

A related problem had to be solved at the same time. Most fish and amphibians release their gametes into the water, where sperm meets egg and fertilization occurs. Once the shell develops around the egg, which happens in the female's reproductive tract, there's no way for the sperm to get inside and fertilize it. The egg must be fertilized before it leaves the female's body. Thus all reptiles employ internal fertilization.

Anatomical differences also make reptiles better adapted to life on land. Their skeletons, for example, provide better support than do amphibian skeletons.

Why is structural support more important for a land-dwelling organism than it is for an aquatic one?

Water is a more buoyant medium than air, so terrestrial animals require more support.

In contrast to the amphibians, reptiles have dry, scaly skin that protects against dehydration. Since the skin is no longer part of the respiratory system as it is in amphibians, the lungs are more efficient. The circulatory, excretory, and nervous systems have also undergone evolutionary changes to adapt reptiles to terrestrial life.

There are approximately 8,600 species of **birds**, which makes them second only to the bony fishes in vertebrate diversity. Yet birds are generally very similar to each other in many ways, because flight requires a very specialized anatomy.

All birds have wings, which are modifications of the front limbs, and hind limbs that may be suitable for walking, perching, swimming, grasping prey, and scratching. All birds have horny beaks and all lay eggs. In addition, all birds have a unique feature shared by no other animals: feathers. Birds evolved from a line of reptiles, and feathers evolved from reptilian scales. Birds still have scales on parts of their bodies, especially the feet. In addition to their essential role in flight, feathers provide insulation from the elements.

A display of photos of bird bills and feet (or actual specimens, if available) would be a useful addition to the lab.

Birds eat a variety of animal foods, including insects and other invertebrates, such as worms, and vertebrates such as fish and frogs, reptiles, and small mammals. They also eat plant foods such as seeds, fruit, and nectar. Their beaks are adapted for their food sources. The long probing bill of the hummingbird, the hooked bill of the hawk, and the thick seed-crushing bill of the cardinal are just a few examples.

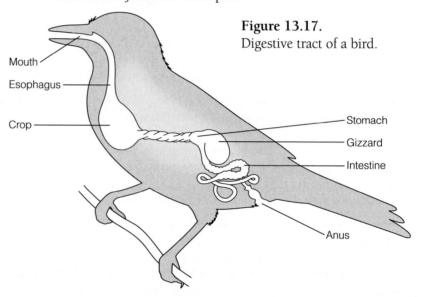

Figure 13.17.
Digestive tract of a bird.

Mouth
Esophagus
Crop
Stomach
Gizzard
Intestine
Anus

Figure 13.17 shows the digestive system of a typical bird. Compare this to the digestive system of the earthworm and the frog. Explain the similarities and differences among the animals in terms of their diets.

Like both the earthworm and the frog, the bird's digestive tract has specialized organs. The esophagus has the same function in birds as in the others. The bird has a crop and a gizzard like the earthworm. The frog lacks these organs. As carnivores, frogs consume more easily digested food. The food of earthworms and most birds needs additional processing, especially by the grinding action of the gizzard. (Many birds eat gravel to assist in this process.) The earthworm does not have a stomach, but the vertebrates do. Presumably this reflects the greater complexity and sophistication of vertebrate systems.

Like their reptilian ancestors, birds have amniotic eggs. The chicken egg is a familiar example. The yolk and the protein in the egg white supply the food for the developing embryo. The egg white also contains a great deal of water, and the whole thing is sealed into a watertight shell. The only substance that enters the egg from the environment is oxygen.

Which came first, the chicken or the egg?

> *The egg. Reptiles were first to have amniotic eggs, and birds evolved from reptiles.*

The **mammals** include approximately 4,500 species ranging in size from under an inch long (the masked shrew) to 100 feet (the blue whale) and adapted for diverse lifestyles. Mammals fly, swim, climb trees, burrow underground, run on four legs, and walk upright on two legs. Note that locomotion in mammals differs from that of reptiles in that the limbs are placed more directly under the body. The characteristics shared by all mammals are bearing fur or hair and nourishing the young with milk. The young of most mammals develop in a uterus inside the mother's body, but the duck-billed platypus and the spiny anteater lay eggs. Marsupial mammals, including opossums, kangaroos, and koalas, do develop their young in a uterus, but only for a brief period. Development is completed in a pouch on the mother's body after the offspring are born.

As in other groups, many of the adaptations of mammals reflect diet. Some use specialized food resources, while others eat a wide range of foods. Herbivores, carnivores, insectivores, and omnivores are all found among the mammals. The digestive tracts of mammals also reflect their diets. In general, carnivores have relatively short digestive tracts, as you saw in the carnivorous frog. Carnivores have separate meals with time between. Herbivores, on the other hand, spend much of their day eating. What might be the reason for this difference? (Hint: Consider the composition of the diets. Do you get more calories from a salad or a hamburger?)

> *Compared to the fat and protein content of meat, the diet of herbivores is relatively low in calories. Plant material contains a lot of cellulose—fiber, to us—that "dilutes" the digestible carbohydrates.*

Like termites, vertebrates lack the enzymes necessary for digesting cellulose. And herbivores have a similar solution to the problem: gut symbionts, mainly bacteria. Herbivores have very long digestive tracts to allow time and space for the extra processing steps needed.

Since almost all mammals have teeth, comparison of teeth is a good way to study feeding adaptations. Mammals have four types of teeth (Figure 13.18). Incisors are used primarily for snipping or biting off a piece of food, the sharp canines are used to pierce, premolars shear or slice, and molars crush and grind. Depending on the animal's diet, certain teeth are emphasized and modified, while other types may be absent.

There are two basic designs. Herbivores such as horses and rabbits have an enlarged cecum—a pouch at the juncture of the small and large intestines—where the microorganisms live. Ruminants such as cows, sheep, goats, buffalo, and deer have a more elaborate system. The huge ruminant stomach (26-gallon capacity in cows) has four chambers for different steps in the processing of plant matter. Ruminants chew cud, regurgitated material, to further break down the plant fibers mechanically.

Figure 13.18.
Types of teeth in a human.

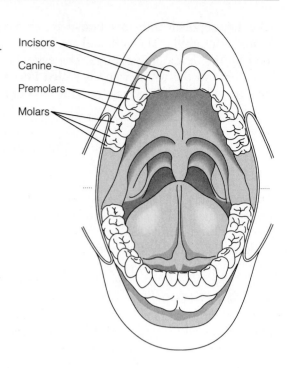

Incisors

Canine

Premolars

Molars

Procedure

View the skulls or photographs on display. For each specimen, describe the teeth and state your inferences about the animal's diet.

Answers depend on the specimens available.

Questions for Review

1. Examine the "unknown" animals on display. Decide which phylum each animal belongs to and explain what characteristics you used to make your decision.

2. Why are sponges named pore bearers?

   Their bodies are full of pores, which are used to bring in the large amount of water needed for suspension feeding.

3. Bivalves have little sensory ability and heavy shells, while cephalopods have well-developed senses and reduced shells. How are these characteristics adaptive for each group?

   Cephalopods live a more mobile life than bivalves, actively seeking prey. Their degree of mobility also requires awareness of their surroundings. Bivalves are filter feeders needing only to find a suitable spot to occupy. In addition, most species are buried in the sediment. Their sedentary habit requires that their bodies have a very effective means of protection; it also means they don't have to carry the weight around with them.

4. Compare the echinoderms with the cnidarians. Consider habitat, symmetry, locomotion, defense, and means of obtaining food.

   Echinoderms and cnidarians are both radially symmetrical. Echinoderms are exclusively marine and most cnidarians are marine. All members of both groups are aquatic. Locomotion in echinoderms is slow and not graceful. Although the medusa forms of cnidarians are quite motile, the polyp forms are sessile and capable only of an ungainly tumbling motion. Cnidarians defend themselves with nematocysts and have no other form of protection. Echinoderms have an endoskeleton of hard plates and may have spines. Cnidarians are carnivorous and use their nematocysts and tentacles to capture prey. Echinoderms may eat other animals or plants or may scavenge organic material. Their food-gathering mechanisms are varied.

5. Name the three phyla that contain animals called worms. Discuss the differences among the digestive tracts of these three phyla.

   Flatworms (Platyhelminthes)—One-opening digestive tract

   Roundworms—Two-opening digestive tract but no specialization of orgams

   Annelids—Specialized organs in digestive tract

6. Concentrating on the evolutionary transition from amphibians to reptiles, discuss the adaptations that allow vertebrates to be successful in terrestrial habitats.

   Reptiles have dry, scaly skin that resists dehydration; their lungs are more efficient to make up for the loss of gas exchange through the skin. The skeleton is adapted to provide more support, since reptiles are not buoyed up by water. And, with the amniotic egg, reptiles are freed from needing water for reproduction. Neither fertilization nor development requires water.

7. Using as examples at least two bilaterally symmetrical organisms and two radially symmetrical organisms, discuss the adaptive value of bilateral versus radial symmetry.

8. Choose three animals from different phyla and describe their adaptations for obtaining food. Explain how these adaptations are suited to each animal's environment.

9. Choose three animals from different phyla and compare their means of digesting food. Include in your answer a comparison of the efficiency of each digestive system.

LAB TOPIC 13

Animal Diversity Instructor Notes

If possible, students should see living examples of as many animals as possible. For example, a saltwater tank is an excellent way for students to see some of the invertebrates. Or perhaps you can put together an exhibit of arthropods. A number of amphibians and reptiles are available from biological supply houses, or you may be able to arrange to borrow some temporarily from a local pet store.

Organization

Introduction (5 minutes)

Briefly review the characteristics of animals and Figure 13.1 (phylogenetic tree).

Prep for Procedures (5 minutes)

Some materials will be available to each pair of students (activities for Porifera, Cnidaria, planaria, and Arthropoda). The others (tapeworm, Nematoda, Annelida, Mollusca, Echinodermata, and Chordata) will be limited to stations or centers so that students share materials. Explain where specimens are located. You may want to assign different pairs of students to start with different activities.

If you have living vertebrates on display, a few words about humane treatment of animals are in order. If the animals are not to be handled, be sure there are signs cautioning students not to touch them. If you do allow students to pick up the animals (for example, to see how dry reptile skin is), instruct your students in proper handling techniques. They should also be told to wash their hands immediately after touching the animals.

Summary (15 minutes)

Have the students share their observations and results from the laboratory activities and answer the Questions for Review. As usual, students should contribute their answers rather than having the instructor deliver the information. If you had students classify unknown animals as suggested at the end of the lab, ask students to give the answers they came up with and their reasons for making their decisions.

For a 2-Hour Lab

You may wish to emphasize some groups at the expense of others. The list of time requirements will help you choose the activities that suit both your interests and your time budget. Alternatively, you may wish to spread the activities over two lab periods, which affords more time for discussion and supplementary diversity displays.

How to Open a Clam

To open a clam, use a knife or strong scalpel to cut away as much as you can of the hinge that holds the two shells together. You should be able to insert the blade into one end of the hinge and make an opening big enough to force in a screwdriver. Insert the screwdriver; then twist it to pry the shells apart slightly. Either leave that screwdriver in place or insert some kind of wedge to hold the shell open. Then insert a screwdriver on the other side of the shell and twist it to open the shell further. If you want to finish opening the clam later, you can now slip a wooden peg between the shells.

At this point you should be able to slide a knife or scalpel between the shells and cut the two muscles that hold it together. Next, open the shell slowly until you can see inside. The mantle will be clinging to the top shell. Carefully run the scalpel underneath it to free it from the shell. You will have to do a little extra cutting where it is attached to the muscles. Finally, bend the hinge back so that the clam can be laid down flat for display.

Digestion

Introduction

Where do the molecules in our bodies come from? Unlike plants, we cannot turn carbon dioxide from the air into carbohydrates, nor can we absorb other necessary nutrients from the environment. We and other animals ingest the raw materials for our molecules in the form of food. In the process of **digestion,** the food items are broken down into their component parts. They are then changed and repackaged as molecules that fit our immediate needs. They may also be stored for later use. This is analogous to the way a car is stripped by professional thieves: Its parts are redistributed and are no longer recognizable as parts belonging to the original car.

Most of the food that we eat consists of the biological macromolecules that were studied in Lab Topic 3 (Macromolecules): proteins, carbohydrates, lipids, and nucleic acids. In this lab topic, you will learn how macromolecules are broken down into their subunits so that they can be absorbed into the cells of the body. You will then design and perform an experiment using one of the methods you have learned for investigating food molecules.

Instructor Overview
In this lab topic, students perform demonstration exercises to learn simple tests to assay three enzymes, trypsin, lipase, and amylase, that are responsible for the digestion of three major types of food molecules, protein, fats, and starch. Each student team will then design its own experiment using one or more of the assays. The lab prompts students to organize their investigations (see Exercise 14.4) and answer a series of questions that leads them to interpret their results (see Exercise 14.5). Alternatively, you may require students to write lab reports.

 Wear safety goggles while performing these experiments.

Outline

Exercise 14.1: Digestion of Proteins by Trypsin

Exercise 14.2: Digestion of Lipids by Lipase

Exercise 14.3: Digestion of Starch by Amylase

Exercise 14.4: Designing an Experiment

Exercise 14.5: Performing the Experiment and Interpreting the Results

Time Requirements
Exercise 14.1: 10–15 minutes (other exercises are performed while waiting for complete results, which take an additional 20–30 minutes)
Exercise 14.2: 20 minutes (same note as Exercise 14.1)
Exercise 14.3: 20 minutes
Exercise 14.4: 20 minutes
Exercise 14.5: ~60 minutes (or available time)

At the beginning of the period, the instructor should make up the gelatin solution for the class. See the Instructor Notes at the end of the lab topic for recipe.

EXERCISE 14.1
Digestion of Proteins by Trypsin

Objectives

After completing this exercise, you should be able to

1. Explain the purpose of protein digestion.
2. Identify the enzymes involved in protein digestion.
3. Explain how the procedure demonstrated in this exercise is used to determine whether protein digestion has occurred.

Introduction

The proteins we eat can be neither absorbed nor used by the body directly. They must be completely dismantled into their subunits, amino acids. We then use these amino acids to assemble our own proteins. Thus, even though we might eat quantities of muscle (meat) containing the proteins actin and myosin, these protein molecules do not necessarily become actin and myosin in the muscles in our bodies. Their amino acids are reassigned to whatever proteins are needed at the moment, which might be keratin for hair, hemoglobin for red blood cells, an enzyme in the aerobic respiration pathway, or any one of thousands of other proteins. Nine of the 20 amino acids that we need to make proteins can be obtained only by breaking down proteins from food. These are called the nine essential amino acids. We can synthesize the other 11 from simpler precursor molecules.

Chemical digestion of protein begins in the stomach. The pH of the fluids in the stomach is extremely acidic, and the primary enzyme for protein digestion in the stomach, pepsin, works best in those conditions. Further protein digestion is carried out in the small intestine by **trypsin,** the enzyme that you will use for the following procedure. Trypsin is produced in the **pancreas,** a gland that secretes hormones as well as digestive enzymes, and then exported to the small intestine. Unlike pepsin, trypsin functions most effectively under the slightly alkaline conditions of the small intestine. The enzymes that digest proteins are called **proteases.**

Method

In the following procedure you will use the enzyme trypsin, a protease present in an extract made from the pancreas, to digest gelatin.

Gelatin is a protein obtained from sources such as animal hooves and hides. It is capable of taking up large amounts of water, which cause it to swell and set in the familiar Jell-O fashion. If the gelatin is digested—for example, if trypsin has begun to break down the protein molecules into their subunits—the gelatin's structure is destroyed. Without the proper structure, the gelatin will not gel. Figure 14.1 illustrates this test.

If you were to design an experiment using this technique, what dependent variable would you be measuring?

Whether or not the gelatin sets.

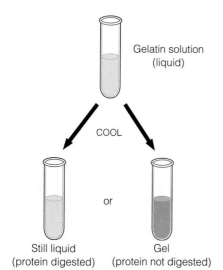

Figure 14.1.
Gelatin test for protease activity. The gelatin will gel only if its protein structure is intact. Digested protein will not gel.

How is this dependent variable related to the activity of trypsin?

Trypsin is an enzyme that digests protein (gelatin). If the trypsin is active, the gelatin will be digested and will not gel.

What factors might be used as independent variables?

Any factor that affects enzyme activity: for example, pH, incubation temperature, amount of trypsin, amount of gelatin.

Predict the outcomes in Tubes 1 and 2. Justify your predictions.

Example: Tube 2 contains protein-digesting enzyme, so the gelatin will not gel. There is nothing to prevent Tube 1 from gelling.

Procedure

1. Get two test tubes and label them 1 and 2 with a wax marker.
2. Use a pipet to measure 2 mL of water into Tube 1.
3. Use a pipet to measure 2 mL of pancreatic extract (trypsin) into Tube 2.
4. Use a pipet or a 10-mL graduated cylinder to measure 10 mL of liquid gelatin into each tube. Swirl to mix.
5. Place the test tubes in ice for 20–30 minutes.

 Go on to the next exercise while you are waiting for the results of this experiment.

6. After 20–30 minutes, place your thumb over the top of each tube and invert it to test whether or not the gelatin has set.
7. Record your results below and explain what happened in each tube.

Tube 1:

This is gelatin with nothing else added. The gel should set.

Tube 2:

Trypsin has begun to digest the gelatin, so the gelatin doesn't set. It is still liquid.

 After you have obtained results for the experiment, be sure to clean your glassware thoroughly.

EXERCISE 14.2
Digestion of Lipids by Lipase

Objectives

After completing this exercise, you should be able to

1. Explain the purpose of lipid digestion.
2. Explain the difference between emulsion and digestion and explain why emulsion of lipids is necessary.
3. Identify the enzyme involved in lipid digestion.
4. Explain how the procedure in this exercise is used to determine whether lipid digestion has occurred.

Introduction

A fundamental characteristic of lipids is that they are insoluble in water. For example, salad dressing made from oil and vinegar will separate into two layers. Why does this property of lipids pose a problem for the digestive system? (Hint: What is the primary ingredient in the beverages we drink?)

> *The digestive processes are water-based. Notice that the procedures in this lab topic use aqueous solutions and the enzymes are also in aqueous solution. The beverages we drink are aqueous. Water is recovered from the large intestine when digestion is complete.*

Lipid digestion must begin by making the molecules more compatible with water so that the digestive enzymes have access to them. This is accomplished by breaking up the lipid into small droplets, which are then distributed throughout the water. This type of mixture is called an **emulsion.** Immediately before pouring out salad dressing, you shake the bottle. The separation problem is solved only temporarily, however. In a few minutes, the oil droplets have again joined up with each other and the oil and aqueous vinegar have separated. A permanent emulsion can be established by using an emulsifying agent to keep the oil droplets dispersed. In salad dressing, mustard may be added to stabilize the oil-and-vinegar emulsion. Egg yolks are also emulsifiers. For example, mayonnaise is an emulsion containing oil, lemon juice, and egg yolks. In the digestive process, **bile,** which is produced by the liver and stored in the gall bladder, emulsifies lipids.

Once lipids are emulsified they are broken down into their subunits by enzymes called **lipases.** The subunits can then be absorbed into the body. Lipases are another type of digestive enzyme that is produced in the pancreas and secreted into the small intestine. Lipid emulsification and digestion then take place in the small intestine, and the subunits are absorbed into the blood.

Fats, which constitute a major subclass of lipids, are composed of a glycerol molecule linked to fatty acids. Most fats found in nature have three fatty acids, so they are called triglycerides. There are also fats that contain one or two fatty acids linked to a glycerol molecule. Enzymatic digestion of fats releases the fatty acids. As weak acids, the fatty acids affect the pH of a solution.

Method

In the following procedure you will use the acidic property of fatty acids to determine whether digestion of fats has occurred. You will mix vegetable oil, a fat, with pancreatic extract, which contains lipases as well as trypsin. If fats are being digested, will the pH of the solution increase or decrease?

Decrease (become more acidic), due to release of fatty acids.

You will use phenolphthalein as a pH indicator. NaOH (sodium hydroxide) is added to make each solution strongly basic. Phenolphthalein turns pink at pH 10. As digestion occurs and fatty acids are released, the phenolphthalein will become colorless, indicating that the solution is now only slightly basic (approximately pH 8).

Examine Table 14.1, which shows how the experiment is set up, and answer the questions that follow.

Table 14.1
Amounts of Reactants Used in Lipid Digestion Procedure

	Tube		
	1	**2**	**3**
Vegetable oil (mL)	1	1	1
Bile (mL)	0	1	1
Water (mL)	2	1	6
Phenolphthalein (dropperful)	1	1	1
Pancreatic extract (mL)	5	5	0

What hypothesis is being tested in this experiment?

Example: Bile increases lipase activity. (Note: Students might also view this as a lipase experiment and hypothesize that lipase is required for fat digestion.)

What dependent variable is measured by the procedure that is used?

Color of the mixture, which indicates its pH.

How is this dependent variable related to the activity of lipase?

As lipase digests fats and releases fatty acids, the pH of the solution decreases (becomes more acidic).

What is the independent variable in this experiment?

Presence of bile (or lipase, if the hypothesis concerns lipase).

What other factors might affect lipase activity?

Any factor that affects enzyme activity—for example, pH, incubation temperature, amount of enzyme, or amount of substrate; also type of substrate (different fats or oils), different types of emulsifiers.

What control treatments are used in this experiment?

Tube 1 contains enzyme but no bile; Tube 3 contains bile but no enzyme.

Predict the outcome of this experiment and justify your prediction.

Example: Tubes 1 and 2 both contain enzyme, so they will both be colorless at the end of the experiment, but Tube 2 will become colorless faster because it also contains bile. Tube 3 will show no change because it doesn't have any enzyme in it.

Procedure

1. Get three test tubes and label them 1, 2, and 3 with a wax marker.
2. Use a pipet to measure 1 mL of vegetable oil into Tubes 1, 2, and 3.
3. Use a pipet to measure 1 mL of bile into Tubes 2 and 3.
4. Use a pipet to measure water into each tube, as shown in Table 14.1.
5. Add 1 dropperful of phenolphthalein from a dropping bottle to each tube.
6. Using a pipet, add 5 mL pancreatic extract to Tubes 1 and 2.
7. Cover each tube tightly with Parafilm, put your thumb over the top, and shake vigorously ten times.

 NaOH is a caustic solution. Avoid contact with your skin or clothing. Inform your instructor immediately if a spill occurs.

8. Remove the Parafilm and put 2–3 drops of NaOH from a dropping bottle into each tube. Replace the Parafilm and shake to mix the solution. Add drops of NaOH until the solution stays pink.
9. Record the color in each test tube in Table 14.2 at time 0.

 Release of fatty acids will change the solution from pink to colorless. However, the bile solution in Tubes 2 and 3 will give them a pale yellow cast throughout the experiment.

Table 14.2
Results of Lipid Digestion Procedure

Time (min)	Tube		
	1	**2**	**3**
	Oil + Lipase	Oil + Bile + Lipase	Oil + Bile
0	Pink	Pink	Pink
1	Pink	Lighter pink	Pink
2	Pink	Much lighter	Pink
3	Pink	Colorless	Pink
4	Lighter pink		Pink
5	Lighter pink		Pink
20	Lighter pink		Pink
30	Colorless		Pink

10. For the next 5 minutes, record the color in each tube at 1-minute intervals in Table 14.2.

11. Record the colors again after 20 minutes.

All of the enzymatic reactions are temperature dependent. If results are not obtained, let the reaction run longer. The Prep Guide instructs the prep staff to have all solutions at room temperature for this lab.

While you are waiting for the results from this procedure, go back to Exercise 14.1 and observe the results. Then begin Exercise 14.3.

12. Record the colors again after 30 minutes, if time permits.

When you have obtained the results of this experiment, be sure to clean your glassware thoroughly. Soaps and detergents are also emulsifying agents, breaking grease up into small particles that are more easily carried away from your dishes. Please keep this in mind when you clean up your glassware.

The indicator range for phenolphthalein is 8.3 (colorless) to 10 (pink).

Explain the color change or lack of color change in each tube.

Tube 1

It takes about 25 minutes or so for this tube to decolorize. Lipase is digesting the fats, but slowly because most of the oil is in one big lump rather than dispersed through the solution, where it will be in contact with the lipase.

Tube 2

This tube decolorizes rapidly. The oil has been emulsified by the bile, permitting more access to the fat by the lipase.

Tube 3

This tube contains no enzyme, so despite being emulsified by the bile, the fat is not digested. The color in the tube remains pink.

E X E R C I S E   1 4 . 3
Digestion of Starch by Amylase

Objectives

After completing this exercise, you should be able to

1. Explain the purpose of carbohydrate digestion.
2. Explain how the procedures in this exercise are used to determine whether carbohydrate digestion has occurred.

Introduction

Sugars are simple carbohydrates. Polysaccharides such as starch and cellulose are complex carbohydrates. What subunits compose these two polysaccharides?

Glucose.

The ultimate function of carbohydrate digestion is to convert all carbohydrate molecules into monosaccharide subunits that can be absorbed into the blood and ultimately into body cells. The giant polysaccharides obviously must be chopped up into their component subunits, but even disaccharides such as sucrose (table sugar) and lactose (milk sugar) must be converted to monosaccharides for absorption.

Cellulose, which composes the cell walls of plants, is a large polysaccharide and a component of many plant foods we eat, but our digestive systems cannot break cellulose molecules down into their component sugars. Because it is indigestible, cellulose is the largest component of dietary fiber. It adds bulk to the intestinal contents and thus helps the remains of a meal pass through the digestive system more quickly.

On the other hand, starch is a major source of nutrients. Digestion of starch begins in the mouth, where salivary **amylase** initiates the process of clipping the huge polysaccharide into smaller strands, then into disaccharide maltose molecules, and finally into its monosaccharide glucose subunits. Like trypsin and lipase, amylase is also secreted by the pancreas into the small intestine. In the small intestine, starch digestion that was begun in the mouth is completed, and the monosaccharides are absorbed into the blood.

Method

For this procedure you will use pancreatic extract as a source of amylase. Since the activity of digestive enzymes is affected by pH, you will add HCl (hydrochloric acid) to one of the experimental tubes to determine its effect on amylase activity.

Starch is added to the experimental tubes as a substrate for amylase. You will use iodine reagent, I_2KI (iodine-potassium iodide), which turns dark blue in the presence of starch, to determine whether starch digestion has occurred. If no starch is present, the solution remains yellow. If the starch has been partially digested, a red-brown color will result.

Examine Table 14.3 to see how the experiment has been set up, then answer the questions that follow.

Table 14.3
Amounts of Reactants Used in Starch Digestion Procedure

	Tube			
	1	2	3	4
Pancreatic extract (mL)	0	1	1	1
Starch (mL)	5	5	5	0
0.5% HCl (dropperful)	0	0	1	0
Water (mL)	2	1	0	6

What hypothesis is being tested in this experiment?

Example: Amylase is active in acidic conditions.

What dependent variable is measured by this procedure?

Presence of starch, as indicated by the iodine test.

How is this dependent variable related to the activity of amylase?

When amylase has digested starch, the iodine test will be negative.

What is the independent variable in this experiment?

pH (addition of HCl to one tube).

What other factors can affect amylase activity?

Any factor that affects enzyme activity—for example, incubation temperature, amount of enzyme, or amount of substrate; also, type of substrate (different sources of starch or different polysaccharides), origin of amylase.

What control treatments are used in this experiment?

Tube 1 doesn't contain enzyme, so it will show whether starch breaks down without enzyme present. Tube 4 contains no starch, so it will show that the enzyme itself doesn't produce a positive reaction to the iodine test.

Predict the outcome of this experiment and justify your prediction.

Example: Tubes 1 and 2 will still test positive for starch at the end because amylase needs acidic conditions to work. (Note: This is not true, but the hypothesis and prediction should be based on what the student thinks, not on what you know.) Tube 3 will test negative for starch after 5 minutes because amylase digests starch in the presence of acid. Tube 4 will still have a negative result because it didn't contain any starch to begin with.

Procedure

1. Get four test tubes and label them 1, 2, 3, and 4 with a wax marker.
2. Use a pipet to measure 5 mL starch solution into Tubes 1, 2, and 3.
3. Add 1 dropperful of 0.5% HCl from a dropping bottle to Tube 3.

Students can use pH paper to measure pH. The pH of Tube 3 is ~4. The pH of the other tubes is ~6.

 HCl is a caustic solution. Avoid contact with your skin or clothing. Inform your instructor immediately if a spill occurs.

4. Using a pipet, add 2 mL water to Tube 1, 1 mL water to Tube 2, and 6 mL water to Tube 4 (see Table 14.3).
5. Using a pipet, add 1 mL of pancreatic extract to Tubes 2, 3, and 4.
6. Cover each tube tightly with Parafilm and invert once to mix.
7. Use a wax marker to label a second set of test tubes 1, 2, 3, and 4. Draw one pasteur pipetful of solution out of each tube in the first set and put it into the corresponding tube in the second set.
8. Put a few drops of iodine reagent from a dropping bottle into each tube in the second set.
9. Record the results as + (starch present) or − (no starch) in Table 14.4.

 A blue color indicates that undigested starch is present. Yellow means that no starch is present.

Table 14.4
Results of Starch Digestion Experiment

	Tube			
	1	**2**	**3**	**4**
Iodine Test for Starch	**Starch**	**Amylase + Starch**	**Amylase + Starch + HCl**	**Amylase**
Initial	+	+	+	−
Final	+	−	+	−

10. Wait at least 5 minutes for digestion to occur in the first set of tubes. While you are waiting, label a third set of test tubes 1, 2, 3, and 4.

11. After 5 minutes, test again for starch in each tube by drawing one pasteur pipetful of solution out of each tube in the first (experimental) set and putting it into the corresponding tube in the third set. Then add a few drops of iodine reagent to each tube in the third set.

12. Record the results in Table 14.4.

 After you have obtained results for the experiment, be sure to clean your glassware thoroughly.

What is a positive reaction to this test?

Starch turns dark blue.

If a solution that contains amylase (pancreatic extract) tests positive for starch after 5 minutes, what does this indicate about amylase activity?

The amylase is not active: The starch has not been digested.

Is your hypothesis supported or proven false by the results? Explain.

Example: Proven false. Starch was not digested in the acidic conditions of Tube 3 (positive iodine test), but it was digested in Tube 2, which had no acid added.

EXERCISE 14.4
Designing an Experiment

Objective

After completing this exercise, you should be able to

1. Design an original experiment to investigate some factor that affects the process of digestion.

The Instructor Notes at the end of the lab topic give you some suggestions for the types of experiments that students can do. You may allow students to choose the method they want to use for their investigation or assign a method to each group in order to make the prep more predictable. Let the prep staff know what additional materials you would like to have available for the students. For example, you might want to have a series of pH buffers, water baths, ice, hot plates, thermometers, and different substances for testing.

In Exercises 14.1, 14.2, and 14.3 you learned techniques for investigating three digestive enzymes that are contained in an extract made from the pancreas: trypsin, lipase, and amylase. In Exercises 14.4 and 14.5 your lab team will design an experiment using one of these methods, perform your experiment, and present and interpret your results. You may review the factors that you listed as possibly affecting the activity of trypsin, lipase, and amylase to help you decide on an independent variable for your investigtion.

The following materials will be provided for your team.

For Trypsin Experiment

test tubes	ice chest with ice
test tube rack	pancreatic extract
10-mL graduated cylinder	2% gelatin solution
5-mL pipets with pi-pump	

For Lipase Experiment

test tubes	0.1N NaOH
small Parafilm squares	vegetable oil
pancreatic extract	5-mL pipets with pi-pump
phenolphthalein solution	

For Amylase Experiment

test tubes	pancreatic extract
1-mL pipets with pi-pump	starch solution
5-mL pipets with pi-pump	iodine reagent

Your instructor will tell you what additional materials will be available.

Describe your experiment below.

Question or Hypothesis

Dependent Variable

Independent Variable

Explain why you think this independent variable will affect the dependent variable.

Control Treatment(s)

Replication

Brief Explanation of Experiment

Predictions

What results would support your hypothesis? What results would prove your hypothesis false?

Method

You may find it helpful to construct a table showing the contents of each reaction tube.

Design a Table to Collect Your Data

.

List Any Additional Materials You Will Require

E X E R C I S E   1 4 . 5
Performing the Experiment and Interpreting the Results

Objectives

After completing this exercise, you should be able to

1. Perform the experiment your lab team designed.
2. Present and interpret the results of your experiment.

Before you do the experiment, be sure that everyone on your lab team understands the techniques that will be used. You may divide up the tasks before you begin work.

Be thorough in collecting data. Don't just write down numbers; record what they mean as well. Don't rely on your memory for information that you will need when reporting on your experiment later! If you have any questions, doubts, or problems during the experiment, be sure to write them down, too.

Results

Before you begin to prepare your results for presentation, decide on the best format to use. Remember, you want to give the reader a clear, concise picture of what your experiment showed. Refer to the data presentation section of Appendix A (Tools for Scientific Inquiry) for help. If you are drawing graphs, use graph paper. Complete your tables and/or graphs before attempting to interpret your results.

Write a few sentences below *describing* the results (don't explain why you got these results or draw conclusions yet).

Discussion

Look back at the hypothesis or question you posed in this experiment. Look at the graphs or tables of your data. Do your results support your hypothesis or prove it false? Explain your answer, using your data for support.

Did your results correspond to the prediction you made? If not, explain how your results are different from your expectations and why this might have occurred.

Describe how your data are supported by information from other sources (for example, textbooks or other lab teams working on a similar problem).

If you had any problems with the procedure or questionable results, explain how they might have influenced your conclusion.

If you had an opportunity to repeat and extend this experiment to make your results more convincing, what would you do?

Summarize the conclusion you have drawn from your results.

Questions for Review

1. As you are happily munching on your Big Mac, your mind wanders back to bio lab and what you learned about digestion. You are moved to describe for your friends the fate of the two all-beef patties, special sauce, lettuce, cheese, pickles, and onions on a sesame seed bun. How is each of these ingredients digested and in what form do they finally reach your bloodstream? (Also include any indigestible parts of the food.)

Two all-beef patties

The protein is digested in the stomach (pepsin) and small intestine (trypsin). The products of digestion are amino acids, which are absorbed into the blood from the small intestine. As we all know, an "all-beef" hamburger contains fat as well as beef. The fat is first emulsified by bile in the small intestine and then digested by lipases. Fats end up as fatty acids and glycerol molecules, which are absorbed in the small intestine. We have not discussed the digestion of cholesterol and other lipids.

Special sauce

At least one component of the special sauce is fat (digestion described above).

Onions, lettuce, pickles

These are all plant materials and so will contain some cellulose, which is an indigestible polysaccharide. Miscellaneous small sugars are also present. Monosaccharides can be absorbed directly. Sucrose must be digested so that the glucose and fructose subunits can be absorbed. This most likely happens in the small intestine.

Cheese

Cheese contains fat and protein, so digestion is the same as for the hamburger.

Sesame seed bun

Bread contains starch. Digestion of starch begins in the mouth (amylase) and resumes in the small intestine (also amylase). Other enzymes besides amylase are required (for example, maltase) in order to reduce starch completely to its subunits (glucose), which are absorbed in the small intestine. The sesame seeds contain cellulose and probably also fats.

2. Fungi lie in contact with their food sources (for example, a slice of bread) as they grow. The fungus secretes enzymes that break down large food molecules, and the fungal cells then absorb the products of

that digestion. Compare this process to the digestive processes described in this lab topic.

Food that we eat is taken to the interior of the body, but since the lumen of the alimentary canal is technically outside of the body, the food doesn't actually enter the body (cross cell membranes) until it is digested. Thus what we do is very similar to what fungi do: secrete enzymes to digest macromolecules and then absorb the subunits.

3. Describe a test to determine whether gelatin has been digested.

    Gelatin is a protein that will set into a gel if its structural integrity is intact. Otherwise it will not gel. So a test of gelatin digestion is to determine whether the gelatin sets.

4. An unknown combination of macromolecules is mixed with pancreatic extract. Samples are tested at the beginning of the experiment and after 30 minutes. The following results are observed.

    Phenolphthalein test: Initially red, colorless after 30 minutes.

    Iodine test: An initial sample is blue. A sample taken after 30 minutes is yellow.

    What macromolecules were present in this sample before digestion? How do you know?

    Fat (because of the phenolphthalein results) and starch (because of the iodine reagent results).

    What molecules are present in the sample after 30 minutes? How do you know?

    Fatty acids (color change in phenolphthalein), monosaccharides (negative starch test).

    Could there have been macromolecules in the sample whose presence you didn't detect? Explain.

    Yes; there was no test for protein or nucleic acids.

Acknowledgments

Procedures for the enzyme assays were adapted from the following sources:

Armstrong, W. D., and C. W. Carr. *Physiological Chemistry Laboratory Directions,* 3rd ed. Minneapolis: Burgess Publishing, 1963.

Dotti, L. B., and J. M. Orten. *Laboratory Instructions in Biochemistry,* 8th ed. St. Louis: C. V. Mosby, 1971.

Oser, B. L., ed. *Hawk's Physiological Chemistry,* 14th ed. New York: McGraw-Hill, 1965.

LAB TOPIC 14

Digestion
Instructor Notes

Organization

The instructor must make up the gelatin solution at the beginning of lab (see 2% Gelatin Recipe on page 14-21).

Time requirements for performing the exercises will vary depending on the laboratory skills of the students. The time estimates assume that students have gained some familiarity with using pipets and following directions for laboratory procedures by performing labs earlier in the course. Students will also be able to design their investigations more readily if they have had at least one previous investigative lab. If you think that your students may not be prepared to perform these exercises in the time alotted without being rushed, eliminate one of the techniques (Exercise 14.1, 14.2, or 14.3).

Introduction (10–15 minutes)

Be sure that students understand what digestion means and why it is necessary for animals. Review the macromolecules and their subunits (Lab Topic 3, Macromolecules). Explain that pancreatic extract contains various digestive enzymes. Explain what each assay (Exercises 14.1, 14.2, and 14.3) tests.

Prep for Procedures (5 minutes)

Tell students to move on to other exercises during the waiting periods.

Prep for Investigations (15–20 minutes)

Review the results of Exercises 14.1, 14.2, and 14.3.

Before you tell students to do Exercise 14.4, give them some idea of extra materials that are available for their investigations (for example, pH buffers, different macromolecules to use as substrates, different proteases to compare with trypsin, means of varying temperature). Writing a list on the board would be helpful. See Possible Investigations on page 14-21 for additional suggestions.

At the previous lab meeting, tell students to bring in ideas and materials for this experiment. A grade incentive helps. For example, you can give them a few points for bringing to lab a list of factors that affect enzyme activity or a list of foods that contain each type of macromolecule.

If your lab schedule permits, you can end the lab after performing the demonstration exercises and give students until the next lab period to design their experiments and bring in any extra materials they will need.

If this is the first investigation, it may initially be difficult to get students started on designing an experiment. One useful technique is to have each group brainstorm ideas before settling on one. They should quickly write down everything they can think of that might affect digestive enzymes. If they can't think of anything at all, you can help out by asking them to review Lab Topics 3 (Macromolecules) and 5 (Enzymes).

Once students have a list of broad ideas, they can go back through it and consider each idea as a starting point for an experiment, eliminating those that don't lend themselves to the kinds of investigations that can be done in lab.

Keep in mind that students are used to having the instructor tell them what experiment to do. Many students will delay independent thinking for as long as possible, hoping that you will eventually give up and tell them what to do. Some students believe that you have an experiment in mind and they are supposed to figure out what you want. They will continually ask if their ideas are correct. Resist the temptation to take the traditional role of the instructor; allow your students to develop their own ideas.

Teams work independently on performing their investigations. For evaluation, students can formalize and hand in their results and answer the questions in the lab manual for Exercise 14.3 or they can write a formal report.

For a 2-Hour Lab

Choose one of the assays to use (do Exercise 14.1, 14.2, or 14.3). Then go ahead with Exercises 14.4 and 14.5.

2% Gelatin Recipe

Each group will use ~20 mL. To dissolve 2 g gelatin powder in 100 mL water, first blend the gelatin powder with a little cold water to wet it. Bring about half of the water to a near boil and add it to the gelatin, stirring to dissolve it. Once the gelatin has dissolved, add the rest of the water, which should be cold, while continuing to stir.

Possible Investigations

The types of investigations that can be done with these techniques are essentially investigations of factors affecting enzyme activity. Amounts of substrate and enzyme and type of substrate are relevant to the digestive process. So is pH, since stomach and intestinal pH differ, but pH is not a suitable independent variable for the assays for lipase and trypsin activity (see following notes). The effects of temperature are less relevant, since body temperature is relatively constant.

Digestion of Gelatin by Trypsin (or other protein-digesting enzyme)

Students can investigate the effects of different enzymes or enzyme sources besides pancreatic trypsin. For example, papain (found in meat tenderizer) is a protease. Jell-O directions caution against using kiwi fruit and fresh or frozen pineapple, which prevent the gelatin from setting because they contain proteolytic enzymes. Canned pineapple is okay, presumably because the enzyme (bromelin) is denatured during the canning process. Students can boil fresh pineapple to attempt to denature the enzyme, too.

Experiments using pepsin or testing the effects of pH on trypsin will produce negative results with gelatin. Background reading should tell students that the pH optimum of pepsin, which is found in the stomach, is very low (about 2). In fact, gelatin will set when incubated with pepsin at a pH above 6 and will remain liquid at a low pH, indicating that pepsin digests protein best in low pH. Unfortunately, gelatin also remains liquid when made up at low pH *without* the pepsin. This makes a good lesson in the importance of using proper controls.

To get gelatin at pH 2–4, add 1–2 mL 0.1N HCl to 10 mL gelatin solution. To get pH 5, add 2 mL 0.5% HCl. To get pH 8, add 1–2 mL 1% $NaHCO_3$. To get pH 10, add 1 mL 0.1N NaOH. A mixture containing just gelatin (10 mL) and the enzyme (2 mL) is approximately pH 6.

Digestion of Fats with Lipase

Type of substrate is a good independent variable for the lipase test. Amount of enzyme and amount of bile are also good independent variables. The effect of temperature can be investigated. Students can also try using different emulsifiers. (Mustard and egg yolk are mentioned in the lab exercise.)

The nature of the assay precludes using pH as an independent variable. For the color change of the phenolphthalein to be observed, the initial pH of the mixture should be approximately 10. Fat digestion results in a decrease of the pH value to about 8, which is the pH value at which phenolphthalein becomes colorless.

Digestion of Starch

The amylase assay is a good technique for investigating substrate concentration. It also gives good results with enzyme concentration or temperature as the independent variable. The iodine test for starch could be supplemented by the Benedict test for reducing sugars (directions are given in Lab Topic 3, Macromolecules).

Salivary amylase is mentioned in the laboratory, but having students collect and investigate their own salivary amylase is *not* recommended because of the risk of disease transmission.

Comparison of Enzyme Characteristics

Students can investigate similiarities and differences among the digestive enzymes by doing a set of experiments using the same independent variable for two or three of the enzymes.

Circulation

Introduction

In order to carry on all the functions necessary for life, animal cells must receive nutrients and oxygen from their environment. They must also be able to rid themselves of metabolic wastes, including carbon dioxide. In the least complex animals, such as sponges, cnidarians, flatworms, and roundworms, most cells either are in direct contact with the external environment or are only one cell away. Therefore, these animals can obtain oxygen and nutrients directly into their cells by diffusion. In more complex animals, however, a circulatory system is needed to transport nutrients and oxygen to all cells in the body and to remove metabolic wastes. Blood is the tissue responsible for these transport functions. In this lab topic you will learn techniques for investigating two aspects of the human circulatory system, which transports oxygen, nutrients, and other substances throughout the body. You will also design and perform your own experiment using these techniques.

Outline

Exercise 15.1: Anatomy of the Mammalian Circulatory System

Exercise 15.2: Measuring Cardiovascular Function in Humans

    Activity A: Pulse Rate and Heart Sounds

    Activity B: Blood Pressure

Exercise 15.3: Designing an Experiment

Exercise 15.4: Performing the Experiment and
               Interpreting the Results

Instructor Overview
In this lab topic students review the structural features of the mammalian circulatory system by using diagrams and a pig heart. They learn two techniques for measuring cardiovascular function in humans, pulse rate and blood pressure, which are explained in relation to the cardiac cycle. Each student team will then design its own experiment using pulse rate and/or blood pressure as the dependent variable(s). The lab topic prompts students to organize their investigations (see Exercise 15.3) and answer a series of questions that leads them to interpret their results (see Exercise 15.4). Alternatively, you may require students to write lab reports.

Time Requirements
Exercise 15.1: 20–30 minutes
Exercise 15.2: 30–40 minutes
Exercise 15.3: 15–30 minutes
Exercise 15.4: 40–60 minutes

EXERCISE 15.1
Anatomy of the Mammalian Circulatory System

Objectives

After completing this exercise, you should be able to

1. Explain the roles of the heart, arteries, veins, and capillaries in the circulatory system.

2. Identify the structures of the mammalian heart and indicate the direction of blood flow into, through, and out of the heart.

3. Distinguish between arteries and veins.
4. Explain how the structure of arteries is related to their function.
5. Explain how the structure of veins is related to their function.
6. Explain how the structure of capillaries is related to their function.

Introduction

Blood is distributed from the heart to the lungs and body tissues by a system of blood vessels (Figure 15.1). **Arteries** carry blood away from the heart. Major arteries branch into smaller arteries called **arterioles** that narrow into capillaries that are in close contact with the tissues. **Capillaries** are thin-walled vessels that permit exchange of gases and other substances between the blood and the tissues. Capillaries merge into **venules,** small vessels that connect the capillaries with veins. The **veins** carry blood back to the heart to complete the circuit.

Figure 15.1.
Components of the mammalian circulatory system.

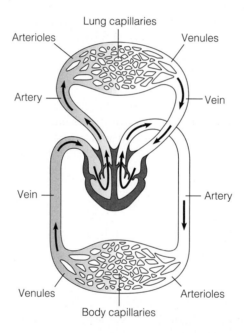

Lung capillaries
Arterioles
Venules
Artery
Vein
Vein
Artery
Venules
Arterioles
Body capillaries

In this exercise you will identify the structures of the mammalian heart using a pig heart as an example. Look at the pig heart provided for this exercise and identify the structures that are listed in the following paragraphs in boldface type. Label these structures on the diagram of the heart (Figure 15.2). Draw arrows to indicate the direction of blood flow into, through, and out of the heart.

Note that in Figure 15.2 you are viewing the heart as if you were facing the animal: The right side of the diagram is the left side of the heart. Hold the pig heart so that it is in the same position as the diagram.

 Wear latex gloves when you are handling the heart.

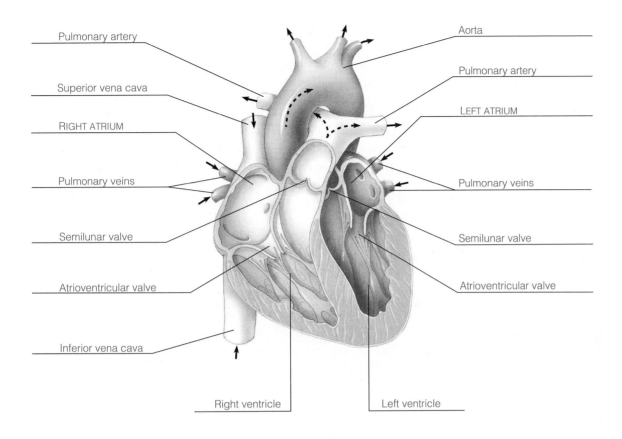

Figure 15.2.
Mammalian heart. The human heart, shown above, is very similar to the
pig heart that you will observe in lab.

Blood returning to the heart from the body enters the **right atrium** via the
two vessels called the **venae cavae** (singular: vena cava). The superior
vena cava carries blood from the upper portions of the body and the inferior vena cava brings blood from the lower body. From the right atrium,
blood passes through the **right atrioventricular (AV) valve** to the **right
ventricle,** which pumps it through the **pulmonary artery** to the lungs.
There is a **semilunar valve** between the right ventricle and the pulmonary
artery. Blood returns to the **left atrium** from the lungs via the **pulmonary
veins.** The blood then passes through the **left atrioventricular valve** to
the **left ventricle,** which pumps it out to the body through the **aorta.**
Notice that the muscle of the left ventricular wall is thicker than the right
ventricular wall. Blood leaving the left ventricle must be pumped farther
than blood leaving the right ventricle. There is another semilunar valve
between the left ventricle and the aorta.

What is the function of the atrioventricular valves?

*They prevent backflow of blood from the ventricles to the atria
when the ventricles contract.*

What is the function of the semilunar valves?

They prevent backflow of blood into the ventricles. One prevents backflow of blood from the pulmonary artery to the right ventricle. The other prevents backflow of blood from the aorta to the left ventricle.

The arteries can be viewed as muscular, elastic tubes that offer little resistance to blood flow and provide pressure to drive blood through the body. When the ventricles of the heart contract, more blood flows into the arteries than leaves them (systole, Figure 15.3). The volume of fluid causes the elastic arterial walls to bulge. When the ventricles relax, the arterial walls recoil and give the blood an additional push (diastole, Figure 15.3).

Figure 15.3.
Movement of blood through arteries.

Systole

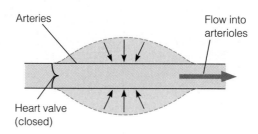

Diastole

By the time blood reaches the venules to begin its journey back to the heart, blood pressure is very low. The veins have large diameters and provide little resistance to blood flow. They also have thinner walls, which are able to stretch more than arterial walls; therefore, veins can hold larger volumes of blood with less increase in pressure than arteries. Venous pressure can be increased by contraction of the smooth muscle in the walls of the veins. It is also increased when veins are squeezed by the surrounding skeletal muscle.

Procedure

1. View the microscope slide of an artery and a vein in cross section (the artery has a thicker wall and is more circular in shape than the vein).

2. Sketch the artery in the margin of your lab manual. Notice the thick middle layer of elastic connective tissue and smooth muscle in the artery. This layer of tissue provides resistance to stretching, thus keeping the blood under pressure.

Student sketch of artery and vein.

3. Sketch the vein in the margin.

What is the principal difference between the two vessels?

> *The wall of the artery is thicker than that of the vein, particularly the middle layer of tissue. Lacking the extra elastic tissue that helps the arteries maintain their shape, the veins tend to collapse during preparation of the slides.*

The function of **capillaries** is to provide an interface between the blood and the other body tissues. The diameter of capillaries is often not much larger than the diameter of a red blood cell, and the cells must travel in single file. Capillary walls are only one cell layer thick. Explain how this structure serves capillary function.

> *The thinness of the capillary walls allows diffusion. The small size of the capillaries also enables them to penetrate all tissues.*

E X E R C I S E   1 5 . 2
Measuring Cardiovascular Function in Humans

Objectives

After completing this exercise, you should be able to

1. Determine pulse rate for an individual.
2. Determine blood pressure for an individual.
3. Describe the cardiac cycle and how it relates to heart sounds and blood pressure.
4. Explain how and why hypertension, atherosclerosis, and arteriosclerosis affect circulation.

Activity A: Pulse Rate and Heart Sounds

The most accessible measure of cardiac activity is **pulse rate.** As the blood is pumped, the intermittent pressure can be felt in the arteries, especially at the wrist and throat. (See Figure 15.4.) Counting the pulse rate tells us the number of heart beats per minute.

Procedure for Pulse Rate

1. Find your pulse as shown in Figure 15.4. Use your fingers rather than your thumb, since there is a slight pulse in the thumb.
2. Count your pulse for 15 seconds _____ .
3. Multiply this figure by four to determine the number of beats per minute: _____ . This is your resting pulse rate.

Figure 15.4.
Measuring pulse rate.

Contraction of the ventricles propels blood through the arteries, creating the pulse. The period from the beginning of one ventricular contraction to the beginning of the next is called the **cardiac cycle.** The ventricular contraction itself is referred to as **systole,** and it is followed by a period of relaxation called **diastole.** You will use a stethoscope to listen to the events of the cardiac cycle.

Procedure for Heart Sounds

1. Clean the ear pieces of a stethoscope with alcohol.
2. Place the chestpiece of the stethoscope to the left of your (or your partner's) breastbone and listen to the heart.
3. You should hear a low-pitched "lub" sound followed by a higher-pitched, more forceful "dup." "Lub" is the sound of the atrioventricular (AV) valves closing. Would this mark the beginning of systole or diastole? (Refer to Figure 15.2 if you have forgotten the location of the AV valves.)

   Systole.

4. Now listen for the "dup" sound. "Dup" is heard when the semilunar valves close. Is "dup" the beginning of systole or diastole?

   Diastole.

Normally you hear only the vibrations made by the valve closures; the blood flowing through the heart is not heard. A **heart murmur** means that the rush of blood is audible. This may occur because the valves fail to close completely, allowing backflow; there may be a small hole in one of the walls that divide the heart chambers; or the valve opening may be abnormally narrow. In any case the murmur is a result of faster, more turbulent blood flow than is normal.

Activity B: Blood Pressure

The flow of blood into, through, and out of the heart is driven by pressure differences created by the cardiac contractions. Once the blood is out in the body, it is pushed along by **arterial pressure** (review Figure 15.3). What we call **blood pressure** is actually arterial pressure. Maximum arterial pressure coincides with ventricular contraction (systole). Minimum pressure (diastole) occurs before the next ventricular contraction. Note that although the pressure is lower during ventricular relaxation, it is well above zero. Blood flow does not stop between ventricular contractions.

You can present blood pressure data either by graphing systolic pressure and diastolic pressure on the same axes (Figure 15.5a) or by plotting the difference between the systolic and diastolic pressures (Figure 15.5b), which is called **pulse pressure.** It is useful to show pulse pressures when comparing one group of subjects to another or when comparing data taken from a subject at different times.

a.

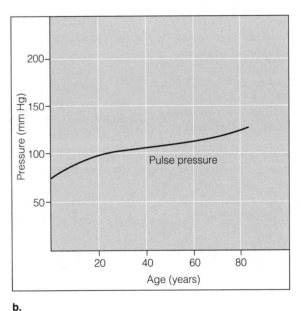
b.

Figure 15.5.
Graphical presentation of blood pressure data. (a) Changes in systolic and diastolic pressures with age. (b) Changes in pulse pressure with age.

Systolic pressure normally ranges from 108 to 140 mm Hg (millimeters of mercury, a measure of pressure) and normal diastolic pressure ranges from 60 to 90 mm Hg for 20–24-year-olds. Average values for males are slightly higher than for females, and blood pressures increase for both sexes with age (see Figure 15.5).

High blood pressure, or hypertension, is an increased resistance to blood flow in the arteries; that is, more force is required to pump blood through the arteries. Generally **hypertension** is defined as blood pressure greater than 140/90 mm Hg. High blood pressure can be dangerous because an artery might rupture under the increased pressure. The two most important types of damage to arteries from hypertension are to vessels in the brain and kidneys. High blood pressure also forces the heart muscle to work harder to pump blood to the body. Like all muscle, the heart requires oxygen to generate the ATP necessary for contraction. Oxygen and nutrients are brought to the heart by the coronary arteries. The increased workload of the muscle must be supported by an increase in oxygen. Angina pectoris (chest pain) results from an oxygen deficiency.

What circumstances might trigger angina in a person who has hypertension?

> *Anything that further increases the heart's workload, such as exercise or stress.*

Why does exercise increase the heart rate?

> *The muscles are using oxygen in the process of aerobic respiration to make the ATP needed to move the muscles.*

A more complete explanation for increased heart rate during exercise is that increased oxygen consumption by the muscles, plus increased concentrations of carbon dioxide and other metabolites, cause the arterioles in the muscles to dilate, thus increasing blood flow to the muscles. Arteriole dilation greatly decreases peripheral resistance to blood flow, which would lead to lowered blood pressure and failure to deliver enough blood to other organs as well as to the exercising muscles if it were not compensated by increased cardiac output. Cardiac output is a rate (liters/min) that equals heart rate (beats/min) × stroke volume (liters/beat). Except in trained athletes, stroke volume increases only a little during exercise; it is primarily increased heart rate that accounts for the increase in cardiac output.

Hypertension also promotes **atherosclerosis,** a disease in which deposits of lipids such as cholesterol are found in the large arteries. How would atherosclerosis affect blood pressure?

> *The narrower the bore of the tube, the more force is required to pump blood through it; hence blood pressure increases.*

Arteriosclerosis (literally, arterial hardening) is a form of atherosclerosis in which calcium depositions accompany the lipids, causing the arterial walls to lose their elasticity. What effect would you expect this condition to have on the circulatory system? (Hint: Review Figure 15.3.)

> *Greater systolic pressure develops because it takes more pressure to distend the walls (the walls resist stretching).*

In addition to reduced blood flow, an atherosclerotic artery is prone to blood clots that plug the artery and block flow completely. A sudden reduction or stoppage of the flow of oxygen in the coronary arteries causes the rapid death of cardiac cells, which is known as a heart attack. The dead muscle cells are later replaced by scar tissue, leaving the heart permanently weakened.

To supplement this brief discussion of cardiovascular disease, you may bring in news articles about the latest research in this area for students to read when they have time during the lab.

In the following procedure you will take your partner's blood pressure using a stethoscope, a blood pressure cuff, and a device that measures the pressure in terms of millimeters of mercury (mm Hg). You will take one blood pressure reading with the subject lying down and one reading with the subject standing. What hypothesis is being tested with these data?

> *Example: Blood pressure is higher in a person who is standing than in a person who is lying down.*

What is the independent variable?

> *Body position.*

What is the dependent variable?

Blood pressure.

Why do you think this independent variable might affect the dependent variable?

Example: The heart has to work harder to pump blood "uphill" so the ventricular contractions must be stronger, which results in higher blood pressure.

Predict the outcome of the experiment in terms of your hypothesis.

Example: If the hypothesis is supported, the blood pressure, expressed as average pulse pressure, will be higher in subjects who are in the vertical position than in subjects who are in the horizontal position.

 Read the entire procedure before you start.

You can demonstrate this procedure during your introduction.

Procedure

Figure 15.6 shows where to place the cuff and stethoscope. It is helpful to locate the brachial artery first by feeling for it with your fingers. Figure 15.7 relates the procedure of taking blood pressure to the events of the cardiac cycle and illustrates the variations in arterial pressure. The circled numbers will help you understand how these variations are related to the cardiac cycle.

To save time, use only one member of each pair as the subject. This will give you data from 10–15 subjects. The instructor should collect all the data for the class to use in evaluating the hypothesis.

 Work with a lab partner. One of you should be the subject and the other should take the blood pressure readings.

1. Take the first blood pressure reading with the subject lying down. The subject should lie still for 2–3 minutes before the blood pressure is taken.
2. Wrap the cuff around the arm just above the elbow. It should be just loose enough so that there is room for two fingers between the cuff and the arm.
3. Close the valve on the bulb by turning it all the way clockwise.
4. Locate the brachial artery (see Figure 15.6) and hold the stethoscope in place over it.

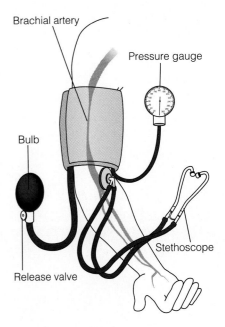

Figure 15.6.
Position of blood pressure cuff and stethoscope.

Figure 15.7.
Relation of arterial and cuff
pressures.

5. Inflate the cuff to a pressure higher than systolic pressure (① on
    Figure 15.7); 160 mm Hg should be adequate. The cuff presses on the
    arteries, shutting off blood flow to the lower arm.

 Do not keep the cuff inflated for more than one minute. If you
can't complete the procedure, deflate the cuff and wait a few
minutes before you try again.

6. *Very slowly* release the pressure by turning the valve slightly counter-
    clockwise. Listen through the stethoscope. When the cuff pressure is
    slightly below systolic pressure, the artery begins to open and blood
    starts to flow through ②. As in a heart murmur, this blood flow is
    more audible because constriction of the artery causes turbulent, high-
    velocity flow. When you hear the first intermittent tapping noises,
    record the pressure: _____ mm Hg. This is the systolic pressure.

7. As the cuff deflates, the sounds are continuous but have become
    muffled ③. When the cuff pressure is slightly below diastolic

pressure, blood flow through the artery is normal and the sounds disappear. Record the point at which you stop hearing the sounds: _____ mm Hg. This is the diastolic pressure.

To summarize: Systolic pressure is recorded when sounds are first heard. Diastolic pressure is recorded when sounds disappear.

 If you are not successful in determining your partner's blood pressure after three attempts, don't try again. Repeatedly collapsing the arteries can be harmful.

---

8. Take the subject's blood pressure again while he or she is standing up. Record systolic/diastolic pressure: _____ / _____ mm Hg.

A table of average pulse pressures is probably the best way of handling these data. A comparison of frequency distributions can also be used.

The instructor will compile class data. Decide on an appropriate method of presentation and interpret your results.

Was your hypothesis proven false or supported by the results? Use data to support your answer.

Example: Supported. The average pulse pressure was higher when the subjects were standing than when they were lying down. (Students should use specific data to support their conclusions.)

What other independent variables might affect blood pressure and/or pulse rate?

Examples: exercise, stress, consumption of caffeine, alcohol or other drugs, smoking, age, sex, race, or physical condition of subjects.

E X E R C I S E   1 5 . 3
Designing an Experiment

Objective

After completing this exercise, you should be able to

1. Design an original experiment to investigate some factor that affects pulse rate and/or blood pressure.

In Exercises 15.1 and 15.2 you learned how to measure pulse rate and blood pressure. In Exercises 15.3 and 15.4 your lab team will design an experiment using one of these methods, perform your experiment, and present and interpret your results. You may review the independent variables you listed at the end of Exercise 15.2 before beginning to design your experiment.

Describe your experiment below. To increase the number of data points you can get, you may need to ask other lab teams to participate in your experiment.

The following materials will be supplied for your group:

apparatus to measure blood pressure

stethoscope

Hypothesis

Dependent Variable

Independent Variable

Explain why you think this independent variable will affect pulse rate and/or blood pressure.

Control Treatment(s)

Replication

Brief Explanation of Experiment

Predictions

What results would support your hypothesis? What results would prove your hypothesis to be false?

Method

Design a Table to Collect Your Data

List Any Additional Materials You Will Require

E X E R C I S E   1 5 . 4

Performing the Experiment and Interpreting the Results

Objectives

After completing this exercise, you should be able to

1. Perform the experiment your lab team designed.
2. Present and interpret the results of your experiment.

Before you do the experiment, be sure that everyone on your lab team understands the techniques that will be used. You may want to divide up the tasks before you begin work.

Be thorough in collecting data. Don't just write down numbers; record what they mean as well. Don't rely on your memory for information that you will need when reporting on your experiment later! If you have any questions, doubts, or problems during the experiment, be sure to write them down, too.

Results

Before you begin to prepare your results for presentation, decide on the best format to use. Remember, you want to give the reader a clear, concise picture of what your experiment showed. Refer to the data presentation section of Appendix A (Tools for Scientific Inquiry) for help. If you are drawing graphs, use graph paper. Complete your tables and/or graphs before attempting to interpret your results.

Write a few sentences *describing* the results (don't explain why you got these results or draw conclusions yet).

Discussion

Look back at the hypothesis you posed in this experiment. Look at the graphs or tables of your data. Do your results support your hypothesis or prove it false? Explain your answer, using your data for support.

Did your results correspond to the prediction you made? If not, explain how your results are different from your expectations and why this might have occurred.

Describe how your data are supported by information from other sources (for example, textbooks or other lab teams working on a similar problem).

If you had any problems with the procedure or had questionable results, explain how they might have influenced your conclusion.

If you had an opportunity to repeat and extend this experiment to make your results more convincing, what would you do?

Summarize the conclusion you have drawn from your results.

Questions for Review

1. When blood returns to the heart from the body, it travels into the heart via the two _____*venae cavae*_____ to the _____*right atrium*_____ of the heart. From there, blood enters the_____*right ventricle*_____ of the heart. Backflow into the atrium is prevented by the _____*atrioventricular*_____ valve. When blood returns to the heart from the lungs, it travels via the_____*pulmonary*_____veins into the _____*left atrium*_____ of the heart. From there, blood enters the_____*left ventricle*_____ of the heart. Backflow into the atrium is prevented by the _____*atrioventricular*_____valve. When the right ventricle contracts, blood leaves the heart via the

_____*pulmonary*_____ artery. The_____*semilunar*_____valve prevents backflow into the ventricle. When the left ventricle contracts, blood leaves the heart via the_____*aorta*_____(blood vessel). The_____*semilunar valve*_____ prevents backflow into the ventricle.

2. The_____*atrioventricular*_____valves close when the ventricles contract. The sound heard when those valves close is_____*"lub"*_____. The ventricles contract during the _____*systolic*_____phase of the cardiac cycle. The_____*semilunar*_____valves close after the ventricles contract. The sound heard when those valves close is_____*"dup"*_____. The ventricles relax during the_____*diastolic*_____phase of the cardiac cycle.

3. Make a schematic diagram of the mammalian heart, showing how the blood flows through the right and left atria, the right and left ventricles, the venae cavae, the aorta, the pulmonary veins, and the pulmonary artery. Draw arrows showing the direction of blood flow.

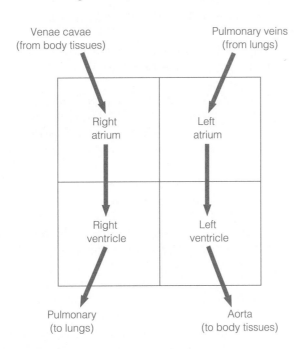

Student sketch of schematic diagram of mammalian heart.

4. If you were shown two microscope slides, how would you be able to tell which one is an artery and which one is a vein?

The layer of smooth muscle and elastic tissue is much thicker in arteries than in veins, so the vessel wall is thicker overall.

5. What are the functions of arteries and how does their structure help them perform these functions?

Arteries carry blood away from the heart and provide pressure to keep blood moving through the body after the initial push from the ventricle. The elastic tissue in the arterial walls helps them resist stretching as they are overfilled with blood, and this provides the arterial pressure.

6. What are atherosclerosis and arteriosclerosis? Why are they harmful to the circulatory system?

    Atherosclerosis is the deposition of lipids in the lumen of the arteries. The diameter of the arteries is diminished by the deposits, causing increased blood pressure (higher pressure is needed to force the blood through the arteries). Also, when the arteries are blocked to a major extent, there may actually be a problem getting enough blood through to supply the organs with oxygen. Blockage of the coronary artery, for example, can deprive the heart tissue of oxygen and thus lead to a heart attack. The lipid deposits can also encourage clotting of the blood; if the clot blocks a major artery, a stroke or heart attack will result.

    Arteriosclerosis adds the further complication of calcium deposits in the atherosclerotic plaques, which reduce the distensibility (capacity to expand) of the arterial walls.

7. What is the function of veins? How does the structure of veins help them perform their function?

    Veins return blood from the capillaries to the heart. They provide even less resistance to blood flow than do the arteries; the diameters of veins are generally greater than the diameters of arteries, and the walls of veins are more distensible (more easily stretched). They serve as a reservoir of blood. The amount of venous return is an important determinant of cardiac output, which is increased if the veins return more blood to the heart.

8. What is the function of capillaries? How does the structure of capillaries help them perform their function?

    Capillaries are the most direct contact between the blood and the tissues. Oxygen and nutrients carried by the blood are delivered to the tissues from the capillaries, and CO_2 and waste products are picked up. The capillary walls are only a single cell layer thick, so diffusion can occur across the cells, and the junctions between the cells are fairly loose, so diffusion occurs between cells, too.

9. What are the two phases of the cardiac cycle?

    Systole (ventricular contraction) and diastole (ventricular relaxation).

Circulation
Instructor Notes

Organization

Introduction (5 minutes)

Give a brief general overview of the importance of the circulatory system.

Explain that students will design their own experiments, so they should be thinking about what to do as they work through Exercises 15.1 and 15.2.

Prep for Procedures (5–10 minutes)

Show where materials are located in the laboratory.

Demonstrate the procedure for measuring blood pressure, including the caution against leaving the cuff inflated for more than a minute and the caution against repeating the procedure more than three times on one subject.

Preparing Students to Design Experiments (15–30 minutes)

Summarize answers and results from Exercises 15.1 and 15.2.

Ask students to brainstorm ideas for independent variables. For this lab topic they will have little difficulty coming up with ideas. Of the ideas suggested, let students know what would be available for their use (for example, coffee, cigarettes, *Psycho* video). If exercise is suggested, use the step test. The "step test" is a standardized exercise for testing heart rate. Perform it by stepping up and down on a low platform or stair approximately 30 times per minute for 3 minutes. Step up with one foot and then the other. Step down with one foot and then the other. Maintain a steady pace of approximately one cycle every 2 seconds.

Once this discussion has been completed, students will only need about 10 minutes to finalize their proposals.

For a 2-Hour Lab

Omit the background exercise (Exercise 15.1) and give a more thorough introduction to Exercise 15.2 so that students understand the heart sounds they are hearing and have some understanding of blood pressure before they begin. To save more time, the entire class should design an experiment collectively. Then each class member will participate as a subject. Compile the data for all students to copy and use for their reports.

Possible Investigations

As with all the investigative lab topics, it helps to introduce the topic to be investigated some time before the lab is performed, either in lecture or in the previous lab. Let students know they will be expected to design their own experiments and encourage them to bring any materials they will need.

Students will have no problem getting ideas for investigations that use pulse rate and blood pressure as the dependent variables. An additional dependent variable you can suggest is recovery time: the length of time it takes for the pulse rate to return to normal after it has been elevated (for example, by exercise).

Collecting data during the laboratory period limits the type of experiments that can be undertaken because the available subjects are limited in type and number. For example, it may not be possible to compare athletes and nonathletes. If possible, have students collect data outside of class to broaden the ranges of investigations that are possible and to allow them to include data from more subjects.

Physical fitness of different groups of subjects, as measured by their performance on the step test, is one possibility for investigations. Directions for the step test are given in Preparing Students to Design Experiments. For the purpose of student investigations, I suggest using a more strenuous version of the step test: The subjects should step up and down for 5 minutes at the fastest rate they can maintain. This will make recovery time a better measure of fitness. Or students may want to use some other standard activity such as running in place, jumping rope, or running stairs. Students should also consider factors that might influence a subject's performance in a single exercise session and decide on a reasonable number of replicates.

Some differences in subject groups that could be investigated include physical conditioning (for example, test athletes from different sports, such as joggers, aerobicizers, or body builders), age, sex, and smoking habits. Discuss the availability of appropriate subjects before approving an investigation. Age may not be a good choice of variable for students who live on campus, since there may not be enough older subjects—there are only so many professors who will be willing to run the stairs for their students. But age might be a good choice for commuters who can draw subjects from their community or on campuses that have a large population of nontraditional students.

Blood pressure is determined by cardiac output (which is partially determined by pulse rate) and peripheral resistance (arterial dilation or constriction); because of decreased peripheral resistance, blood pressure increases only slightly as a result of exercise.

Many factors can speed up pulse rate, including stimulants (for example, caffeine) and stress (for example, selected scenes from a Freddy Krueger movie or the final moments before an exam). Students may want to devise stressful situations for an investigation. Investigations that involve manipulating subjects will work better than comparisons of subjects who have different characteristics if the data must be collected during the lab period.

Under no circumstances should students be permitted to carry out investigations that involve subjects consuming alcohol, illegal or prescription drugs, or doses of over-the-counter drugs (such as cold medicines) higher than recommended on the label. Students who have heart or respiratory problems should not participate in *any* activity that may stress the cardiovascular system.

The Sensory System

Introduction

Animal survival depends on the ability to sense and respond to stimuli in the environment. As you studied animals in Lab Topic 13 (Animal Diversity), you learned that even the simplest animals have some rudimentary means of perceiving their surroundings. For example, the planarian, a flatworm, has light-sensitive eyespots in the head region. The hydra, a Cnidarian, can chemically sense the presence of food in the water. Coupled with the ability to sense a stimulus is the need to be capable of response. The planarian avoids light in order to minimize its exposure to predators. When it senses light it uses the muscles of its body to turn away. The hydra, sensing food, waves its tentacles to capture it.

Vertebrates have much more complex nervous systems. The **sensory system,** which is part of the **peripheral nervous system,** perceives environmental stimuli. It sends information about the stimuli to the central nervous system, which consists of the brain and spinal cord. A message is then sent from the **central nervous system** to the other component of the peripheral nervous system, the **motor system,** directing a response.

The sensory system, which will be studied in this lab topic, is composed of specialized neurons, or nerve cells. These **sensory receptors** are capable of detecting specific types of stimuli such as light, touch, sound, or chemicals. They convert physical stimuli, which are one form of energy, into **action potentials,** which represent electrochemical energy. That is, the stimulus causes a change in the receptor cell's membrane that is communicated from the receptor through a chain of nerve cells to the central nervous system. The sensory receptor thus acts as a transducer, changing energy from one form such as light, touch, or sound into the electrochemical energy of an action potential. For example, sound energy that is perceived by auditory receptors in the ear is transduced into the electrochemical energy of an action potential that travels to the brain. Analogously, a loudspeaker transduces electrical energy from a radio or TV into sound energy.

Sensory receptors are specialized to receive a particular type of stimulus. In these laboratory exercises you will study **chemoreceptors,** which detect chemicals; **photoreceptors,** which sense light; and **tactile receptors,** which are sensitive to pressure and temperature.

Outline

Exercise 16.1: Chemical Senses

    Activity A: Taste Discrimination

    Activity B: Individual Differences in Taste

    Activity C: Smell Discrimination

    Activity D: Influence of Smell on Taste

Instructor Overview
In this lab topic students explore aspects of their own senses of taste, smell, vision, and touch. For the sense of taste, students distinguish the sweet, sour, salty, and bitter tastes and discover umami. They learn about variability of taste perception among individuals by tasting PTC, thiourea, and sodium benzoate on taste papers. Smell discrimination is tested with a variety of scents, and the relationship of olfactory receptors to taste is demonstrated by tasting a piece of candy while the sense of smell is blocked. Students examine the structural features of the eye by dissecting a sheep eye. They also learn how rods and cones function, observe their blind spot, and determine their accommodation distance. The skin senses are investigated using the two-point discrimination test. Students also perform a temperature adaptation experiment. The Instructor Notes include additional background information and suggestions for instructors who wish to have their students design and perform their own experiments using the procedures learned in this lab topic.

Time Requirements
Exercise 16.1:
    Activity A: 20 minutes
    Activity B: 5 minutes
    Activity C: 15 minutes
    Activity D: 5 minutes

E X E R C I S E   1 6 . 1
Chemical Senses

Objectives

After completing this exercise, you should be able to

1. Name the taste organ and the specific taste receptors and explain how taste is perceived by humans.
2. Name the five basic tastes and give examples of substances that elicit them.
3. Explain the adaptive significance of the five basic tastes.
4. Explain how it is possible that different people perceive the same substances as different tastes (or having no taste).
5. Explain how smells are perceived by humans.
6. Compare the complexity of the sense of smell with taste.
7. Discuss the influence of smell on taste.

Introduction

The ability to sense chemicals in the environment arose early in the evolution of animals. Although in most animals the chemosensors are concentrated in specific areas of the head, there are variations as well. House flies have chemosensors on their feet, enabling them to "taste" a potential food source as soon as they land on it. Besides taste receptors on the whisker-like barbels or "feelers" around the mouth, catfish also have taste buds all over their bodies. Humans receive chemical stimuli in two different ways. Molecules that are carried by water enter our mouths and are sensed by taste buds. Airborne molecules are sensed by cells in the nasal cavity.

Why is it beneficial for us to have a separate system for detection of taste and smell? What is the "specialty" of each?

> *Smell allows long-distance detection. The presence of food, danger, or potential mates, for example, can be detected by airborne molecules. Taste requires actual contact, and can be used to determine whether food is safe.*

The human tongue contains specialized receptor organs, the taste buds, to detect dissolved chemicals. The bumps on the tongue, called papillae, are separated by trenches, and the taste buds are located along the walls of the trenches. Particles of food that become lodged between the papillae can rot and contribute to bad breath, so oral hygienists recommend that people gently brush their tongues along with their teeth. Adults have about 10,000 taste buds; children have more. As you have probably observed, the tongue and the mucous membrane of the mouth also have receptors for pressure, temperature, and pain. These receptors are responsible for the "mouth feel" that contributes the sense of pleasure or distaste aroused by particular foods.

Figure 16.1 shows a taste bud. There are 20 to 30 taste receptor cells in each taste bud. The receptor cells are continually being replaced; their life span is about 10 days. One end of the receptor cell projects into a pore of the bud so it can contact chemical substances directly. The other end is in contact with a neuron. When the receptor is stimulated by a chemical, it transduces chemical energy into an action potential, which is transmitted to the adjoining neuron and eventually to the brain.

The olfactory receptor cells, which are responsible for the sense of smell, are located in the mucous membrane in the upper part of the nasal cavity. The receptors have cilia on one end that extend into the mucous layer (Figure 16.2). For a substance to be experienced as an odor, it must release molecules that diffuse through the air to the olfactory receptors. The molecules then dissolve in the mucous layer and interact with the receptors. In the human nose, there are as many as 40 million olfactory receptors in a space the size of a dime. That sounds like a lot, but our sense of smell is rather poor compared to other animals. Dogs, for example, have a much larger area devoted to olfactory sensors and have ten times as many receptor cells as we do. As a result, they can detect scents at much weaker concentrations.

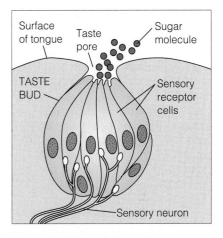

Figure 16.1.
Human taste bud. When sensory cells in the taste bud are stimulated by sugar molecules, sensory neurons transmit action potentials to the brain.

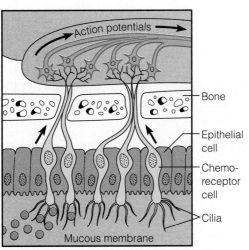

Figure 16.2.
Olfactory epithelium. The smell receptor cells sense molecules that dissolve in the mucous layer. Information from the receptor cells is sent to the brain via nerve fibers. The olfactory epithelium also includes nonreceptor epithelial cells that surround the receptors.

Activity A: Taste Discrimination

In this activity you will determine the tastes that you can identify. Each student should perform the entire procedure, but you should share materials with others working nearby.

Procedure

1. Get a small amount of the five testing solutions labeled 1, 2, 3, 4, and 5 in separate disposable cups.

2. Get 3 cotton swabs and a cup of water for rinsing.

3. Dip a cotton swab into Solution 1 and touch it on your tongue until you taste it. If you don't taste it immediately, try the swab on a different area of the tongue.

There are regions of the tongue where taste buds are more dense and areas where there are no taste buds, so students should just move the swab around until they taste the solution. An article that appeared in the March 2001 issue of Scientific American *stated that the famous "taste map" of the tongue is not accurate—there are no specific regions where the receptors for specific tastes are located.*

 If you need to get more of the same solution, use the clean end of the cotton swab. Do not double dip!

---

4. Record the taste of Solution 1 in Table 16.1 below.

5. Rinse your mouth thoroughly with water.

6. Repeat steps 3–5 for Solutions 2, 3, 4, and 5 and record the results in Table 16.1. Be sure to rinse your mouth after each tasting.

 If you are sensitive to MSG, do not taste Solution 5.

Table 16.1
Results of Tasting Solutions

Solution	Taste
1	*salty*
2	*sour*
3	*bitter*
4	*sweet*
5	*"meaty"*

Check with your classmates. Did everyone identify the solutions with the same descriptions?

Probably all students identify the familiar four basic tastes, salty, sour, bitter, and sweet. Solution 5 will get a variety of responses.

The taste of Solution 5 has been recognized in Japan for nearly a century. Some Western researchers have recently concluded that it is a fifth basic taste, though not all agree. It is called umami (ooh mah me), a Japanese word without an exact English equivalent. *Savory, essence, pungent,* and *meaty* have been suggested to describe this taste. The umami taste is elicited by glutamate, an amino acid, as well as by some other chemicals. The solution you tasted was monosodium glutamate (MSG), which is used as a flavor enhancer. Glutamate is the most abundant amino acid and is present in many protein-containing foods including meat, seafood, and aged cheeses such as Parmesan. We can't taste it when it is bound with other amino acids in protein molecules, but cooking or other processing such as fermentation or aging releases the free amino acids.

A sweet taste can be caused by numerous chemicals besides sugar. Much effort has been devoted to the search for a low-calorie substitute for sugar. Aspartame (brand name NutraSweet) is a combination of amino acids that has been very successful as a sweetener. Some sweet-tasting substances are even poisonous, including some lead compounds, glycols (for example, antifreeze), and chloroform. However, those are not substances one would naturally encounter as food.

The salty taste is caused by various ionized salts. Table salt (NaCl), which produces sodium and chloride ions when dissolved in water, is the most familiar. The taste can also be elicited by related salts such as $CaCl_2$ (calcium chloride), KCl (potassium chloride), and NaI (sodium iodide).

Are sweet, salty, and umami tastes ones that you enjoy?

> *Presumably students will answer yes.*

Why do you suppose our taste buds can sense sweet, salty, and umami tastes? (Hint: In what foods do these tastes occur?)

> *The sweet taste drives our appetite for carbohydrates; umami encourages us to eat protein; and the salty taste ensures that we get sodium and minerals in our diets. If you consider naturally occurring foods (not the junk food we've invented), these three tastes encourage healthy diets.*

Sourness is really a measure of pH. The stronger the acid, the more sour the taste. A 5% solution of acetic acid (the same as vinegar) was used for the sour taste in this activity.

Is the sour taste one you enjoy or one you avoid? Why might it be adaptive to avoid food or drink that tastes sour?

> *We do enjoy some sourness, for example in dill pickles and vinegar-and-oil salad dressings. In naturally occurring foods (or those left in the refrigerator too long) however, a sour taste indicates that the food or beverage has spoiled. Also, many fruits are sour before they are ripe.*

Like the sweet taste, the bitter taste can be caused by various chemicals, especially alkaloids. Alkaloids are complex molecules that occur naturally in many plants. Some have powerful physiological effects, including death, in humans. Some have been used or modified for their medicinal properties. Caffeine, strychnine, nicotine, cocaine, morphine, and LSD are alkaloids. So is quinine, which was used as the tasting solution. Quinine is used in the treatment of malaria and gives the bitter flavor to tonic water.

Many plants contain alkaloids as deterrents to herbivores.

Is the bitter taste one you enjoy or one you avoid? Why might it be adaptive to avoid food or drink that tastes bitter?

With rare exceptions such as tonic water, people avoid the bitter taste. In the natural world, anything bitter is likely to be harmful, so this is an adaptation for distinguishing poisonous plants from edible ones.

Table 16.2 shows the minimum concentrations of different substances needed to stimulate the sweet, sour, salty, and bitter taste receptors. (Information on the umami taste is not available.) Explain why these differences make sense in terms of human survival.

We are extremely sensitive to the bitter taste (that is, it is sensed at a very low concentration), and it is most likely to be harmful. Sensitivity to "sour" is also much greater than sensitivity to sweet and salty, which indicate desirable foods.

Table 16.2
Minimum Concentrations for
Taste Receptor Stimulation

Sucrose (sweet)	0.01M
NaCl (salty)	0.01M
HCl (sour)	0.0009M
Quinine (bitter)	0.000008M

Why is it important for animals, including humans, to have sense organs for taste in the mouth rather than elsewhere in the digestive system?

Again, taste is a way of determining what is safe or unsafe to consume. Once the substance gets past the mouth it may be too late to prevent harmful effects.

 Before you do the next activity, rinse your mouth out thoroughly with water.

Activity B: Individual Differences in Taste

Food preferences may be attributed to various factors such as texture that have nothing to do with taste, or to the number and sensitivity of a person's taste buds. In this exercise, you will investigate individual differences in the ability to taste certain chemicals.

Procedure

1. Get one piece each of the four color-coded taste papers: control (which has no chemical on it), PTC (phenylthiocarbamide), thiourea, and sodium benzoate.

2. Taste the control paper first by placing it on your tongue and allowing your saliva to dissolve any chemicals present in the paper. Record the taste in Table 16.3. Discard the control taste paper.

3. Taste each of the remaining papers in the order listed in Table 16.3 and record what you taste.

4. Discard each paper in the trash after you have tasted it.

Table 16.3
Results of Testing Taste Papers

Paper	Color	Taste
Control	White	
PTC	Blue	
Thiourea	Yellow	
Sodium benzoate	Pink	

Compare your responses to the taste papers with the responses of your classmates. Did everyone experience the same tastes? Explain any differences.

Probably not. The ability to taste PTC is a dominant genetic characteristic. Some people do not taste sodium benzoate, while others taste it as salty, sweet, bitter, or sour.

The ability to taste the chemical PTC is determined by a single gene. Tasters have at least one copy of the dominant allele and nontasters are homozygous recessive. In the natural world, would there be an advantage or disadvantage to being a "taster"?

"Tasters" can sense the bitter taste of this chemical, which would be an advantage since bitterness indicates toxicity. Nontasters could consume a poisonous plant.

Activity C: Smell Discrimination

In this activity you will explore your ability to identify odors. You should have several opaque containers that are covered tightly with lids. Each one is numbered so that you can record your observations, but it is not necessary to take them in order. Work with a lab partner. One person should be the subject and the other the experimenter. When you have finished, switch roles and repeat the procedure.

Procedure

1. The subject should close her or his eyes. Without looking into the container (because he or she will be the subject later), the experimenter should raise the lid of a container and hold it up to the subject's nose.

 Be sure to close the lid after each trial to keep the materials fresh.

2. The subject should identify the odor. Without telling your lab partner what you smelled, write this observation in Table 16.4 (you may open your eyes to do this!).
3. Repeat the procedure until the subject has identified the odor in each container.
4. Switch roles and repeat the procedure.

Results will depend on the sources used. The Prep Guide offers some suggestions. There should be some familiar scents (for example, pine, lemon) and some that are more difficult to identify and therefore must be described. For example, students may recognize the scent of fresh herbs but not be able to name them, so they will have to use descriptive terms.

Table 16.4
Results of Smell Discrimination Testing

Jar	Smell
1	
2	
3	
4	
5	
6	
7	

Compare your results to your lab partner's results. Did you agree on identifying the odors? Did other students agree with your results? Explain.

Depending on the sources used, some scents will be very definitely identifiable and others won't. For example, everyone agrees that an orange smells orangy, but a less familiar scent such as rosemary might be called piney, woodsy, sharp, or some other adjective. Students will learn that we do not have such precise categories for smell as we do for taste.

In contrast to the four or five basic tastes we can identify, we can distinguish perhaps 10,000 odors. Scientists do not understand exactly how this is accomplished. Some believe that smells, like tastes, can be classified in several basic categories, but there is no agreement as to what those categories are. One proposed scheme groups smells as fragrant, burnt, "goaty," or acid. Another scheme groups them as fruity, spicy, flowery, foul, or resinous.

Activity D: Influence of Smell on Taste

 This test should be done after the other taste tests.

Most foods are complex mixtures of chemicals that stimulate our sensory receptors with the sweet, salty, bitter, sour, or umami taste, or some combination of them. In addition, we can perceive texture and temperature. Since the mouth also contains air, scent molecules from food enter the nasal cavity and stimulate the olfactory epithelium. Thus the sense of smell plays a large role in what we "taste."

Procedure

1. Close your eyes while your lab partner gets a Life Saver.
2. With your eyes still closed, pinch your nostrils closed while your partner puts the candy on your tongue.
3. Keep holding your nose while the candy dissolves in your mouth. Guess its flavor:

4. Release your nose. Does the Life Saver taste different? Guess its flavor again.

5. What is the actual flavor of the candy?

Did you think the flavor was the same when you could taste and smell as you did when you could only taste? Explain your results.

> *Usually students can't guess the flavor of the candy without using the sense of smell. The taste receptors give us the information that the candy is sweet or possibly sweet and sour, but the "flavor," such as cherry or grape, is the result of olfactory stimulation. A subject who has a cold, however, won't be able to tell the difference.*

A similar demonstration can be done using any foods that have the same texture. For example, if you chew on a piece of raw onion while you hold your nose, you'll find its taste indistinguishable from the taste of raw potato.

EXERCISE 16.2
Vision

Objectives

After completing this exercise, you should be able to

1. Identify the lens, iris, pupil, retina, sclera, cornea, fovea, and choroid coat and give the function of each.
2. Describe the functions of the rods and cones.
3. Explain how eyes become adapted to the dark and to bright light.
4. Explain the physical basis of red-green color blindness.
5. Explain the presence of a blind spot.
6. Explain accommodation.

Vision is a result of our ability to sense and interpret light energy. When photoreceptors are stimulated by light, they transmit an action potential, which is carried to the brain. In this exercise you will learn about the properties of the photoreceptors, color vision, and some structural features of the eye that affect vision.

Activity A: Structural Features of the Eye

The eye is constructed so that images (patterns of light) will be focused on light-sensitive photoreceptors in the retina. A camera has the same function: The photographer wishes to focus an image on photosensitive film. A comparison of the eye and the camera (Table 16.5) will help you understand the functions of the various parts of the eye.

In this exercise you will examine the external features of the human eye. You will then dissect a sheep eye, which is structurally very similar to the human eye, to observe the internal features. Figure 16.3 illustrates the structures you will see.

Procedure

1. Look at your lab partner's eye and identify the sclera, cornea, iris, and pupil.

The white of the eye is the **sclera,** a tough coat that extends around the eyeball.

Part of the front of the sclera, the **cornea,** is transparent to permit passage of light. The cornea can be seen best by looking at a side view of the eye. Notice that it bulges out slightly. The cornea plays a role in bending the light that enters the eye; it acts as a fixed lens. If the curvature of the cornea is too great, myopia, or nearsightedness, results.

The cornea is subject to injury from particles that lodge in the eye. Corneal transplants are used to replace damaged or defective corneas. This type of transplant is unique in that the recipient will not reject the transplanted tissue, because there is no blood supply to the cornea and thus the immune system is not involved. About 10,000 corneal transplants are performed each year. In addition, there is a procedure called radial keratotomy in which the shape of the cornea is surgically altered to correct myopia.

Table 16.5
Comparison of the Eye and a Camera

Part of Eye	Part of Camera	Function
Lens	Lens	Focuses light
Iris	Diaphragm	Regulates size of light opening
Pupil	Light opening	Allows light to enter
Retina	Film	Senses light
Sclera	Camera body	Protects the eye and helps maintain shape of the eyeball
Cornea	Colorless filter	Covers and protects light opening; bends light
Choroid coat	Black paint on inside of camera	Absorbs extra light to prevent internally reflected light from blurring the image

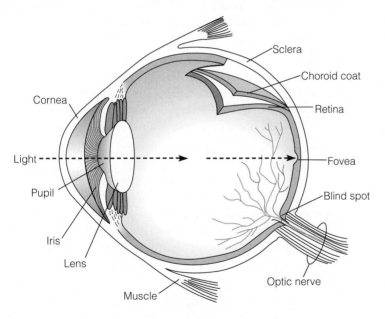

Figure 16.3.
The human eye.

The pigmented structure is the **iris.** It contains muscles to regulate the size of the **pupil,** which is simply an opening to allow light to enter.

2. Get a sheep eye in a dissecting pan and a scalpel. You may wear latex gloves while you perform this procedure.

3. Examine the external tissues. Notice that there is a large amount of fat. This cushions the eye in its socket.

4. As you move aside the fat, look for the muscle tissue that held the eye in place. The muscles are attached to the sclera.

The color of the iris depends on the amount of pigment present. Black eyes have the greatest amount and blue eyes the least. If the amount of pigment is small, the color results from the light scattered by unpigmented parts of the iris. Newborn caucasian babies typically have blue or gray eyes because the pigment has not yet completely developed.

> Handle the scalpel with caution. When cutting, be sure to cut away from your body and keep your fingers out of the way. The scalpel blade is extremely sharp.

5. Cut away the fat and muscle as best you can. You will see a thick white cord emerging from the back of the eye. This is the **optic nerve,** which transmits information from the sensory cells of the eye to the brain.

6. Locate the pupil, iris, sclera, and cornea of the sheep eye. The preservative caused the cornea to become cloudy; the cornea of any living eye is clear.

7. Place the eye in the dissecting pan with the cornea down. Make a circular incision about 5 mm ($^1/_4$ inch) above the cornea all the way around the eye. The small cavity between the cornea and the iris contains a watery liquid that helps maintain the shape of the eye. This liquid will drain out as you cut the eye.

Diabetics may develop cataracts as a result of the disease. There may also be a link between smoking and UV exposure and cataract formation. Lenses that have cataracts may be replaced with artificial lenses.

8. Locate the lens, which is a hard white disk. Like the cornea, it is transparent in the living eye. A **cataract** is a clouding of the lens that prevents light from passing through it unobstructed. The image formed is then distorted, and the person may even become blind. A tendency to form cataracts may be inherited, but many cataracts are the result of aging.

9. To observe the retina, first remove the jellylike material that fills the large cavity of the eyeball. In the preserved eye, this substance has usually dried up to some extent. Its function is to help maintain the shape of the eyeball. The retina, which is a yellowish layer, contains the rods and cones—the actual sensory cells of the eye.

10. Notice that the inner surface of the eye is lined by the choroid coat, a black coating that absorbs stray light rays that might otherwise interfere with clear vision. In the sheep eye, as in many mammals (but not humans), stray light that has not been absorbed in the retina bounces off the **tapetum lucidum,** a reflective layer located between the choroid coat and the retina. This gives the retina a second chance to absorb the light. The tapetum is responsible for the iridescent appearance of the inner eye. It is also the reason that the eyes of some animals glow in the beam of a flashlight or car headlights.

    Which would be more likely to have a tapetum lucidum, nocturnal animals (those that are active primarily at night) or diurnal animals (those that are active during the day)? Explain your answer.

    Nocturnal animals, because they must function during low-light conditions. The tapetum is an adaptation that allows them to more completely absorb the light entering the eye.

11. Locate the optic nerve again and follow it to its origin inside the eye. The gathering of neurons in this place results in a blind spot where no image is formed on the retina. The blind spot will be studied in Activity C.

Activity B: Photoreceptors

Light entering the eye falls on the retina, which contains neurons that are modified into photosensitive cells, the **rods** and the **cones.** The human eye contains about 125 million rods and 7 million cones.

The rods and cones contain pigments that absorb light energy, causing rapid (within nanoseconds) breakdown of the pigments. This change stimulates the nerve fiber leading to the brain. The final product of the breakdown is a form of vitamin A, which can be stored in the retina and later reconverted to photosensitive pigments.

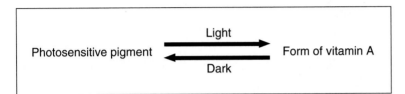

Figure 16.4.
Forms of photosensitive pigments. The pigments undergo a molecular change when they absorb light. After a short period of darkness, the change is reversed.

Night blindness is a condition in which people see well in the daytime but very poorly at night. Using Figure 16.4, suggest a mechanism and a cure for this affliction.

If there is a deficiency of vitamin A, the photosensitive pigments and their precursors are present in lower amounts than normal, so the eyes are not as sensitive to light. In the daytime there is enough light that this isn't noticed, but the person has trouble seeing at night. This is not a common affliction because vitamin A is stored in the body and it takes months to develop a deficiency. The cure, vitamin A, can work rapidly to restore normal sensitivity to light.

If your retinas are exposed to strong light over a long period of time, most of the photosensitive form of the pigment is converted to vitamin A and your eyes become less sensitive to light. If you then walk into a dark theater, for example, your eyes will not be sensitive enough to distinguish the light spots from the dark areas. In photographic terms, your image is underexposed—everything looks dark. In the dark, however, the photosensitive pigments are reconstructed from vitamin A, and within 1 minute sensitivity to light increases tenfold. After 40 minutes sensitivity increases 25,000 times. This is called **dark adaptation.**

When you walk out of the theater into the bright sunlight, the image you see is overexposed. The dark areas of the image look bright too, because your retina is sensitive even to dim light. Adaptation to light, though, is more rapid than dark adaptation, and will be complete in a few minutes.

You can demonstrate "local" adaptation of the retina by fixing your eyes on the white dot in the center of the peace symbol in Figure 16.5 for 30 seconds. Then look at the black dot at the right of Figure 16.5.

Figure 16.5.
Adaptation demonstration.

What do you see when you stare at the black dot?

A "negative" of the image in Figure 16.5.

Explain why.

The photosensitive cells in the part of the retina that is exposed to the dark (black) part of the diagram have converted their pigment from a form of vitamin A to a photosensitive form and are now more sensitive to light. In other words, that part of the retina is dark adapted. At the same time, the part of the retina that was exposed to the bright (white) part of the diagram became light adapted. When the entire retina is now exposed to the white page, the dark-adapted area looks brighter than the light-adapted area. That is, the dark-adapted area of the retina has just walked out of a dark theater. The light-adapted area looks dark because the photoreceptors in that area of the retina are less sensitive to light. That is, the light-adapted area of the retina has just walked into a dark theater.

Why don't you see an afterimage like this when you are reading black print on a white page?

If you really are reading, the pattern of light and dark on the retina is constantly shifting, so no adaptation occurs. However, most of us have probably had the experience of finding ourselves staring fixedly at some particularly boring text. In that case, you do see a negative afterimage if you look at a blank part of the page.

Besides adaptation of the photoreceptors, what other mechanism helps the eye adjust to changes in light and dark? (Hint: Review Table 16.5.)

Pupillary contraction and dilation.

Usually during a cave tour the guide will turn off all lanterns and flashlights to let everyone experience total darkness. No matter how long you wait, you will not be able to see anything. Why not? Why don't your eyes adapt to the dark?

> *Dark adaptation is actually adaptation to very low levels of light. If there is absolutely no light, the photoreceptors will never be stimulated no matter how long you wait. This situation is, of course, very unusual; no matter how dark it seems there is almost always some small amount of light. That's why it's so interesting to experience total darkness in a cave.*

Cones are responsible for bright-light vision and for the perception of detail and color. Cones are concentrated in the center of the retina in an area called the **fovea,** which is directly in line with the center of the cornea and lens. The human fovea has approximately 150,000 cones per square millimeter, but some birds have up to a million cones in the same area.

There are three types of cones, called blue, green, and red, which contain different photosensitive pigments. The red cones are sometimes called yellow because the photopigment they contain also absorbs yellow light. Figure 16.6 shows the range of wavelengths of light absorbed by each type of cone.

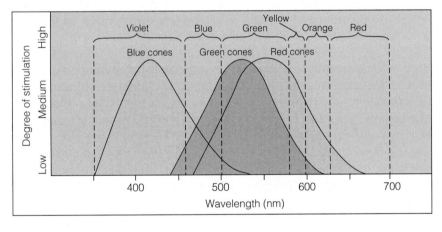

Figure 16.6.
Cone sensitivity to light. Each type of cone contains a different photopigment that absorbs light. When light is absorbed, the cone is stimulated to produce an action potential.

When a photopigment absorbs light, the cone is stimulated and sends an action potential to the brain. As Figure 16.6 shows, the blue cones are maximally stimulated in response to light of 420 nm. The blue cones can also absorb light at other wavelengths. For example, when light of 530 nm is received by the retina, green cones are stimulated maximally. But both the blue cones and the red cones are also stimulated to some degree. All three curves overlap.

We see white when all the cones are stimulated and black when none of them is stimulated.

The reason that we see so many gradations of colors is the overlap in the curves. The brain interprets the *ratio* of stimulation of the three cones. For example, light with wavelengths from 580 to 595 nm stimulates both the red and the green cones, and the brain interprets that color as yellow. This can be demonstrated by using red and green colored filters to make red and green light. The red light stimulates the red cones and the green light stimulates the green cones.

Procedure

1. Aim two light sources so that the beams overlap.
2. Hold a red filter in front of one light source and a green filter in front of the other.
3. Position the lights so that all three colors (red, green, and a third color where red and green intersect) can be seen, and project the beams onto a sheet of white paper.

What color do you see in the area where the lights overlap?

> *Yellow.*

You can supplement this exercise by using the Ishihara charts, which test for color blindness. Ordering information is in the Prep Guide.

Two percent of all men lack red cones. Using Figure 16.6, estimate which colors these people have the most difficulty seeing. Explain your answer.

> *Yellow, orange, and red, because there is little or no overlap between the red and green cones for those wavelengths of light.*

Six percent of all men lack green cones. Using Figure 16.6, estimate which colors these people have the most difficulty seeing. Explain your answer.

> *They can still see all wavelengths of light because the absorption curve for the green cones overlaps the other two. However, green light stimulates their red cones alone instead of producing a ratio of stimulation of green and red. Therefore green is seen as red. Green and blue are difficult to distinguish.*

Both of these conditions are called red-green color blindness. A lack of blue cones is also possible, but it is quite rare. Other types of color-vision deficiencies result when individuals have all three types of pigments but one is defective. The only truly color-*blind* people, though, are those who lack two of the three types of cones; they see everything in shades of gray. Less than 0.01% of the population has this affliction. Because the color genes are located on the X chromosome, meaning that the trait is sex-linked, color blindness in females is also rare.

Rods are more numerous on the periphery (edges) of the retina, away from the fovea. These photoreceptors are 50 to 100 times more light-sensitive than the cones. They function in dim-light vision and are insensitive to colors.

Some animals are so specialized for diurnal or nocturnal existence that their retinas contain only one type of photoreceptor.

Gray squirrels and some birds are purely diurnal. Their retinas contain only ___*cones*___.

Bats and owls, which are strictly nocturnal, have only ___*rods*___.

If you are trying to see an object in dim light, should you look at the object directly so that the image falls on the cones or should you look at it out of the side of your eye so that the image falls on the rods? Explain your answer.

> *If you look directly at an object, the image falls on the fovea (cones). The rods are more sensitive, so you want the image to fall on them.*

Activity C: The Blind Spot

The retina is supplied with blood vessels and neurons, which are in front of the photoreceptors. As you saw when you dissected the sheep eye, the neurons all converge at one point, called the optic disk, where the optic nerve exits the eye and carries impulses to the brain. There are no photoreceptors at the optic disk, so any light that falls on this area cannot be perceived; there is a **blind spot.** You can demonstrate the presence of the blind spot with the following procedure.

Procedure

1. Hold up your lab manual with Figure 16.7 about 12 cm (5 inches) in front of you.

Figure 16.7.
Demonstration of blind spot.

2. Close your left eye and focus the right eye on the plus sign in the figure.
3. Very slowly move the lab manual away from you until the circle disappears.
4. Repeat the procedure with the right eye closed: Focus on the circle with the left eye and move the book until the plus sign disappears.

When you locate your blind spot, the circle or the plus sign disappears, but you still see something. What do you see?

> *The brain fills in the graph paper background because that's what the rest of the retina perceives. Students may say at first that they see nothing because the circle or plus sign is gone; tell them to look again.*

Even though you have a blind spot in each eye where there are no rods or cones, you don't perceive a blind spot in your vision under normal circumstances. Why not? (Hint: You usually use both eyes to look at an object.)

> *The blind spot in the left eye blocks out a different part of the image than the blind spot in the right eye (so the left eye sees what the right eye is missing, and vice versa). Also, as just demonstrated, the brain fills in missing information.*

Figure 16.8.
Measuring accommodation distance.

Activity D: Accommodation

Muscles holding the lens change the shape of the lens in order to focus on objects at different distances. The ability to focus on a close object, which requires the lens to become almost spherical, is called **accommodation.** During the aging process, physical changes in the lens itself affect its ability to change shape. A 10-year-old can focus on a point 9 cm from the eye, but a loss of accommodation is usually noticeable by age 40 and typically must be corrected with reading glasses when a person reaches the mid-40s. In the following procedure, you will measure your accommodation distance.

Procedure

1. Cover one eye and hold out a pencil at arm's length.
2. Focus on the point and slowly bring the pencil in toward the eye until it is no longer in focus, as shown in Figure 16.8.
3. Have your lab partner measure the distance between the eye and the pencil. Record your measurements.

   Right eye:_____cm.

   Left eye:_____cm.

   If you wear glasses, repeat the procedure without your glasses.

   Right eye:_____cm.

   Left eye:_____cm.

This makes a good graphing exercise. Have students collect data from individuals of a wide range of ages and construct a frequency histogram.

Your instructor will collect data from everyone in the class. Explain any similarities and differences you see among the members of the class.

If there is a range of age, students will see a related range of accommodation distances. Students who wear glasses to correct farsightedness will have longer accommodation distances than students who don't wear glasses. If everyone is of a similar age and wearing any corrective lenses necessary for their eyes, they will have very similar accommodation distances.

EXERCISE 16.3
Skin Senses

Objectives

After completing this exercise, you should be able to

1. Explain adaptation and distinguish between rapidly and slowly adapting receptors.
2. Discuss how receptor adaptation affects our perception of temperature.

We receive a great deal of information about the external environment through sensory receptors localized in different places in the skin. A variety of receptors contribute to your tactile sensitivity, or sense of touch. There are also thermoreceptors that sense temperature.

Activity A: Sensitivity to Pressure

One measure of sensitivity of touch is the **two-point threshold,** the minimum distance that must be between two points before they are felt as separate points. In a very sensitive area, this distance may be as little as 1 mm. That is, if you are touched by two points that are 1 mm apart, you feel two touches. On a less sensitive part of the body, two points 1 mm apart will feel like one touch.

Procedure

1. Get a compass and a ruler.
2. Using the ruler to measure point-to-point distance, set the points of the compass 5 mm apart.
3. While your lab partner has his or her eyes closed, touch both points simultaneously to the palm of his or her hand.

 Be careful not to stimulate pain receptors!

4. If your partner feels two separate points, reduce the distance between the points. If he or she feels only one point, increase the distance. Your goal is to find the smallest distance at which your partner can still distinguish two separate touches. Record this distance in Table 16.6.
5. Repeat the procedure for the other parts of the body listed in Table 16.6.

Table 16.6
Touch Discrimination

Location	Two-Point Threshold (mm)
Palm	
Fingertip	
Lips	
Outer forearm	
Outer shoulder	
Calf	

 While experimenting, your point-to-point distance should range between 1 mm and 30 mm. Touch your partner with only one point from time to time as a test.

Why might certain points of your body have lower two-point thresholds than other parts?

> *The number of sensory receptors varies in different parts of the body.*

Compare your results with those of your classmates. Are there differences among individuals for the two-point threshold?

> *Most students should have the same rankings for the various areas of the body, but the actual distance in millimeters may vary. Although the threshold of sensitivity is lower for women than for men, there is no difference in two-point discrimination.*

Activity B: Perception of Temperature

There are at least four types of nerve endings involved in sensing temperature. Figure 16.9 shows the temperature range that stimulates each type. Note there is some overlap between cold and warmth receptors. As is the case with the cones and color vision, our perception of temperature results from the relative degrees of stimulation of each type of receptor. Imagine that you get into a tub filled with water that is 35°C (95°F, slightly below body temperature). Both cold and warmth receptors are stimulated and we perceive this combination as lukewarm. On the other hand, if the water is 25°C (77°F) only cold receptors are stimulated, and we definitely perceive cold.

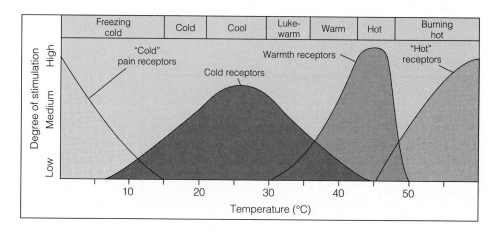

Figure 16.9.
Response of thermal receptors.

Our perception of temperature is complicated by the fact that most areas of the body have three to four times as many cold receptors as warmth receptors. The number of receptors per body part also varies, from 15 to 25 cold receptors per square centimeter in the lips to fewer than 1 cold receptor per square centimeter in broad surface areas such as the back.

Certain chemicals can also stimulate the temperature receptors. Mint can give a sensation of cold, while mustard can give a feeling of warmth. An old folk remedy for colds was to use a mustard plaster on the chest. The sensation of heat from rubbing compounds such as Ben-Gay is also a result of chemical stimulation of temperature receptors.

Nearly all receptors have the ability to **adapt;** that is, they stop responding to the stimulus. When a receptor has adapted completely, it no longer sends a message to the central nervous system that a stimulus is occurring. (We have already discussed the adaptation of photoreceptors to light.) Some tactile receptors adapt rapidly. They react while a change is taking place, so they transmit a signal when light pressure begins and another one when the pressure stops. In this way they detect movement. You are usually not aware of the clothes on your body because the touch receptors on the skin have adapted. However, if someone tugs at your sleeve, you feel the movement of the cloth against your skin.

Other receptors adapt slowly: The receptor continues to transmit information to the brain as long as the stimulus is present. Tight-fitting clothes stimulate slow-adapting pressure receptors that make you uncomfortably aware that something is constricting your body.

The nerve endings that are specialized as pain receptors do not adapt (or adapt very little). Why?

> *Pain is a signal that something is wrong; the signal must continue until the problem is corrected.*

The thermal receptors, on the other hand, can adapt to a great extent, as the following procedure shows.

Procedure

1. Working with a lab partner, prepare three beakers of water, with temperatures approximately 20°C, 33°C, and 44°C. Label them A, B, and C, respectively.

2. Dip a different finger briefly into each beaker and record your perception of the temperature (hot, indifferent, etc.).

   A: ___*Cold*___ B: ___*Lukewarm*___ C: ___*Hot*___

3. Measure the temperature in each beaker and record the actual temperatures.

   A: ___*20°C*___ B: ___*33°C*___ C: ___*44°C*___

4. Immerse the index finger of your right hand in Beaker C and your left index finger in Beaker A. Record the time it takes for each finger to adapt.

   Right:_____ Left:_____

 This will take several minutes. Don't abandon your lab partner! Examine Figure 16.9 and make predictions about the outcome of the experiment while you wait.

---

5. After adaptation has occurred, move both fingers to Beaker B. Record how each finger feels.

   Right:_____ Left:_____

Explain what has happened in terms of receptor adaptation: What happens when the cold receptors adapt? What happens when the warmth receptors adapt? Why do your fingers feel different when they are moved to the second beaker?

When a finger is immersed in 20°C water, stimulation of the cold receptors is very strong at first. Adaptation is rapid for the first few seconds, then progresses more slowly to near-completion over several minutes. When adaptation is complete, the cold receptors are no longer firing. When the finger is then placed immediately in 33°C water, it feels hot because for a short time only the warmth receptors are firing. At 44°C, only the warmth receptors are firing. Adaptation of the warmth receptors followed by immersion in 33°C water results in perception of cold because only the cold receptors are stimulated.

Thus at 33°C, which feels neither hot nor cold, both cold and warmth receptors are stimulated. If one type of receptor doesn't fire (is completely adapted), the other type furnishes the sole input to the brain. This is analogous to what happens in color blindness.

As a consequence of adaptation, the thermal receptors respond noticeably to changes in temperature. When the cold finger in Beaker A moved to water that was warmer than the finger, the temperature of the finger began rising rapidly. This change in temperature made your finger feel much warmer than it would if it were already close to the temperature of the water. Thus, our experience of warmth or cold is strongly influenced by changes in temperature. List a few examples of situations when a change in temperature exaggerates the sensation of warmth or cold:

For example, coming into an air-conditioned room from a hot day or vice versa, getting out of a warm shower, or jumping into the pool on a hot day.

Your ability to perceive a rapid temperature change is proportional to the area of skin involved. If your whole body is subjected to a change as small as 0.01°C, you may be able to detect it. If a very small area of the body is involved, you might not notice a temperature change 100 times that great (1°C).

Considering all you have learned about perception of temperatures, would you advocate the cautious slow-wading/splashing approach to getting wet in the cold ocean or the quick dunk? Defend your choice.

This is still a matter of personal preference, but students can use it to review what they have learned about temperature perception: It depends on the current state of adaptation of the temperature receptors and on the parts and total area of the body subjected to a change.

Questions for Review

1. Explain how information about the environment gets to the brain.

   Sensory receptors receive stimuli such as chemical substances, light, or mechanical energy and change (transduce) those stimuli into action potentials, which are then conducted to the brain by nerves.

2. If you close your eyes and pinch your nostrils closed, is it easier for you to distinguish between lime Kool-Aid and lime juice or between lime Kool-Aid and cherry Kool-Aid? Explain your answer.

   Without the sense of smell you can only distinguish by taste. Sweet and sour will be easier to distinguish than two sweet substances.

3. Explain how the substances that stimulate taste and smell receptors are similar to each other.

   Both require that molecules of a substance be water-soluble, since the chemical must be dissolved in saliva or mucus to be perceived.

4. Dogs and horses are only able to see images in black and white. Give a reason why this might be the case.

   Their retinas do not contain the cones necessary for color vision.

5. Explain the scientific basis for the saying "All cats are gray in the dark."

   The rods, which are responsible for low-light vision, register only shades of gray. In the dark there is not enough light to stimulate the cones, which are responsible for color vision.

6. Which photoreceptors are primarily responsible for dark adaptation? Explain how dark adaptation works.

   Rods. Under conditions of low light, the photosensitive pigment is regenerated from vitamin A, which increases the light sensitivity of the rods.

7. How does the retina of a person who is red-green color-blind differ from the retina of a person with normal color vision?

   A normal retina has three types of cones: red, green, and blue. In red-green color blindness, either the red or green cones are lacking.

8. Explain why older people complain that their arms seem to be getting shorter when they pick up something to read.

   The minimum distance at which the eye can focus (accommodation) increases with age, so they have to hold reading material farther away in order to focus on it.

9. Compare two-point touch discrimination with the resolving power of the microscope. (Recall from Lab Topic 4 that resolving power is the ability to distinguish two separate points.)

   In a microscope, resolving power is the ability to detect two images as separate (for example, two lines that are close together). In the touch test, resolving power is the ability to detect two stimuli as separate.

10. How does the ability of tactile receptors to adapt slowly or rapidly affect the kind of stimuli they sense?

    A tactile receptor that adapts slowly continues to send action potentials as long as the stimulus continues; slow-adapting receptors sense pressure. A fast-adapting receptor receiving the same touch would sense movement; it fires when the stimulus starts and again when it stops.

11. You've been working outside on a cold day and when you come inside you wash your hands. Even though the water is lukewarm, you feel like your hands are burning. What happened?

    The cold receptors have adapted and now any water warmer than 30°C stimulates only the warmth receptors. Without the modulating effect of the cold receptors, your hands feel very warm.

LAB TOPIC 16

The Sensory System
Instructor Notes

Organization

Introduction (10 minutes)

Have students list categories of stimuli that animals must sense and respond to for survival. For example, animals must be able to locate and identify food, avoid predators, locate shelter, find mates, and so forth.

Briefly define receptor types, which are named for specific stimuli: chemo = chemical (taste and smell); photo = light.

Prep for Procedures (5 minutes)

Students should work in pairs. They may start with any exercise, but once they begin an exercise students should do all the activities in order before going to the next exercise. (This is especially important with the taste tests.)

Summary (15 minutes)

Review results of taste and smell discrimination (Exercise 16.1). Poll the class on differences in tastes of taste papers (Exercise 16.1, Activity B) and on accommodation distance (Exercise 16.2, Activity D). Discuss results of the two-point touch test (Exercise 16.3, Activity A).

Discuss questions students have had difficulty answering during lab. These will probably include how local adaptation works, color blindness, why we don't normally see a blind spot, and adaptation of thermal receptors.

For a 2-Hour Lab

Have only one lab partner be the subject for smell discrimination (Exercise 16.1, Activity C) and the two-point touch test (Exercise 16.3, Activity A) instead of having each partner collect data on the other. This saves ~15 minutes.

Also, eliminate either Exercise 16.2, Activity A (dissection of sheep eye, Objective 1) or Exercise 16.2, Activity B (photoreceptors, Objectives 2, 3, and 4). Either way saves 20–30 minutes. Omission of Activity B also saves explanation time during the summary. Activity B and the corresponding review questions (4, 5, 6, and 7) can be assigned for homework or extra credit, since everything except the red + green = yellow light demonstration is in the lab manual.

Investigative Extensions

Following are some suggestions for types of experiments that students can design and perform if you want to extend this lab topic in an investigative format. Background references are included for your information. If you decide to have students do their own investigations, you should briefly present each of these ideas (or as many as you want to use). It is not likely that students will have enough background information to come up with the independent variables for meaningful investigations on these topics by themselves. They can, however, design the rest of the experiment.

Concentration of Tasting Solution

Concentration of a solution can affect its taste. For example, at very low concentrations, both KCl and NaCl taste sweet (Ludel, 1978). At slightly higher concentrations, NaCl tastes salty but KCl tastes bitter. KCl only tastes salty at very high concentrations (so if KCl is used as a substitute for table salt, larger amounts must be used). Similarly, saccharin is sweet in low concentration but bitter in high concentration. Investigating these differences is a good exercise for teaching about concentration and dilutions.

Use the cotton swab technique and apply solutions in turn to the tongue, rinsing well between applications.

Temperature

Taste also varies according to temperature. Sensitivity to salty and bitter tastes increases as the temperature decreases. Sensitivity to the sweet taste increases with increasing temperature. Sensitivity to the sour taste is not affected by temperature.

Students can use the standard solutions that have been made up for the lab topic and vary the temperature using an ice bath and hot plate (test hot solutions before applying to the tongue). Since the temperature of a solution will change very rapidly on a cotton swab, the subject should be given premeasured mouthfuls of the test solutions. Be sure to rinse the mouth thoroughly between tests. An investigation of this type requires students to deal with the problem of subjectivity. They will have to establish a qualitative scale for subjects to use in rating the test solutions (for example, 0 = not salty, 1 = mild, 2 = average, 3 = strong, 4 = unpalatable).

Comparison of Various Chemicals

The lab manual mentions that various chemicals can elicit the same taste. Students can research the ingredients of artificial sweeteners and compare the intensity of sweetness of different brands. Use a sucrose standard for comparison, and be sure to use equal concentrations. Saccharin is 675 times sweeter than sucrose! Other comparisons of substances that cause the same taste are also interesting—for instance, a series of sugars. Fructose is almost twice as sweet as sucrose, glucose is less sweet, and lactose has only $\frac{1}{3}$ the sweetness of sucrose (Guyton, 1981).

Students will have to determine a qualitative scale for ranking sweetness. The cotton swab method is best for these investigations, and careful rinsing of the mouth between samples is necessary.

DO NOT let students do tasting experiments unless you are certain that the chemicals involved are completely harmless.

Taste Papers

Students can investigate the prevalence of "taste blindness" to PTC (phenylthiocarbamide) or thiourea by taking home packets of taste papers and obtaining results from a large sample of people; 15–30% of the population cannot taste PTC and related thiourea compounds (Guyton, 1981).

People who taste PTC as bitter and sodium benzoate as salty generally like the taste of buttermilk, sauerkraut, and turnips, while those who taste both PTC and sodium benzoate as bitter tend to dislike those foods. Sodium benzoate is also used as a food preservative. Students can get very interesting correlations between the ability to taste these chemicals and food preferences.

Sensitivity to Pressure

Students can investigate further by testing more body parts at home and constructing a map similar to the one here. They should gather data from several people and obtain an average.

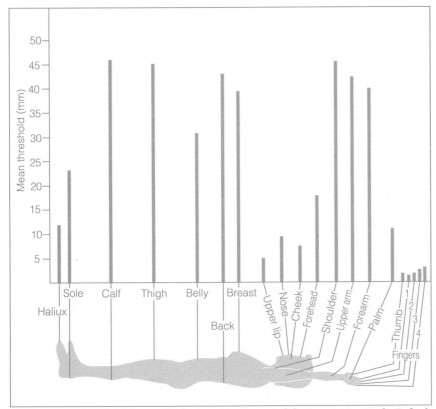

Map of sensitivity to two-point test.

Source: Brown, E. L., and K. Deffenbacher. *Perception and the Senses*. New York: Oxford University Press, 1979, fig. 2.7, p. 56. © 1979 Oxford University Press.

References

Brown, E. L., and K. Deffenbacher. *Perception and the Senses*. New York: Oxford University Press, 1979.

Guyton, A. C. *Textbook of Medical Physiology*, 6th ed. Philadelphia: W. B. Saunders, 1981.

Ludel, J. *Introduction to Sensory Processes*. San Francisco: W. H. Freeman, 1978.

Marieb, E. *Human Anatomy and Physiology*. Redwood City, CA: Benjamin/Cummings, 1992.

Schmidt, R. F., ed. *Fundamentals of Sensory Physiology*. New York: Springer-Verlag, 1978.

Plant Structure

Introduction

As discussed in Lab Topic 13 (Animal Diversity), all birds have structural similarities that allow us to identify certain animals as birds, whether they are owls or ostriches. Although different species of birds may be quite diverse in appearance and behavior, they also have many structural features in common, including feathers, wings, scaly feet, and horny beaks. Similarly, despite the diversity of adaptations in the approximately 280,000 species of plants, most plants are recognizable as plants because of certain features that they share. Stems, leaves, and roots are included in our concept of "plant," and indeed almost all plants have these structures. In addition, almost all plants contain the green pigment chlorophyll, which captures the sun's energy for photosynthesis. All of these characteristics enable us to look at an organism and immediately classify it as a plant.

In this lab topic you will study plant structure by examining the cells and tissues of several different angiosperms, or flowering plants. As you may recall from Lab Topic 12 (Plant Diversity), angiosperms are vascular plants. That is, they contain tissues that carry food and water throughout the plant. Gymnosperms (for example, pines) are also vascular plants, as are ferns and their relatives. Lower plants such as mosses lack vascular tissue, which restricts them to moist habitats. However, plant cells are similar in all types of plants, and because of the presence of chlorophyll, we have no trouble recognizing mosses and liverworts as members of the plant kingdom along with flowering plants.

Outline

Exercise 17.1: Plant Cells and Tissue Systems

Exercise 17.2: Plant Organs

    Activity A: Leaves

    Activity B: Roots

    Activity C: Stems

Exercise 17.3: Dermal Tissue

    Activity A: Dermal Tissue in Stems

    Activity B: Dermal Tissue in Roots

    Activity C: Dermal Tissue in Leaves

Exercise 17.4: Ground Tissue

    Activity A: Parenchyma

    Activity B: Collenchyma

    Activity C: Sclerenchyma

Instructor Overview
In this lab topic students examine plant cell and tissue types. They cut and stain their own stem cross sections of Coleus (a dicot) using a nut-and-bolt microtome. Prepared slides of dicot root and leaf cross sections are also used. Students first examine the dermal tissue system, then the ground tissue system, and finally the vascular tissue system in stems, roots, and leaves. The structure of the lab topic allows students to see the similarities and differences among the tissues in different organs. As they work through the exercises, students add detail to their sketches of root, leaf, and stem cross sections so when finished they have a complete, labeled drawing of the cells and tissues in each organ. The cross sections are supplemented by freshly prepared slides of root hairs, an epidermal peel from a leaf, and pear sclereids. Although this lab topic is traditional in its presentation of the material, students learn both a technique for preparing their own plant sections and enough background information to interpret a wide range of plant tissues. The exercises thus provide a good foundation for student-designed investigations. Suggestions for investigations are given in the Instructor Notes.

Time Requirements
Exercise 17.1: 10–15 minutes (covered during the instructor's introduction)
Exercise 17.2: 30–40 minutes
Exercise 17.3: 20 minutes
Exercise 17.4: 30–40 minutes
Exercise 17.5: 15 minutes

Exercise 17.5: Vascular Tissue
Activity A: Vascular Tissue in Leaves
Activity B: Vascular Tissue in Roots
Activity C: Vascular Tissue in Stems

E X E R C I S E 1 7 . 1
Plant Cells and Tissue Systems

Objectives

After completing this exercise, you should be able to

1. Describe the characteristics of primary and secondary cell walls.
2. Define the three tissue systems.

In this exercise you will review some features of plant cells that distinguish one cell type from another. A living cell has a protoplast. The protoplast consists of the cytoplasm, or cell contents, and the nucleus. The cytoplasm contains the enzymes and organelles that carry out the metabolic processes of the cell. The nucleus contains the genetic material. The protoplast of a plant cell is bounded by the cell membrane, as it is in animal cells. In addition, plant cells have a cell wall outside the membrane. As you may recall from Lab Topic 4 (Using the Microscope), plant cells contain vacuoles that take up water. The cell wall, though permeable to water, serves to limit expansion of the protoplast as the vacuole swells.

Living cells, those that are carrying on metabolic processes or are actively dividing, have primary cell walls. Primary cell walls are composed mostly of cellulose, a polymer made up of glucose subunits, and are somewhat stretchable to accommodate growth. As you will see in this lab topic, primary cell walls may be thin and delicate or they may be thick and tough.

Certain cells also have a **secondary cell wall,** which is laid down after the cell matures. The protoplast often dies after depositing the secondary wall, so no further metabolic activities take place in the mature cell. Secondary walls are strong and rigid and are mostly found in specialized cells that have a support function. Another polymer, lignin, preserves the secondary cell walls of many cells as well as strengthening them. We take advantage of the strength and longevity of secondary cell walls when we use wood, which consists entirely of lignified secondary cell walls.

Besides the cell wall, plant cells are also distinguished from animal cells by the presence of chloroplasts. Chloroplasts are the site of photosynthesis and are one of the organelles located in the cytoplasm. Since they contain the green pigment chlorophyll, they are readily apparent by light microscopy of unstained materials.

Compared to the bodies of animals such as humans, plant bodies are marvels of simplicity. The organs are few and are relatively similar throughout the higher plants: roots, stems, leaves, and reproductive structures, which are specialized leaves. Plant bodies are all made up of the same cell

Plant tissues arise either from primary growth (growth from the tips of the roots and shoots) or from secondary growth (an increase in girth). Meristems and growth patterns are not discussed in this lab topic, and all of the structures described are from primary growth.

and tissue types, which are very similar throughout the plant. Because of this continuity, three tissue systems can be identified in the plant body (Figure 17.1).

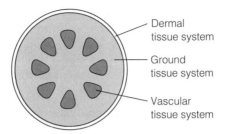

Figure 17.1.
The three tissue systems in a typical dicot stem.

In a simple dicot stem the dermal tissue system consists of the epidermis. In the plants you will examine for this lab topic, the epidermis is a single layer of cells.

The vascular tissue system consists of the xylem (water-conducting) and phloem (food-conducting) tissues. In stems, leaves, and some roots the vascular tissue is found in distinct clusters called vascular bundles.

Vascular tissues are arranged differently in the roots, stems, and leaves. The arrangement also differs between monocotyledons (one seed leaf, monocots) and dicotyledons (two seed leaves, dicots), which are the two large subdivisions of flowering plants. The basic organs, tissues, and cells are the same in both monocots and dicots. The examples used in this laboratory will all be dicots. Other characteristics of the two subdivisions will be studied in Lab Topic 18 (Flowers, Fruits, and Seeds).

The ground (fundamental) tissue system consists of everything that isn't vascular tissue or dermal tissue. Parenchyma, sclerenchyma, and collenchyma cells make up the ground tissue. You will learn the characteristics of these cells in this laboratory.

In primary plant growth the epidermis is generally a single layer of cells, although there are plants that have multiple layers of epidermis. When plants undergo secondary growth, they develop a more complex dermal layer.

E X E R C I S E   1 7 . 2
Plant Organs

Objectives

After completing this exercise, you should be able to

1. Describe the primary functions of leaves, roots, and stems.
2. Briefly describe how the microscope slides that you viewed were prepared.

Activity A: Leaves

The primary function of leaves is photosynthesis. Their structure is designed to provide surface area to intercept sunlight. The leaf surface also has openings, called **stomata**, to permit entry of the CO_2 that is needed for photosynthesis. Internally, the cells and tissues of leaves are either involved in carrying out photosynthesis, in bringing water and minerals to the cells that carry out photosynthesis, or in exporting the sugars produced by photosynthesis to other parts of the plant.

In this procedure you will begin your examination of a prepared slide of a thin section of a leaf. The slide was made using a **microtome,** an instrument that slices tissue so thinly that light can pass through it. You may recall from Lab Topic 4 (Using the Microscope) that tissues are also stained so that various structures can be seen. In the prepared slides that you will view for this lab topic, primary cell walls are stained green and secondary cell walls are stained red. In most cells you will not be able to distinguish any features of the protoplast.

Procedure

1. Get a prepared slide of a typical dicot leaf cross section.
2. Place the slide on the stage of a compound microscope and secure it with the stage clips.
3. Using the low-power objective (10×), focus on the center of the leaf.
4. Use one of the blank pages at the end of this topic to make a sketch of the leaf cross section. For this sketch, simply draw an outline of the leaf. You will add details later as you learn about the cells and tissues.

Activity B: Roots

Roots serve to anchor the plant in the soil, where they absorb water and minerals that are then conducted to the rest of the plant. Many plants also store food in the root until the next growing season. We take advantage of the stored sugar or starch in roots such as carrots, radishes, sweet potatoes, and turnips.

In this procedure you will begin your examination of a prepared slide of a thin section of a root. The root section was made using the same procedure that was used for the leaf section.

Procedure

1. Get a prepared slide of a typical dicot root cross section.
2. Place the slide on the compound microscope and focus on the root section using the low-power (10×) objective.
3. Sketch an outline of the root cross section following your sketch of the leaf. You will add details later as you learn about the cells and tissues of the root.

Where in the root are the cells that have secondary walls located?

In the center.

Activity C: Stems

Stems support the leaves, enabling them to be positioned advantageously for photosynthesis. They also serve as a conduit for the exchange of materials between the leaves and roots. Some stems, for example cactus stems, are able to photosynthesize. Stems may also be a significant food storage site, as in asparagus and Irish (white) potatoes.

Instead of using a prepared slide, you will make your own cross sections of a *Coleus* stem. Your instructor has prepared short stem pieces that are supported by paraffin to make slicing easier. A crude microtome was made by a nut-and-bolt assembly. It will help you make the very thin slices that are desirable for viewing on the compound microscope. After obtaining thin sections, you will apply a stain called toluidine blue to help you distinguish the different cell types.

Procedure

 Before beginning, read through all of the instructions and set up the three dishes for staining as directed. Be careful with the toluidine blue, since it can stain your skin and clothing.

Cutting the Sections

1. Get a nut-and-bolt microtome from your instructor. It contains a short piece of *Coleus* stem embedded in paraffin.
2. Hold the head of the bolt flat on the table with one hand (your left, if you are right-handed). Holding the razor blade in your other hand, remove the excess wax on top by slicing it down even with the top of the nut (Figure 17.2).

For complete instructions on preparing the tissue blocks, see the Prep Guide and the section Preparing the Tissue Blocks in the Instructor Notes.

 Slice as illustrated in Figure 17.2 to keep your fingers out of the way of the razor blade.

Figure 17.2.
Using the microtome.

3. Twist the bolt just a little so that a wafer-thin section sticks up out of the nut.

4. Use a slicing motion to cut this section down to the nut. When slicing, use as much of the razor blade as possible. That is, start on one end of the blade and slide down to the other with each slice, as shown in Figure 17.2. This technique results in a cleaner cut than you would get using a chopping motion.

5. As each slice is cut, transfer it to a small dish of 50% ethanol.

6. Continue to twist up thin sections of the block and slice them off, trying to get the thinnest possible slice each time. It is better to get a crescent of a thin slice than a thick slice that is entirely round. A paraffin block of this size will yield several sections. The last few millimeters of the block will fall out of the nut and can't be sliced.

Staining

1. Leave the stem sections in 50% ethanol for 5 minutes. Free the plant tissue from the paraffin surrounding it if necessary.

2. Using forceps, move the sections to a dish of toluidine blue stain and leave them there for 5–10 minutes.

3. Rinse the sections in a dish of distilled water.

4. Mount each section in a drop of 50% glycerine on a slide. You can put two or three sections on each slide. Mount all of your sections so you can find the best one.

5. Add a coverslip and observe your sections on the compound microscope.

6. Following your leaf and root sketches (Activities A and B), draw an outline of a wedge of your best section. You will add details later as you learn about the cells and tissues of the stem.

Students' results sometimes show little correspondence of color to cell wall type, but the walls will still be quite vividly stained.

In general, toluidine blue stains secondary cell walls green to blue-green or blue. Primary cell walls stain red or purplish colors.

Where are the cells with secondary walls located in the stem?

Either in groups or in a ring about a third of the way into the stem.

The cross sections of the stem and the leaf are clearly different shapes. Considering the function of each organ, why might their forms be different?

The main function of the stem is support. The main function of the leaf is photosynthesis, which requires light and carbon dioxide. The leaf has a large area to intercept light, and it is flat to facilitate diffusion of carbon dioxide to the cells that need it.

E X E R C I S E   1 7 . 3
Dermal Tissue

Objectives

After completing this exercise, you should be able to

1. Describe the general functions of dermal tissue.
2. Identify the dermal tissue on a slide or diagram of a stem, root, and leaf cross section.
3. Explain how and why the epidermis of roots and stems differs.
4. Identify guard cells and stomata and explain their function.

The **epidermis** is the outermost layer of cells; it serves as a covering for the plant body. As the plant's interface with the environment, the epidermis regulates the passage of substances to and from the environment. Epidermal cells have primary cell walls and retain their protoplasts. For the above-ground plant parts, the epidermis provides protection against water loss, which is a major problem for land-dwelling plants. Some epidermal cells deposit a coating called a **cuticle** to cover the outer wall. The cuticle consists of lipid substances that are impermeable to water, which helps the plant retain water. The cuticle is especially important to plant parts such as fruits that have a high water content and to plants that inhabit very dry regions.

Activity A: Dermal Tissue in Stems

In this procedure you will examine the dermal tissue system of the *Coleus* stem. As an above-ground organ, the stem must be protected against water loss. This is the primary function of the stem epidermis. It is typically only a single layer of cells, which may be covered with a cuticle. Many stems and leaves also have specialized epidermal outgrowths such as hairs that, in some plants, aid in preventing water loss or discourage predation by insects. You should see abundant hairs on your *Coleus* stem section.

If secondary growth occurs, the epidermis will be replaced by the periderm, which constitutes part of the bark on trees. Some of the cells in the periderm have secondary cell walls. However, all of the specimens examined in this lab topic have only primary growth.

Procedure

1. Identify the epidermis in your *Coleus* stem section.
2. On the stem outline that you sketched previously, draw and label several representative epidermal cells.

Activity B: Dermal Tissue in Roots

The epidermis of young, active roots is very permeable to water. It is the function of roots to absorb water and minerals into the plant. In these procedures you will examine the dermal tissue system of the root and root hairs in germinating seeds.

Procedure for Examining Root Dermal Tissue

1. Return your dicot root slide to the microscope and identify the epidermis.
2. On the root outline you sketched previously, draw and label several representative epidermal cells.

In addition to the epidermal cells that you labeled on your root cross section, part of the root has specialized epidermal cells with extensions called root hairs. Root hairs are found just behind the growing tip of a root, and they are responsible for most of the absorption of water and minerals into the plant. Since they are thin, the root hairs collectively have an extremely large surface area for absorption. Gardeners refer to these active, growing roots as feeder roots and take care not to damage the delicate hairs during transplanting.

Example of student sketch of root hairs.

Procedure for Observing Root Hairs in Germinating Seeds

1. Get a germinated seed from your instructor and place it on a clean microscope slide. No coverslip is needed.
2. Observe the root hairs using either a dissecting microscope or the scanning-power objective of a compound microscope.
3. Sketch the root hairs of the germinated seedling in the margin of your lab manual.

The lining of the small intestine in animals has many microscopic projections that create a vast surface area. What do you think is one role that the small intestine plays in the digestive system? Explain by relating that role to the large surface area.

Absorption of nutrients. A large surface area, whether in roots or the small intestine, indicates an absorptive function.

Activity C: Dermal Tissue in Leaves

Because leaves have such a large surface area, they are particularly vulnerable to water loss. The epidermal cells of leaves fit together tightly and, as in stems, may derive additional protection from a cuticle. However, since the primary function of leaves is photosynthesis, the leaf must be at least partially open to the atmosphere in order to take in the carbon dioxide that is needed. Stomata (singular: stoma) are the openings that allow CO_2 to enter, but they do not remain open continually. Instead, they are regulated by a pair of guard cells that border the opening. In general, the stomata are open during daylight and when the CO_2 concentration in the leaf is low. They close in the dark and when the CO_2 concentration in the leaf is high. Stomata also close to conserve water if the plant is in danger of wilting.

In these procedures you will examine the dermal tissue system of the leaf by cross section and by making an epidermal peel.

Procedure for Examining Dermal Tissue Cross Section

1. Place the leaf cross section on the microscope.
2. Locate a stoma, which is a gap in the epidermis with one guard cell on each side of it. Stomata are more abundant on the lower surface of the leaf.
3. On your previous drawing of the leaf cross section, draw and label a pair of guard cells and several other epidermal cells.

Procedure for Examining Epidermal Tissue

Epidermal tissue can also be examined by peeling that layer of cells from a leaf. When you view your slide, you will be looking down on the cells, much like examining the squamous epithelium from your cheek in Lab Topic 4 (Using the Microscope), rather than viewing a cross section as you have done for the other slides viewed thus far in the lab.

1. Get a leaf from your instructor.

2. Put a drop of water on a microscope slide.

3. Fold the leaf over so that its lower surface is broken. Using forceps, grasp the edge of the epidermis that is exposed along the break and pull gently away from the break as shown in Figure 17.3. A thin sheet of tissue should peel off. If you're not successful, try again.

4. Float the epidermal peel on the drop of water on the slide. If your peel includes a thick section of tissue, cut that off with a razor blade before you add the coverslip.

5. Add a coverslip and view the tissue under the compound microscope. Draw several cells, including at least one pair of guard cells, in the margin of your lab manual.

How are the guard cells different from the other epidermal cells?

They contain chloroplasts. They are also a different shape.

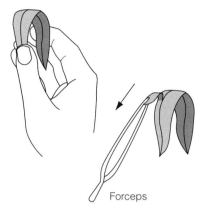

Figure 17.3.
Making an epidermal peel.

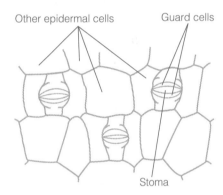

Example of student sketch of epidermal peel showing stomata and guard cells. Shape of the cells differs in different species.

E X E R C I S E   1 7 . 4

Ground Tissue

Objectives

After completing this exercise, you should be able to

1. Identify the areas of ground tissue on a slide or diagram of a stem, root, and leaf cross section.

2. Identify parenchyma cells on a slide or diagram of a stem, root, and leaf cross section.

3. Describe the general functions of parenchyma tissue in stems and leaves.

4. Describe the general function of collenchyma and sclerenchyma tissue.

5. Identify collenchyma cells on a slide or diagram of a stem and leaf cross section.

6. Identify sclerenchyma fibers on a slide or diagram of a stem cross section.

7. Identify a slide or diagram of sclereids.

Plant ground tissue consists of parenchyma, collenchyma, and sclerenchyma cells. Ground tissue serves more than one function. Depending on its composition and location, ground tissue may serve for storage, photosynthesis, structural support, or protection.

Activity A: Parenchyma

The most abundant plant cell type is parenchyma. These cells are relatively unspecialized and retain their protoplasts throughout their existence. They have primary cell walls and make up the mesophyll of leaves (where most photosynthetic activity takes place), the flesh of fruits, the pith of stems, and parts of the root and stem. In roots, stems, and fruits, many parenchyma cells are used to store food. Most of a sweet potato (root), white potato (stem), and the edible portions of a banana, for example, consist of parenchyma cells that are filled with starch grains.

Leaf Parenchyma

The ground tissue of the leaf, called the mesophyll, consists mainly of parenchyma cells. Your leaf cross section includes two distinct types of parenchyma in the mesophyll. The cells in the upper layer are elongated and loosely packed. Most of the photosynthesis in the leaf is carried out in these cells. The rest of the parenchyma consists of cells that are irregularly shaped and widely separated by spaces.

Procedure

1. Place the leaf cross section on your microscope.
2. Focus on an area of the mesophyll.
3. On your previous drawing of the leaf cross section, sketch and label several mesophyll cells. Include both types of parenchyma.

Why are the leaf parenchyma cells so loosely packed? (Hint: The spaces between the parenchyma cells are continuous with the stomata.)

> *Carbon dioxide can circulate freely among the parenchyma cells via the intercellular spaces.*

Root Parenchyma

In the root cross section that you are studying, the ground tissue consists of the thin-walled parenchyma cells that surround the central axis. This area is called the cortex. The cells contain stored starch and are loosely packed to allow water and oxygen to pass through the spaces between the cells. Because they have primary cell walls, the parenchyma cells are stained green in this prepared slide.

Procedure

1. Place the root cross section on your microscope.
2. Focus on the cortex.
3. On your previous drawing of the root cross section, sketch and label several representative parenchyma cells.

Why do root cells require oxygen?

> *They need it for aerobic respiration, just like the rest of the plant. When soil becomes packed or water-logged, roots don't get enough oxygen to metabolize.*

The innermost layer of cells in the root cortex is the endodermis. These cells fit tightly together and parts of their walls are impregnated with a lipid substance that is impermeable to water, much like the cuticle. At this point, water entering the plant is forced into the cells rather than being permitted to flow between them.

Why is there no cuticle on the root epidermis?

> *Because roots must take up water. The cuticle on the stem and leaves keeps water in the plant; in the root, it would prevent water from entering the plant.*

Stem Parenchyma

On your *Coleus* stem section, parenchyma is located in the center of the stem, which is called the pith. The cell walls are thin and, because they are primary cell walls, they should be stained a reddish-purple color by the toluidine blue.

Procedure

1. Place your *Coleus* stem section on your microscope.
2. Focus on the pith area (center).
3. On your previous drawing of the stem section, sketch and label several representative parenchyma cells.

Where else in the stem can you identify parenchyma cells?

> *There are parenchyma cells between the vascular bundles. Some sections may have parenchyma beneath the epidermis and within the vascular bundles.*

What similarities can you observe between the parenchyma cells in the root cross section and in the stem cross section?

> *Parenchyma cells in both cross sections are thin walled and have irregular shapes. Although different stains were used, the parenchyma cells in both slides have taken up the color for primary cell walls (green in the root, reddish-purple in the stem).*

Activity B: Collenchyma

Many young stems and leaves contain collenchyma cells for support. These cells have primary cell walls and their protoplasts remain alive. An identifying characteristic of collenchyma is the noticeable thickening at the corners of the cell walls, which strengthens the wall.

Stem Collenchyma

Coleus, like other members of the mint family, has a square stem. You may have noticed this feature when you made the stem cross sections in Exercise 17.1. The stiffness of the stem, particularly at the four ribs, is due to collenchyma cells.

Procedure

1. Place the stem cross section on your microscope.

2. Focus on the layers of cells just below the epidermis. Look for the characteristic thickened corners that identify collenchyma cells.

3. On your previous drawing of the stem cross section, sketch and label several collenchyma cells.

Leaf Collenchyma

Collenchyma cells are also found as support tissue in leaves. When you look at a leaf, you will notice the veins that contain the vascular tissue. The midrib, or central vein, is especially prominent. In many leaves, including the one you are examining in this lab, collenchyma cells support the midrib. Collenchyma cells are also used for support in the petiole, which is the part of the leaf that connects the blade to the stem. Celery stalks are enlarged petioles, and the strings in celery are collenchyma cells.

Procedure

1. Place the leaf cross section on your microscope.

2. Locate a vascular bundle. Look for collenchyma cells below the vascular bundle.

3. On your previous drawing of the leaf cross section, sketch and label several collenchyma cells.

Activity C: Sclerenchyma

Like collenchyma, the function of sclerenchyma is to support and give strength to plant parts. However, sclerenchyma cells typically have hardened secondary cell walls; *skleros* is Greek for hard. Mature sclerenchyma cells may not have a living protoplast. There are two distinct varieties of sclerenchyma cells, fibers and sclereids. Your stem cross section may show fibers, but neither the root slide nor the leaf slide shows either type of sclerenchyma.

Fibers

Fibers are elongated sclerenchyma cells; some are as long as 25 cm (almost 10 inches). Fibers may be found in leaves, stems, and fruits. Cotton fibers grow on the surface of the seed, but fibers are more commonly associated with vascular tissue. Besides cotton, other plant fibers that are used to make textiles include flax, which is used to make linen, and ramie, which is sometimes blended with cotton or linen. Other plant fibers are used to make rope, twine, burlap, canvas, mattress stuffing, and many other products. The luffa, a fruit related to cucumber and squash, is another example of plant fibers. Once the fruit is dried and the fiber skeleton is exposed, it is used as a scratchy bath sponge.

You might want to ask the prep staff to have a luffa available in the lab.

Procedure

1. Place the *Coleus* stem cross section on your microscope.

2. Locate a vascular bundle. Not all stem sections have fibers, but if fibers are present, they will be found in a group just outside the vascular bundle. Fibers can be identified by their thick secondary cell walls and geometric shape.

3. On your previous drawing of the stem cross section, sketch and label a group of fiber cells.

Older stems are more likely to have fibers than younger stems.

Collenchyma cells and sclerenchyma fibers are similar in form and function. What are their differences, and why might these differences be important to a plant?

> *Collenchyma cells have a primary cell wall and a living protoplast. Fibers have secondary cell walls and no protoplast. Therefore collenchyma cells can still metabolize and grow, while sclerenchyma cells are actually dead. Collenchyma cells are also more flexible, since they retain the stretchable primary cell wall.*

Sclereids

Sclereids are relatively short cells with very thick, typically lignified cell walls. Although they may be found in any part of the plant, you will most readily recognize them as components of fruits, nutshells, and other hard coverings such as the peach pit. Because they are so hard, they are sometimes called stone cells.

Pears contain clusters of sclereids that give them their characteristic gritty texture. In this procedure you will isolate pear sclereid cells and observe them microscopically.

Procedure

1. Get a bit of pear flesh on the end of a teasing needle and put it on one end of a microscope slide.

2. Use the dissecting scope to locate a cluster of sclereids. They are large, round, white structures.

3. With the teasing needle, push the sclereid cluster to the other end of the slide to separate it from the pear parenchyma.

4. Add a drop of water and a coverslip to the slide.

5. Apply gentle pressure to the coverslip to flatten the sclereids.

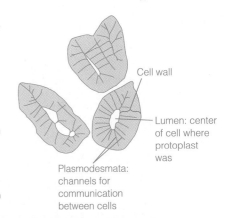

Example of student sketch of pear sclereids.

 It takes a fair amount of pressure to break open a cluster of sclereids, but be careful not to smash the coverslip.

6. Observe the sclereids on the compound microscope and, in the margin of your lab manual, sketch what you see.

Why do you think you found collenchyma and sclerenchyma cells in stem and leaf tissue, but not in the root?

> *Above-ground plant parts need support, while roots do not.*

EXERCISE 17.5
Vascular Tissue

Objectives

After completing this exercise, you should be able to

1. Describe the general function of xylem and list its cell types.
2. Identify xylem on a slide or diagram of a root, stem, and leaf cross section.
3. Describe the general function of phloem and list its cell types.
4. Identify phloem on a slide or diagram of a root, stem, and leaf cross section.

Xylem and phloem, the components of vascular tissue, are usually found together. Vascular bundles, discrete areas of vascular tissue separated by ground tissue, are a common arrangement in stems and leaves. Roots generally have a central vascular cylinder where the xylem and phloem are located.

Xylem is a tissue composed of different cell types that interact to conduct water and minerals through the plant. The cells that actually carry the water were misnamed "tracheary elements" in the seventeenth century, when it was thought that they served a function similar to that of the air ducts (tracheae) in animals; no one bothered to correct the name when the truth was discovered. The earliest tracheary elements to evolve in plants were similar to sclerenchyma fibers with thick, lignified secondary walls. These cells, the **tracheids,** are arranged with overlapping ends and a series of membrane-covered pits for passing water from one cell to the next. Angiosperms, which are the most highly evolved plants, have **vessel elements** in addition to tracheids. Since they are shorter, have larger diameters, and have holes in the cell walls that are not covered by membranes, vessel elements offer less resistance to water flow than do tracheids. Neither vessel elements nor tracheids contain protoplasts at maturity. Xylem tissue may also contain parenchyma and fibers.

Wood is xylem. We take advantage of the structural strength of its lignified secondary cell walls by using wood for building and making furniture. The softer woods, especially from conifers, are used to make paper.

Phloem is a tissue consisting of different cell types that interact to distribute the food made in the leaves to other parts of the plant. The phloem of flowering plants contains **sieve-tube members,** which have primary cell walls and living protoplasts at maturity. Associated with the sieve-tube members are the smaller **companion cells,** which play an important role in the transfer of substances from cell to cell. Phloem tissue may also contain sclerenchyma fibers and parenchyma.

Activity A: Vascular Tissue in Leaves

In the leaf cross section, xylem and phloem are found in vascular bundles that are scattered throughout the mesophyll. In a dicot leaf, veins run in all directions, so most of the vascular bundles have been cut at an angle. The best view of a cross section of a vascular bundle is the midrib, the major vein that runs through the center of the leaf. The most common arrangement of leaf vascular tissue is for the xylem to be located on the upper side of the vascular bundle and the phloem on the lower side. In this prepared slide, the xylem elements are stained red and have thick secondary cell walls, while the phloem cells are stained green due to their primary cell walls.

Procedure

1. Place the leaf cross section on your microscope.
2. Locate the vascular bundle at the midrib of the leaf.
3. On your previous drawing of the leaf cross section, sketch and label representative xylem and phloem tissue.

Activity B: Vascular Tissue in Roots

In the root cross section that you have been studying, the vessel elements are the thick-walled cells that form a cross in the center of the root. Since they have secondary cell walls, they are stained red.

The sieve-tube members and companion cells on the root cross section are stained green, since they possess primary cell walls. The phloem is located between the "arms" of xylem within the vascular cylinder delineated by the endodermis.

Procedure

1. Place the root cross section on your microscope.
2. Locate the central vascular cylinder. Identify the xylem and phloem.
3. On your previous drawing of the root cross section, sketch and label representative xylem and phloem cells in the vascular cylinder.

Activity C: Vascular Tissue in Stems

The vascular tissue in your *Coleus* stem cross section may be found in vascular bundles or it may be in a more or less continuous ring. The vessel elements are much larger than any other cells on the slide, so they are easily identifiable even if they are not differentiated from other cells by the stain. Notice again the thickness of the cell walls. The cells surrounding the vessel elements are also part of the xylem. The phloem is located just exterior to the xylem, toward the outside of the stem.

The differential staining properties of toluidine blue actually depend on the chemical composition of the cell wall. If the tracheids and vessel elements are not lignified, they will not be differentiated from primary cell walls by color. However, they can still be identified by the thickness of the cell wall and the large, hollow cross-sectional area.

Procedure

1. Place the stem cross section on your microscope.

2. Locate a vascular bundle or, if the vascular tissue is continuous, an area of the slide that contains vascular tissue. Look for the large, thick-walled vessel elements. The phloem tissue is to the outside of the xylem. Unless you have a very good section, however, you may find it difficult to identify the phloem, which is easily crushed in the sectioning process.

3. On your drawing of the stem cross section from Exercise 17.2, sketch and label the xylem and phloem.

Notice that the vessel elements you have observed are hollow, since the protoplasts have died, and their cross-sectional area is very large compared to that of other cells. How are these structural features related to the function of the xylem?

> *The vessel elements serve as pipes to conduct water. The larger the diameter of the pipe, the less resistance it offers to water flow. Similarly, the absence of the protoplast removes an obstruction.*

In the stem, phloem is located exterior to the xylem. Explain why it is then logical that phloem in the leaf should be located below the xylem.

> *The leaf grows from the stem. Try to visualize how the vascular tissue of the stem is continuous with the vascular tissue of the midrib of the leaf.*

As the stem gains in girth, the vascular bundles are joined into a continuous ring, and the width of the band of xylem increases. The band of phloem, however, remains approximately the same width as old phloem cells are crushed by the new growth. Explain this difference between xylem and phloem. (Hint: Consider the cell walls.)

> *Xylem elements have thick, hard secondary cell walls. The cells of phloem tissue, on the other hand, have primary cell walls that are easily crushed by the pressure of the expanding xylem.*

Questions for Review

1. If you are looking at a slide of a thin section of plant tissue, what clues might you have to tell you whether a cell has a primary or secondary cell wall?

> *Thickness of the cell wall or differential staining. If the cell has a protoplast (which is often but not always apparent), it has a primary cell wall.*

2. Why do botanists use microtomes and stains when they want to look at plant tissues on the compound microscope?

    Slides are viewed on the compound microscope by passing light through the specimen, which therefore must be very thin. Stains help visualize the cells and their component parts, which otherwise would appear transparent. Stains may also be used to differentiate among cellular features (for example, staining primary and secondary cell walls different colors).

3. What are the general functions of the following?
    Dermal tissue:

    Regulation of passage of substances to and from environment; protection against water loss; mechanical protection.

    Vascular tissue:

    Conduction of food and water through the plant (xylem is also a support tissue).

    Ground tissue:

    Storage; support; photosynthesis.

4. How do the epidermal tissues of roots differ from those of stems and leaves? Suggest a reason for the differences.

    The epidermal tissues of roots have little or no cuticle. In some parts of the root there are also root hairs. Stems and leaves are covered with a cuticle. The differences are a result of the different functions of the organs: Roots absorb water from the ground and need a large surface area and permeable epidermis; stems and leaves are above ground and susceptible to water loss.

5. Figure 17.4 shows the weight change in three apples over time. Which apple had its cuticle removed? Explain your answer.

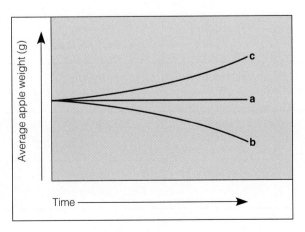

Figure 17.4.
Weight change in three apples.

The cuticle was removed from apple b. Once this protective barrier was gone, the fruit lost weight by dehydration.

6. How does the function of the parenchyma located in leaves differ from the function of the parenchyma located in the roots?

   Parenchyma cells in the leaves carry out photosynthesis. Parenchyma cells in the roots store sugar or starch produced by photosynthesis.

7. Describe how vessel elements are adapted for conducting water.

   Vessel elements have large diameters and holes in the walls to lessen resistance to water flow. They also have thick secondary walls so they can withstand the pressure of the water traveling through them.

8. Summarize your knowledge of plant cells and tissues by filling in Table 17.1.

Table 17.1
Summary of Plant Cell and Tissue Types

	Living or Dead at Maturity	Primary or Secondary Wall	Location(s)	Function(s)
Epidermis	*Living*	*Primary*	*Outer layer of plant*	*Protection; regulation of substances entering and leaving plant*
Parenchyma	*Living*	*Primary*	*Throughout plant*	*Food storage; photosynthesis*
Collenchyma	*Living*	*Primary*	*Beneath epidermis, along midrib*	*Support*
Sclerenchyma fibers	*Dead (usually)*	*Secondary*	*On outside of vascular bundles*	*Support*
Sclereids	*Dead (usually)*	*Secondary*	*Could be anywhere in plant*	*Support or protection*
Tracheids	*Dead*	*Secondary*	*Xylem (in vascular bundle)*	*Carry water (also give support)*
Vessel elements	*Dead*	*Secondary*	*Xylem (in vascular bundle)*	*Carry water (also give support)*
Sieve-tube members	*Living*	*Primary*	*Phloem (in vascular bundle)*	*Conduct food through plant*
Companion cells	*Living*	*Primary*	*Phloem (in vascular bundle)*	*Conduct food through plant*

Acknowledgment

The idea for using a nut and bolt for a microtome was suggested to me by Mrs. Rita Ouzts, a life science teacher at Irmo Middle School, Campus 1, Irmo, South Carolina.

LAB TOPIC 17

Plant Structure Instructor Notes

Organization

Introduction (10–15 minutes)

Review the background information that is given in Exercise 17.1. Emphasize the cell wall types, which are referred to throughout the lab. Explain the three tissue systems. Students will study each tissue system in leaves, roots, and stems. You should also explain how the drawings are to be made. Students first make an outline of the leaf, root, and stem cross sections. They then return to this outline and fill in details as each tissue system is studied.

Prep for Procedures (5 minutes)

You may have students work alone or in pairs for these exercises. They should begin with Exercise 17.2 and work through the lab topic in order.

Caution students about using the razor blades safely. Also caution them about the toluidine blue, which stains skin and clothing.

During the Lab Session

Be very active during this lab session. Even when students aren't asking for help, walk by and glance at the drawings they are making to be sure that they are seeing the right things. If many people seem to be having a similar problem, set up a demonstration microscope that shows what they should be looking at. A videocamera on a microscope, if available, is ideal for this purpose.

Summary (15–30 minutes)

If there is time at the end of the lab period, draw an outline of a leaf, root, and stem cross section on the overhead or chalkboard. Ask different students to come to the front and add a particular cell type to each outline in the same way that students constructed their own drawings while working through the lab exercises. If time is short, show students an overhead that already has the cells drawn in and labeled.

For a 2-Hour Lab

Time can be saved by giving students a prepared slide of a stem cross section rather than having them make their own. However, that would eliminate one of the best activities in this lab topic. Alternatively, you can eliminate looking at either the root or the leaf cross sections, so students are comparing just stems and leaves or stems and roots.

Preparing the Tissue Blocks

It takes about 30 minutes to prepare the paraffin blocks. If you are not doing it immediately before the lab begins, be careful to seal the top of the stem with paraffin or the tissue will dry out.

1. Put small chunks of paraffin in a beaker. (If you are using histological paraffin, it is already in pellet or flake form.) Put that beaker inside a larger beaker that contains water. Place this "double boiler" on a hot plate to melt the paraffin.

 Paraffin is flammable. Do not set the beaker containing paraffin directly on the hot plate.

2. Barely thread the nuts onto the bolts to leave the maximum depth inside.
3. Cut pieces of stem tissue approximately $\frac{1}{2}$ inch long, making the cuts as perpendicular to the stem axis as possible. Any straight section of the stem can be used, including the nodes.
4. Set each stem piece upright in the center of a nut.
5. Pour melted paraffin into each nut. The stem section will stay upright. The paraffin must cover the top of the tissue or it will dry out.
6. The paraffin hardens in about 15 minutes. Do not attempt to hasten the hardening with ice or cold water or the paraffin may crack.

Investigative Extensions

This lab topic is presented as a traditional demonstration of plant anatomy. Once students recognize the typical cells and tissues present in plant sections and have practiced the technique for making plant sections, they can do comparative studies. Investigations can include examining different organs within a plant, looking at different stages of development of an organ, comparing hydrophytic, xerophytic, and mesophytic plants, or looking at the differences between monocots and dicots.

Keep in mind that since the paraffin simply surrounds the tissue rather than infiltrating it, firm tissue slices better than soft tissue. We have successfully sectioned and stained corn roots to examine monocot root tissue. The leaves of corn seedlings emerge rolled and have parallel veins, so they also yield excellent sections. It is much more difficult to obtain a good section of a dicot leaf since it is hard to position the tissue in the block.

It is also possible to obtain longitudinal sections to give students an idea of the three-dimensional structure of cells. To do this, pour a little paraffin into the bottom of the nut before you place the tissue in it, because the last few millimeters of the block will be unusable. Place a stem or root section on its side. You won't get many sections per block, especially with thin roots, so set up a lot.

The references cited contain numerous suggestions for plant tissues that stain well with toluidine blue. Rock (1981) used toluidine blue staining on sections cut freehand from bean, pea, goldenrod, ragweed, and the inflorescence stalks of grasses. He recommends avoiding plants with viscous or milky sap. O'Brien et al. (1964) report using the toluidine blue stain on freehand sections of numerous tissues in addition to *Coleus*, including carrot and potato storage tissues, oat and corn coleoptiles, celery petioles, *Elodea* stems, and epicotyl, stem, root, and cotyledons of peas.

References

O'Brien, T., N. Feder, and M. McCully. "Polychromatic staining of plant cell walls by toluidine blue O." *Protoplasma*, 1964, 59:367–373.

Parker, A., E. Haskins, and I. Deyrup-Olsen. "Toluidine blue: A simple, effective stain for plant tissues." *American Biology Teacher*, 1982, 44:487–489.

Rock, B. "Three-dimensional plant anatomy via hand sectioning and differential staining." In J. Glase (ed.), *Tested Studies for Laboratory Teaching*, Vol. II (pp. 1–15). Dubuque: Kendall/Hunt, 1981.

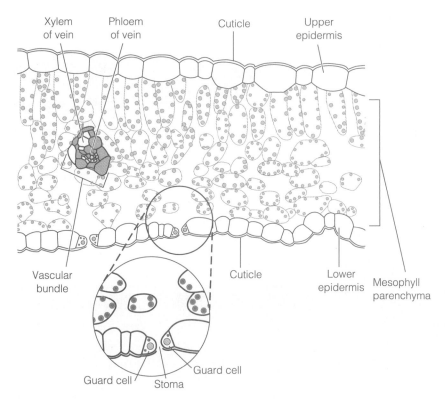

Typical dicot leaf cross section. Example of student sketch after completing the exercises in this lab topic.

Typical dicot root. Example of student sketch after completing the exercises in this lab topic.

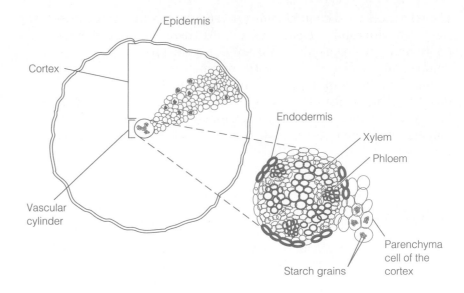

Typical *Coleus* stem. Example of student sketch after completing the exercises in this lab topic.

Flowers, Fruits, and Seeds

Introduction

In seed plants, pollination is the mechanism by which sexual reproduction is carried out. The pollen produced by the "male" structures on one plant must somehow be transported to the "female" structures on another plant so that the sperm produced by the pollen grains can fertilize the eggs. The earliest seed-bearing plants, gymnosperms, were wind-pollinated, as are present-day pine trees. Pollen grains from male pine cones are carried by the wind to the female cones, where they land on a sticky resin. Insects feeding on other resinous parts of gymnosperms discovered this food source and began exploiting it; pollen stuck to their bodies as they fed. As they went from cone to cone and from plant to plant, pollination was also achieved. Since this method of pollen transfer was more effective than just wind, plants that were more attractive to insects had a better chance of being pollinated and reproducing. This is how the angiosperms evolved.

The flower, which is characteristic of the angiosperms (flowering plants), is an adaptation for effective pollination. The diversification of angiosperms and of insect pollinators such as bees and butterflies occurred, not surprisingly, during the same period of evolutionary history.

Figure 18.1 on the next page illustrates how the life cycle proceeds after fertilization. The fertilized egg and other tissues associated with it develop into a seed, while the ovary walls mature to become a fruit. The term angiosperm means "seed container." Angiosperms are not the only seed plants, but they are the only plants whose seeds are contained in an ovary. The fruit that develops may be dry, like peanut shells, or fleshy, like an orange. Besides offering protection, the fruit may have special modifications that help the seeds disperse to a suitable habitat.

In this lab topic, you will learn about different kinds of floral characteristics that attract different kinds of pollinators; you will examine various types of dispersal mechanisms used by plants; and you will look at the parts of a seed and the initial development of seeds into plants.

Outline

Exercise 18.1: Flower Structure

Activity A: Flower Dissection

Activity B: Pollination

Activity C: Pollen Germination

Exercise 18.2: Fruits and Seeds

Activity A: Classifying Simple Fleshy Fruits

Activity B: Seed Dispersal

Instructor Overview
This lab topic features three stages of the angiosperm life cycle. In the first exercise students dissect a flower to learn its basic structures. Then they study flowers as adaptations to attract pollinators. By examining a variety of flowers, students can correlate certain floral characteristics with pollinators' sensory abilities and preferences. Students also see some of the diversity that is found among the four flower parts in different species. By germinating pollen in vitro, students learn that pollination is not fertilization: The pollen must grow before the egg is reached by the sperm. In the second exercise students learn to identify simple fleshy fruits by using a dichotomous key, which compels them to closely observe the similarities and differences among fruits and thus to consider the reasons that fruits have similar and diverse characteristics. Students then explore some of the modifications of fruits and seed coats that assist in seed dispersal. In the third exercise a peanut is used as a tasty introduction to the parts of a seed. Students then examine bean and corn seeds and observe how the seeds develop into seedlings.

Time Requirements
Exercise 18.1: 50–60 minutes (can be shorter or longer depending on the number of examples used for Activity B)
Exercise 18.2: 40–50 minutes (can be shorter or longer depending on the number of examples used for Activity B)
Exercise 18.3: 30 minutes

Figure 18.1.
Life cycle of a flowering plant.

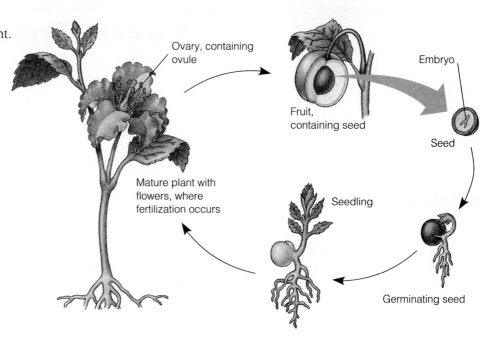

Ovary, containing ovule

Embryo

Fruit, containing seed

Seed

Mature plant with flowers, where fertilization occurs

Seedling

Germinating seed

Exercise 18.3: Seeds and Seedling Development
Activity A: Seed Structure
Activity B: Seedling Development

EXERCISE 18.1
Flower Structure

Objectives

After completing this exercise, you should be able to

1. Identify and describe the function of each part of a flower.
2. Identify flowers as complete or incomplete.
3. Distinguish between pollination and fertilization.
4. Discuss flowers as adaptations for effective pollination and name the floral characteristics preferred by the major pollinators.

Flowers are leaves that have been modified in various ways for reproduction. The fertile parts, which are directly involved in reproduction, are the stamens and the carpels. The sterile parts consist of the sepals and the petals. In general, the sepals protect the fertile parts and the petals attract pollinators. All of these parts are attached to the **receptacle,** a platform at the top of the flower stalk.

A flower that has all four parts (stamens, carpels, sepals and petals) is said to be **complete.** However, as you will see in Activity B, different species display diverse modifications of flower structure. Some flowers lack one or more of the four basic parts; they are called **incomplete.**

Activity A: Flower Dissection

The parts of the flower are arranged in concentric rings or rows around a structure called a receptacle at the top of the stem. From the outside to the inside, the rings are the sepals, petals, stamens, and carpels (Figure 18.2). In this activity you will dissect a flower and identify its parts.

The Prep Guide gives suggestions for flowers to use for dissection.

Procedure

1. Get a flower from your instructor. Examine the sepals, petals, stamens, and carpels, as described in the following steps.

2. The outer ring of **sepals** protects the developing flower when it is a bud. When the flower opens, the sepals protect the fertile parts of the flower. The sepals are usually green, but in some flowers they have been modified to attract pollinators. In most members of the lily family, for example, the sepals are indistinguishable from petals except by their position in the outermost ring.

Some flowers also have showy leaves that are neither sepals nor petals. When you look closely at a dogwood "flower," you'll see that what appear to be large white petals are actually leaves surrounding a cluster of small yellow-petaled flowers. The poinsettia is another example; the red "petals" are actually leaves.

The leaves referred to here are actually bracts, if you want to introduce that term.

Remove the sepals after you have noted their position in the flower.

3. The next ring consists of the **petals,** which are generally colorful to attract pollinators. Petals will be discussed in more detail in Activity B.

Remove the petals after you have noted their position in the flower.

4. The male flower structures called **stamens** produce pollen grains, which are immature male gametophytes. As you may recall from Lab Topic 12 (Plant Diversity), the actual gametes (sperm nuclei) are not produced until after pollination (transfer of pollen to the female organ). The pollen is produced in the **anthers,** which are supported on a stalk.

Remove the stamens after you have noted their position in the flower.

5. The female flower structures are called **carpels,** and their function is to produce the female gametophytes (ovules), which then produce the female gametes (eggs). A flower may have one or more carpels, and if there is more than one they may be either fused or separate. At the top of the carpel is the **stigma,** which receives the pollen. The **ovary,** which encloses the ovules, is at the base of the carpel. If an ovule is fertilized, it matures into a seed. The ovary itself will become a fruit. Gymnosperms, which are also seed plants, have no ovary protecting the seeds, which is why they are called gymnosperms ("naked seeds").

6. Cut across the ovary with a razor blade as shown in Figure 18.2. The ovules are white and egg-shaped. How many carpels do you see?

The answer depends upon the species of flower that is used.

7. Sketch the flower in the margin of your lab manual and label its parts.

Figure 18.2.
Structures found in a typical flower.

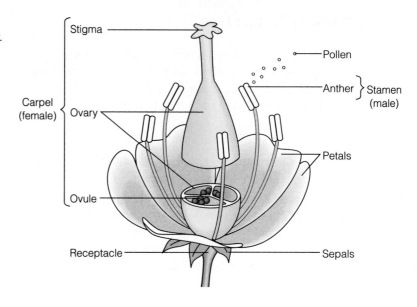

The purpose of this exercise is to have students really look at flowers. It is nice to be able to represent all the different types, but it is not necessary. When we do this lab topic in the spring in South Carolina, we have at least 30 species of flowers available (and students are only able to really examine 6 or 8). When we do it in the fall, there is very little available except the odd composite, so we borrow plants from the greenhouse. This exercise can be done in the dead of winter with slides or pictures, but if no flowers at all are available, it might be better to have students fill in Table 18.1 and spend more lab time on another exercise. See the Instructor Notes for more information about collecting flowers or using flowers from a greenhouse or florist.

Activity B: Pollination

Animals visit flowers to obtain food, which may be pollen or nectar, a sugar-water. Most species of flowers have adaptations that induce pollination by particular animal species that, in turn, have adaptations tuned to the flowers they frequent. In this activity you will examine some general characteristics of the major pollinator and flower types.

Procedure

1. Read the following information about pollinators and complete Table 18.1 on page 18-6.

Beetles

Beetles were probably the first insect pollinators. From the plant's point of view, they are not the best pollinators, because they eat a lot of the pollen and chew on other plant parts as well. However, there are plant species that are solely beetle pollinated. The successful beetle flower has its ovules buried deep, out of harm's way. Because beetle vision is not very good, many beetle-pollinated flowers are white or dull-colored. Some flowers are large and single (magnolia, lily, California poppy, wild rose); others are small and clustered (dogwood, elderberry, spirea). Beetles can smell much better than they can see, so the flowers may have strong odors, usually fruity, spicy, or fetid rather than sweet. Some beetles feed on nectar, but it must be easily accessible because beetles lack the long sucking mouth parts of other pollinators. Color Plate 7 shows a beetle pollinator at work.

Bees

Nearly all of the 20,000 known species of bees visit flowers for food. They quickly learn to recognize colors, odors, and shapes so they can consistently use particular kinds of flowers. Bees can distinguish colors at the ultraviolet end of the spectrum that are invisible to us, but they cannot see red. Bee flowers are usually bright yellow, blue, or purple and they often have distinctive markings to lead bees to the nectar reward. Some of these "nectar guides" absorb light in the ultraviolet range and thus can only be revealed to the human eye by special techniques. Bee flowers usually supply nectar, which is made accessible only to insects with long sucking mouth parts such as bees have, and a landing platform for the bee is often provided by specially adapted petals, as shown in Color Plate 7. The flower odor is mild and sweet. Typical bee flowers are thistle, larkspur, snapdragon, violet, apple, heather, and the various mints.

Flies

There are two types of flies that visit flowers: long-tongued and short-tongued. The long-tongued flies are similar in adaptations and flower preferences to the bees, so we will only consider the short-tongued flies here. These insects have no particular specializations for feeding on flowers, and most have other sources of nourishment. They are attracted to flowers that smell like these other food sources, such as carrion, dung, humus, and blood. Flies do not have keen sight, so fly flowers tend to be dull colors, as illustrated in Color Plate 7. Often there is no nectar for them, and some flowers, such as Dutchman's-pipe, not only don't feed the fly, they trap it for a day or two while the fly receives a load of pollen. This pollen is transported to the next flower, where the fly will likewise be trapped so it can thoroughly pollinate the flower. Examples of fly-pollinated flowers are *Rafflesia* (said to smell like putrefying flesh), black arum (which smells like human dung), and a lily whose odor is compared to fish oil.

Butterflies

Many species of butterflies can see the entire color spectrum visible to us; butterfly flowers may be bright yellow, blue, red, or orange. Flowers are visited for their nectar, and the petals of some butterfly flowers form a long thin tube to store the nectar. Butterflies are successful in obtaining nectar from such a tube because their tongues are about 1–2 cm long (see Color Plate 7). Nectar guides are also common, and landing platforms are provided either by the single flower or by the closeness of a group of flowers. Butterflies are known to pollinate wild geranium, forget-me-not, pinks, and butterfly weed (an orange milkweed).

Moths

Moths often resemble butterflies and also have long tongues for taking up nectar. However, unlike butterflies and other pollinators (except bats) that are active during the day, most moths are nocturnal. Moths have a good sense of smell and typical moth flowers give off a strong, sweet odor, sometimes only at night. Many moth flowers remain closed during the bright daylight hours. In color, shades of white or other colors that stand out in the dusk are most common (see Color Plate 8). Tobacco, Easter lilies, and tuberoses are some moth- pollinated species.

Birds

In tropical and subtropical regions, pollination by sunbirds, hummingbirds, and honeysuckers is common (Color Plate 8). Birds have a poor sense of smell but good color vision, so bird flowers are odorless and red or yellow. Because birds have high metabolic requirements the nectar reward must be high. The abundant nectar is kept in tubes that are inaccessible to most smaller animals, and a red, odorless flower is effectively camouflaged from many insects. Examples of bird flowers are fuschia, hibiscus, cardinal flower, and bird of paradise.

Bats

Residents of temperate regions may be surprised to learn that many tropical species of flowers are bat-pollinated (Color Plate 8). Bat flowers are similar to bird flowers in being large and having abundant nectar. Bats, however, are nocturnal; the flowers they visit are white or dull-colored and may only open at night. Bat flowers have strong fruity or fermenting odors to attract bats. Some examples are calabash, sausage tree, candle tree, and organ-pipe cactus.

Wind

There are numerous wind-pollinated angiosperms that evolved from insect-pollinated ancestors. Wind flowers generally lack the accessories of attraction: They have no nectar, no odor, no bright colors, and are often without petals (Color Plate 8). The stamens are usually exposed to facilitate launching the masses of pollen, and the stigmas are likewise exposed to enhance chances of intercepting the pollen. Many wind-pollinated species have separate male and female flowers. Grasses, cottonwoods, oaks, and walnuts are among the groups of wind-pollinated flowering plants.

Table 18.1
Major Pollinators

Pollinator	Sensory Abilities	
	Visual	**Olfactory**
Beetles	*Poor*	*Good; prefer spicy or fruity odors*
Bees	*Good; recognize shape and color; see UV but not red*	*Attracted to sweet odors*
Flies	*Poor*	*Attracted to rotten odors*
Butterflies	*Good; can see entire visible spectrum*	*Good*
Moths	*Poor*	*Good; attracted to sweet odors*
Birds	*Good*	*Poor*
Bats	*Poor*	*Attracted to fruity or fermenting odors*

Table 18.2

Flower Modifications

Flower	Complete or Incomplete?	Beetles?	Bees?	Butterflies?	Birds?	Wind?	Other

2. Examine the flowers on display to see whether they are complete or incomplete, and record your observations in Table 18.2. Also comment on whether the pollinating agents listed in Table 18.2 could pollinate each flower, and explain why or why not.

Many species of flowers develop a special relationship with one type of pollinator, and pollinators may have reciprocal adaptations that enhance their "faithfulness" to a particular species of flower. Why would it be adaptive for plants to encourage exclusive pollinator service? (Hint: What happens if a pollinator visits a violet, then a daisy, and then an apple blossom?)

Pollen that is transported to the wrong species of plant is wasted.

If students are permitted to dissect the flowers, it helps to use a dissecting microscope for the very small ones. The Prep Guide has instructions for preparing flower holders from rubber stoppers. The rubber stopper has a slit across one end. Squeeze the stopper to open the slit and position the flower stem in it. When the slit is closed, the flower is held upright for examination under a dissecting scope.

Activity C: Pollen Germination

As you could see when you dissected a flower in Activity A, the stigma is located at the other end of the carpel from the ovules. In some flowers this distance is as much as an inch or more. The microscopic pollen grain must germinate and produce a pollen tube that penetrates the stigma and grows through the entire length of the stalk that leads to the ovary. The pollen itself carries no food supply to provide the energy needed for this growth. Instead, cells on the surface of the stigma secrete a sugar-containing substance to provide nourishment. For many different types of plants, it is possible to germinate pollen in a drop of sucrose solution on a microscope slide.

Procedure

1. Using forceps, pluck a stamen from the flower provided. Hold the stamen by its stalk and brush the anther, face down, in the well of a concavity slide. (You may want to do this under a dissecting scope.)

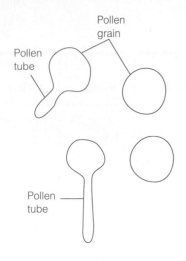

Student sketch of pollen.

2. Put a drop or two of sucrose solution in the well of the slide with the pollen. *Do not* use a coverslip.

3. Examine the slide using the scanning objective of the compound microscope. Sketch a few pollen grains in the margin of your lab manual.

4. Check your slide in 30 minutes to see if any germination has occurred. If the slide is drying out, add another drop of sucrose. Work on other exercises while you wait.

For the flower that you used for this activity, how far will the pollen tube have to grow in order to reach an ovule?

> *Approximately the distance from the stigma to the swelling in the ovary. To reach ovules that are at the bottom of the ovary, of course, the pollen tube will have to be longer.*

Why doesn't pollen carry its own food supply?

> *Pollen needs to be light, since it is airborne. A food supply would add weight.*

EXERCISE 18.2
Fruits and Seeds

Objectives

After completing this exercise, you should be able to

1. Define simple fruit, multiple fruit, aggregate fruit, pericarp, exocarp, mesocarp, and endocarp.

2. Using the key included in this exercise, classify a fruit as a berry, hesperidium, pepo, drupe, or pome.

3. Explain how fruits differ from vegetables.

4. Explain how fruits and seed coats are adapted to carry out seed dispersal.

5. Describe the characteristics of fruits and seeds that are dispersed by animals and by wind and give examples of each type.

Like flowers, fruits are unique to the angiosperms (flowering plants). The ovary first protects the female gametophyte and later the developing seeds. The ovary then develops into a fruit, which serves to protect and disperse the mature seeds. In some plants such as peaches the ovary wall is soft and fleshy when the fruit is mature. In others such as peanuts and grains the ovary wall is dry, papery, or hardened.

Activity A: Classifying Simple Fleshy Fruits

A **simple fruit** is a single ovary from a single flower. The ovary may, however, have more than one carpel, which is the chamber enclosing the seeds. For example, you can usually see three carpels in a cucumber slice (Figure 18.3).

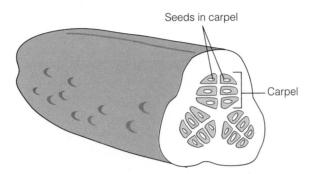

Seeds in carpel

Carpel

Figure 18.3.
Cucumber fruit showing three carpels.

A few fruits such as pineapple and fig develop from numerous flowers that grow very close together; their ovary walls fuse during maturation. Fruits that develop from multiple flowers are called **multiple fruits.** An **aggregate fruit** is a group of simple fruits that are joined together. Aggregate fruits develop when a single flower has numerous ovaries attached to the same flower. Each ovary develops into a separate fruit, but since the fruits are all attached to a single receptacle (the platform at the tip of the stem) they appear to be a single fruit. Blackberries and raspberries are examples of aggregate fruits: each "knob" of a blackberry contains a seed. Strawberries are also aggregate fruits: each "seed" is actually a dry ovary (simple fruit) containing a seed.

As the ovary of fleshy fruits ripens and grows, it also differentiates into three layers. Collectively, the layers are called the **pericarp.** "Carp" means "fruit," so the outer layer of the fruit is called the **exocarp,** the inner layer is the **endocarp,** and the middle layer is the **mesocarp.** Figure 18.4 illustrates the pericarp of a peach. The fuzzy skin is the exocarp, the juicy middle is the mesocarp, and the stony pit is the endocarp. The seed is inside the endocarp, which protects the seed as it passes through an animal's digestive tract. The differences among the pericarps of various fleshy fruits are useful for classifying them.

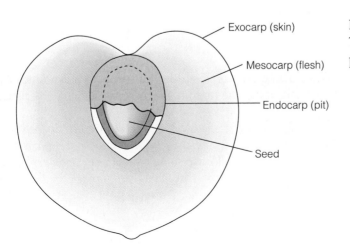

Exocarp (skin)

Mesocarp (flesh)

Endocarp (pit)

Seed

Figure 18.4.
The layers (pericarp) of a peach fruit.

When biologists want to identify an organism, they often turn to a classification scheme known as a **key.** There are keys to identify trees, mushrooms, seashells, mammals, insects, and most other groups of organisms. Keys that are used by experts may include thousands of species, but less complex keys have also been constructed for use by nonscientists who simply want to know more about nature. A key is constructed to lead you through a series of comparisons. At each step, you choose the statement that best describes the sample you are trying to identify. As this process is repeated, the classification is increasingly narrowed until you obtain an answer.

The key that you will use in this exercise includes only fleshy fruits. Dry fruits such as peanuts, beans, grains, and nuts cannot be identified using this key.

To use the key, examine each fruit and choose between the two alternatives for number 1: Is the fruit better described by 1a or by 1b? Continue with the choices listed below the alternative that you selected. For example, if you selected 1a, your next choice is between 2a and 2b below 1a. A peach would key out as follows: The endocarp is not fleshy, since it is the stony pit: choose 1b. The endocarp is stony rather than papery: under 1b, choose 4a. This identifies the fruit as a drupe.

As another example, think about a pumpkin. The endocarp is fleshy (when you scoop out the inside of a pumpkin, the seeds are not separated from the rest of the fruit by stony or papery tissue): choose 1a. The exocarp is thickened: choose 2b. The exocarp is not easily peeled away from the rest of the fruit (have you ever tried to peel a pumpkin?): choose 3b. The pumpkin is a pepo.

Key to Simple Fleshy Fruits

1a. The endocarp is fleshy or pulpy. . . . . Go to 2

  2a. The exocarp is thin and skinlike. **Berry**

  2b. The exocarp is thickened. . . . . Go to 3

      3a. The exocarp and mesocarp are leathery and easily peeled away from the rest of the fruit. The carpels contain large juice-filled hairs. . . . . **Hesperidium**

      3b. The exocarp is thick but not leathery and cannot easily be separated from the rest of the fruit. . . . . **Pepo**

1b. The endocarp is not fleshy. . . . . Go to 4

  4a. The endocarp is stony. . . . . **Drupe**

  4b. The endocarp is a tough, papery membrane. . . . . **Pome**

What characteristics do all the fruits that you examined have in common? Speculate about the reasons for these shared characteristics.

Sometimes students mistakenly key out pepos and hesperidiums as berries, since these fruits are specialized types of berries. Once students have keyed out several fruits, they may begin to catch on to certain relationships. For example, all citrus fruits (and only citrus fruits) are hesperidiums. All members of the squash family are pepos. The only single-seeded fruits in our key are drupes.

They all have seeds because they are all ripened ovaries. They are all fleshy, which students may have various hypotheses about. They all have some type of skin (exocarp) over the softer mesocarp, which is probably for protection. Students may also mention edibility; the reason for this (dispersal) is discussed in the next activity.

What are the major differences among the fruits that you examined?

Color, size, number of seeds, composition of endocarp; others are possible.

Students will be asked to speculate about the reasons for these differences after completing the next activity.

What Is a Vegetable?

Some of the fruits you keyed out are better known to you as vegetables. There is no botanical definition of a vegetable.

What is your definition of a vegetable?

"Vegetable" is not a botanical term; it generally refers to any edible plant part and is sometimes defined as a plant part eaten with a main course.

List some other vegetables.

Which of the vegetables you listed could be classified as fruits?

Based on the botanical definition of fruit, what's the best way to distinguish a fruit from a vegetable?

A fruit is a ripened ovary, so it contains seeds (though there are parthenocarpic fruits, such as bananas, which develop without fertilization and thus have no seeds).

Activity B: Seed Dispersal

Many fruits and seed coats (the layer of tissue covering the seed) are adapted for ensuring that the seeds are carried away from the parent plant, a process known as seed dispersal.

The modifications for seed dispersal parallel the modifications of flowers for pollination. Animal agents may be employed, and plants use color and nutritional rewards to attract the animals to their fruits. Fleshy fruits such as peaches, for example, are meant to be eaten. The animal digests the fruit's flesh, but the seed, in this case protected within the endocarp, passes unharmed through the gut and is deposited some distance from the parent plant. In other fruits, animals consume some seeds while others are buried and forgotten. Name some fruits that use animals to disperse their seeds.

Generally, fleshy fruits that provide a substantial amount of nutrition are good examples, such as apples and berries of various kinds. Nuts such as acorns and hickory nuts also utilize animal dispersal.

Animals are more passive participants in another scheme of seed dispersal, in which the seed coats are adapted to latch onto the fur of passing animals. These plants have names that describe the seed coats, such as sticktights and beggar-ticks.

In a few species of plants, the seeds must pass through an animal's digestive tract in order to germinate. You can tell your students about the tree whose seeds were adapted to such a relationship with the dodo bird on the island of Mauritius. The seeds have a thick seed coat and will not germinate unless they have been subjected to the digestive processes of the dodo bird. Sadly, the dodo bird has been extinct for 300 years, driven to extinction by humans and their domesticated animals. Several of the trees remain and they produce abundant seeds, but no new trees have grown since the day of the dodo. Fortunately, a scientist may have found a way to save the trees by passing their seeds through the guts of turkeys.

Other plants depend on dispersal by the wind. Typically they produce many light seeds, and the seed coat forms some sort of structure that can easily be picked up and carried along by the breeze. Name some examples of seeds that are dispersed by wind.

Dandelion, maple, milkweed; there are many others.

A few seeds are also adapted for dispersal by water, and the fruits of others are simply dropped and carried away passively by rainwater or wind without any special mechanisms.

If your campus and the time of year are suitable, a scavenger hunt is a good way to build and maintain a collection of seeds for this exercise. Give students a list of seed types they must find as part of the grade for this lab topic (for example, five different kinds of wind-dispersed seeds).

Procedure

1. Examine the fruits and seeds on display and determine the probable means of dispersal.
2. In Table 18.3, record the name and probable dispersal method of each plant, and list the characteristics that helped you decide on the dispersal method.

Table 18.3

Seed Dispersal

Plant	Dispersal Method	Characteristics

Why is it advantageous for plants to disperse their offspring?

So they don't compete with the parent and so they have a chance of finding a suitable place to germinate.

In Activity A you listed the similarities and differences that you observed among the fleshy fruits you examined. Suggest reasons for those characteristics.

Differences in color, size, and composition of the fruit's flesh may be adaptations for attracting animal dispersal agents. The nature of the endocarp might also be related to animal dispersal. For example, drupe seeds are protected inside a stony endocarp, while cucumber and apple seeds are not. The fleshiness of all the fruits is a nutritional reward for the dispersing agent.

EXERCISE 18.3
Seeds and Seedling Development

Objectives

After completing this exercise, you should be able to

1. Identify the parts of a seed and describe the function of each part.
2. Identify the parts of a seedling.
3. Compare the development of corn and bean.

Activity A: Seed Structure

The ovules inside the ovary are the female gametophytes. Each ovule produces an egg. If the egg is fertilized by a sperm nucleus that has been produced by the pollen (male gametophyte), it develops into an embryo. Recall from Lab Topic 12 (Plant Diversity) that flowering plants have double fertilization. A second sperm nucleus fertilizes two additional nuclei to form the **endosperm,** which becomes a food source for the embryo when it germinates. The ovule is also surrounded by a layer of tissue that becomes a **seed coat** to protect the seed. In this activity you will look at the three parts of a seed.

Procedure

1. Get a peanut fruit from your instructor.
2. Crack and remove the fruit (shell) from the seeds. As you do, note that the peanut shell can easily be split in half lengthwise. It is a pod, similar to the fruits of other members of the bean family.
3. Remove the red, papery covering, which is the seed coat. Layers of tissue surrounding the ovule thicken and harden into this protective covering.
4. Open the **cotyledons** (seed leaves) gently to avoid damaging the rest of the embryo. The halves of the peanut are the cotyledons. The peanut cotyledons have absorbed the endosperm and will now serve as the food supply. In some plants, for example corn, the endosperm remains separate from the cotyledon(s).

5. Observe the **embryo.** The cotyledons are actually part of the embryo. The rest of the embryo is the root/shoot axis; the plant will grow from the root tip and the shoot tip. Notice the first "foliage" leaves apparent in the embryo.

6. In the margin of your lab manual, sketch and label the parts of the seed.

Activity B: Seedling Development

The molecular structure of the seed coat makes it especially effective at taking up water. As water is absorbed, enzymes are activated that convert the stored starches or fats in the endosperm or cotyledons into usable food molecules. The initial growth of the embryo, which is called germination, depends on the metabolic energy derived from this food. Later, photosynthesis will provide energy for the plant to continue its growth.

Flowering plants are divided into two subgroups, monocotyledons and dicotyledons, depending on whether there is one cotyledon or two in the seed. In this activity you will compare the seeds and early development of a monocot and a dicot.

Procedure

1. Get a bean seed, a corn seed, and a plastic cup containing bean and corn seedlings from your instructor.

2. Dissect the bean seed by removing the seed coat and separating the cotyledons to reveal the embryo inside. Beans are in the same plant family as peanuts.

Are beans monocots or dicots? How do you know?

Dicots (there are two cotyledons).

3. In the margin of your lab manual, sketch the seed and label the seed coat, cotyledons, and embryo.

4. Make a sketch of the bean seedling in the margin of your lab manual.

 Since the number of seedling sets is limited, please do not dig up and destroy these plants!

5. Label the cotyledons on your sketch of the bean seedling.

6. Also label the **hypocotyl** (the part of the shoot below the cotyledons) and the **epicotyl** (the part of the shoot above the cotyledons) on your drawing of the bean seedling. The two leaves at the top of the epicotyl are known as the **first foliage leaves,** in contrast to the cotyledons, or seed leaves.

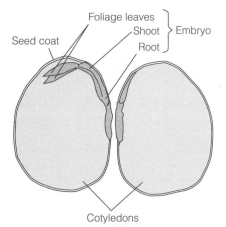

Student sketch of a bean seed.

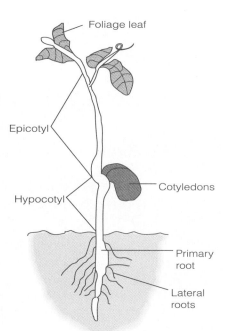

Student sketch of a bean seedling.

7. Label the **primary root,** which is the largest root. The primary root will become a taproot, growing directly downward through the soil. You can see lateral roots growing out of the taproot; they will branch out in all directions to utilize a greater area of the soil beneath the plant, thus increasing the surface area for absorption of water and minerals.

For a light-hearted addition to the lab, get a chia pet and use the germinated seedlings to supplement this exercise. Students can identify the same features that they can see on the bean seedling.

The bean cotyledons may be somewhat shriveled when you observe the seedling. Why are the cotyledons no longer as fleshy as they were in the seed?

> *The starch stored in the cotyledons has been used by the developing plant.*

If you observe seed germination, you will notice that the root emerges from the seed coat first, followed by the shoot. Why should the root develop more quickly? (Hint: What are the functions of the root?)

> *The root serves to anchor the plant in the soil and to absorb water and minerals. The plant thus establishes a position in the soil and begins bringing in the water and minerals that will be needed for further growth.*

8. Cut open a corn seed longitudinally with a razor blade.

 Be careful not to let the razor blade slip. Even though the corn has been soaked in water to soften it, it is very hard to cut.

Corn, like other grains, has a dry fruit that is fused with the seed coat inside, so it is not possible to distinguish the ovary wall from the seed coat.

9. Compare the corn seed you have dissected to Figure 18.5.

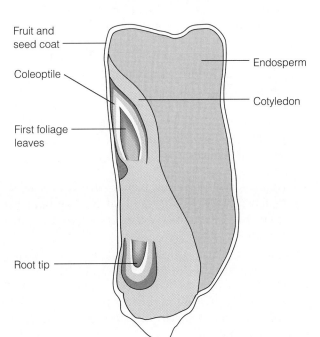

Figure 18.5.
Corn seed. The endosperm in corn is not absorbed by the cotyledon. Notice the coleoptile, which is a sheath enclosing the leaves.

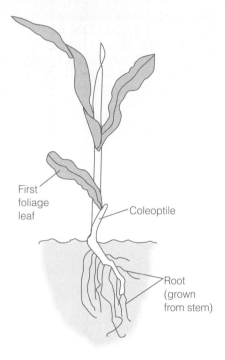

First
foliage
leaf

Coleoptile

Root
(grown
from stem)

Student sketch of a corn seedling.

10. Examine the corn seedling and sketch it in the margin of your lab manual.

11. Label the **first foliage leaf**; it should be apparent. The emerging leaves are encased in a sheath called the **coleoptile.** The coleoptile is found in all members of the grass family (which includes corn and the other grains), but it is not a characteristic of monocots in general.

12. Examine the roots and label them on your drawing.

The primary root lasts for only a short time in monocots; the entire root system develops from the end of the stem. (This is similar to how roots arise from stem cuttings placed in water or moist soil.) If you have ever grown corn in a garden, you have probably noticed how the roots grow from the stem. Mounding soil high up on the stalks stimulates root growth, thus improving the plant's support. Since there is no taproot to penetrate deep into the ground, the roots of monocots tend to be rather shallow and spread out.

Which type of plant do you think is easier to transplant, a dicot or a monocot? Explain.

In transplanting, it's important to bring along as much of the root system as possible. A monocot probably requires a smaller ball of soil since its roots don't penetrate as deeply.

Questions for Review

1. Some angiosperm species have returned to the wind-pollinated mode of gymnosperms (for example, pines). There are benefits and costs associated with both strategies. It costs (energy and materials) to make pollen. It also costs to use animal go-betweens. What must wind-pollinated species "pay" for that animal-pollinated species don't?

They have to produce large amounts of pollen to improve the chances that some will land in the right place. Animal-pollinated plants have a higher success rate, so they make relatively little pollen.

What must animal-pollinated species pay for that wind-pollinated species don't?

They have to make themselves attractive to pollinators (e.g., petals, odor) and offer a reward (nectar, pollen). Wind-pollinated species don't need this.

2. Animals and flowering plants have coevolved in some ways that are beneficial to both parties. That is, they have complementary adaptations. Explain the two examples of coevolution that were seen in this laboratory.

Pollination: Insects, birds, and bats carry the male gametophytes of one plant to the female part of another plant. In return they receive food. Dispersal of seeds from fleshy fruits: Various animals eat fruits, thus receiving nutrition and at the same time dispersing the plant's seeds.

3. An imaginary animal has a poor sense of smell, a long tongue (1–2 cm), and sees well at the short-wavelength end of the spectrum (ultraviolet and blue) but not at the other end of the spectrum (red). Design a flower that attracts this pollinator *and* excludes other pollinators.

    The flower could have little or no odor, nectar at the bottom of a long tube that is formed by the petals, and absorb UV light. Beetles, flies, moths, and bats are attracted by odors, so an odorless plant may not attract their attention. Putting the nectar at the bottom of a long tube excludes beetles and bees, whose tongues are not long enough to reach it. The best petal color would be a UV-absorbing pigment. This color would not be visible to most animals and thus would exclude butterflies and birds as pollinators. If the flowers were blue (the other color the imaginary animal can see), they would attract butterflies.

4. Many fruits are green when they are immature and turn red or orange when they are ripe. What is a possible reason for this?

    Green fruits are camouflaged against green foliage. The bright colors advertise that the fruit is ripe (edible); presumably the seeds are also fully mature at this time and thus ready to be dispersed.

5. Keys like the one you used to classify fruits are based on characteristics that distinguish one group from another. In this case, only the layers of the pericarp were used. What other characteristic of simple fleshy fruits would be useful in a key?

    The number of seeds and whether the seeds are usually eaten as part of the fruit.

6. How do you know that zucchini is a fruit as well as a vegetable?

    It has seeds.

7. Occasionally in nature a mutant plant that lacks the ability to produce chlorophyll occurs. How long could such an "albino" plant survive, if at all? Explain.

    Initial growth depends on the food supply in the seed rather than photosynthesis, so the plant would survive as long as that food lasts. Once it has been exhausted, the plant will be unable to continue growth or maintenance activities since it can't photosynthesize without chlorophyll.

8. How might the amount of food supply that a seed contains be related to its dispersal mechanism?

    Wind-dispersed seeds cannot have a very large food supply since they must be light. Seeds that are adapted for water dispersal or passive dispersal by animals might have the same constraint. Seeds that are eaten or carried by animals can afford to be heavier.

9. What seed characteristic should be considered in deciding how deep to plant a seed? Explain.

   Size of the seed. A very small seed, such as a carrot seed, clearly does not have a large food supply. If planted too deep, it would exhaust its resources before the shoot reached the light needed for photosynthesis. A large seed like a pea, on the other hand, could be expected to survive planting at a greater depth.

LAB TOPIC 18

Flowers, Fruits, and Seeds
Instructor Notes

Organization

Time requirements for this lab topic vary depending on how many specimens you require students to examine for the pollinator, fruit key, and seed dispersal exercises.

Introduction (5–10 minutes)

Review Figure 18.1 to explain how the parts of the lab are connected. Flowers have evolved to facilitate pollination, which is necessary for sexual reproduction. The result of sexual reproduction is the seed, and in angiosperms seeds are contained in fruits. The fruit or the seed coat may be adapted to help disperse the seeds. The seed itself carries a food supply and the embryo of a new plant, protected by a seed coat. When seeds reach a suitable place, they germinate and develop into new plants.

Prep for Procedures (5–10 minutes)

- Explain what to do for the pollinator exercise.
- Tell students which flowers may be dissected.
- Demonstrate how to use the fruit key.
- Caution students to be careful when cutting the corn seed with a razor blade.

If you are collecting flowers, look especially for the ones people usually overlook, like grasses, trees, and small weeds that grow between sidewalk cracks. It certainly isn't necessary to have all the flowers identified, but if you have a botanist handy it's nice to do (use common names, though). It also isn't necessary for you to know what pollinates what, or even what parts are what. Some flowers are very difficult to "interpret"; try figuring out an iris.

If you use flowers from a greenhouse or florist, try to avoid those that have been bred for show, since their colors and multiple petals may be very misleading.

Members of the composite family (daisies, asters, etc.) are very common, so it is worthwhile to know that these "flowers" are actually groups of flowers arranged on a common receptacle. Many composites have two types of flowers. In a daisy, for example, the yellow disk in the center contains one type of flower. The white rays surrounding the disk are the second type of flower. If you examine these flowers closely you can see the reproductive parts in each one.

Summary (10–15 minutes)

- Review some of the flowers and seeds (as time permits).
- Review answers to fleshy fruits.
- Have students draw seedlings with parts labeled on overheads or on the board.

If you wish, you can eliminate most of the summary by having answers for the flower, fruit key, and seed dispersal exercises to show students when they finish.

For a 2-Hour Lab

Eliminate either Exercise 18.1, Activity B, or Exercise 18.2, whichever is more difficult for you to obtain materials for.

Investigative Extensions

Pollen Studies

Students can investigate factors affecting pollen germination. It is first necessary to identify an available source of pollen that is known to germinate *in vitro*. The species used in Exercise 18.1, Activity C, is the best candidate (see the Prep Guide and Schimpf, 1992, for suggestions). Pollen germination is affected by sucrose concentration. Although most species require 5–15%, some require a much higher concentration (especially the composites). Some types of pollen will germinate without sucrose, but adding sucrose enhances the percentage of germination. Pollen may also be able to use sugars other than sucrose for germination, including glucose, galactose, lactose, and raffinose. In some species, sucrose may not produce the best germination percentage.

Boron is known to stimulate germination, though not all species appear to require it. Boric acid can be used to supply the boron; typical stimulatory concentrations are around 0.001%. Calcium may also stimulate germination. Potassium and sodium at $0.01M$ and barium and magnesium at $0.0015M$ have been found to be inhibitory to pollen germination in some species. pH may also affect germination.

If students have the opportunity to collect pollen from different sources, interesting studies can be done comparing the morphology of pollen from wind-pollinated plants and animal-pollinated plants. In general, pollen designed to be carried by the wind has a smooth exterior so the grains won't stick together. There is also an optimum size range, 12–40 μm. Smaller particles tend to carry on past the stigma, and larger particles drop too quickly. Conifer pollen is very large (up to 130 μm) but each grain has two "floats" attached to compensate (conifer pollen looks like it has cartoon mouse ears). On the other hand, pollen that is designed to be carried by insects tends to have sculptured exteriors and lipid exudates, which help the grains stick together.

Pollen can be stained with a drop of aniline blue or acetocarmine for better visibility.

Seed Studies

The topic of seed germination was also introduced in this lab and is a rich source of ideas for student investigations. Investigations of factors such as light, temperature, and various chemicals on seed germination are probably already familiar to you and the students. Other investigations can involve determining how seed size affects germination time and the length of time a seedling can survive without photosynthesis.

Seeds should be surface sterilized before use to discourage the growth of microorganisms. Immerse the seeds in household bleach (5% sodium hypochlorite) for 5–10 minutes, swirling the container occasionally. Rinse thoroughly with water.

References

Clay-Poole, S. T., and I. Slesnick. "The beauty and biology of pollen." *American Biology Teacher,* 1983, 45: 366–370.

Morholt, E., and P. Brandwein. *A Sourcebook for the Biological Sciences,* 3rd ed. New York: Harcourt Brace Jovanovich, 1986.

Schimpf, D. J. "Rapid germination of pollen *in vitro.*" *American Biology Teacher,* 1992, 54: 168–169.

Stanley, R. G. "Pollen Chemistry and Tube Growth." In *Pollen: Development and Physiology,* ed. J. Heslop-Harrison. London: Butterworth, 1971, pp. 131–155.

Animal Behavior

Introduction

In Lab Topic 16 (The Sensory System), you learned about how animals perceive the environment. Sensory input is processed by the central nervous system, which then directs a response. **Behavior** is the term given to those responses. Behavior may be simple and reflexive, such as rapid withdrawal from a source of pain, or it may be complex, such as detecting, approaching, and courting a potential mate. In short, behavior includes everything animals do.

Human behavior is governed by emotion as well as intellect and reasoning skills. Since we are capable of understanding our own behavior, we often find it difficult to adopt a scientific perspective on animal behavior. For many of us, association with pets has created a most unscientific tendency to anthropomorphize, or attribute human thoughts and feelings to animals. For example, when a kitten pounces on a ball of yarn and tumbles it around, we might characterize its behavior as recreation, as if the kitten intentionally decides to relax and have a bit of fun. In fact, this play behavior is probably a rehearsal for the cat's predatory habits later in life. The ball of yarn is a stand-in for a field mouse, and the kitten is learning to chase, capture, and subdue its prey.

As in any scientific field of study, we must be able to formulate testable hypotheses in order to answer questions about behavior. Hypotheses about the causes of behavior can address two general types of questions: What triggers the behavior? and What is the adaptive significance of the behavior? For example, the cue, or trigger, for the kitten's attack on the ball of yarn is whatever stimulus or stimuli caused the ball to be perceived as prey. It might be the ball's shape, color, smell, or movement. The evolutionary reason, or adaptive significance, for the behavior explains why an animal engages in a particular behavior. For example, juvenile rehearsal of prey capture may have provided an evolutionary advantage for the wild ancestors of our domestic cats.

In this lab topic you will investigate a relatively simple behavior, kinesis in pill bugs, and a more complex behavior, agonistic encounters between bettas (Siamese fighting fish). The methods employed allow you to study the immediate causes of these behaviors, but you should also consider their adaptive significance.

Instructor Overview
After a brief introduction to the study of behavior, students perform a demonstration exercise to learn a method for investigating kinesis in pill bugs (Exercise 19.1). In Exercise 19.2, agonistic behavior is introduced. Students observe a betta at rest and displaying to its mirror image. Each student team then designs its own experiment. Students can test a hypothesis either about environmental preferences in pill bugs or about the releasing mechanism for display behavior in bettas. The lab topic prompts students to organize their investigations (Exercise 19.3) and answer a series of questions to interpret their results (Exercise 19.4). Alternatively, the instructor may require students to write lab reports.

Outline

E X E R C I S E   1 9 . 1
Kinesis in Pill Bugs

Objectives

After completing this exercise, you should be able to

1. Define kinesis.

2. Describe how a choice chamber is used for determining pill bug environmental preferences.

3. Explain how the behavior of pill bugs contributes to their survival.

Most humans have sense enough to come in out of the rain, but even lower animals engage in behaviors that will place them or keep them in favorable environments. In one such behavior, called **taxis,** animals move directly toward or away from a stimulus. In another type of behavior, called **kinesis,** animals move randomly in response to environmental stimuli. Their movements are relatively rapid when the environment is unfavorable and slow down when the environment is favorable. How could this random movement be useful to an organism in finding a suitable place?

As long as the animal keeps moving, it could wander into a more favorable environment. When movement slows down in response to a favorable environment, the animal will be staying where conditions are good.

What environmental cues might cause kinetic behavior?

Examples: Light, color of the background, humidity, temperature.

What might be the adaptive significance of kinetic behavior?

Animals that locate a favorable environment are more likely to survive.

Figure 19.1.
The pill bug. Other common names for the pill bug and related species are roly-poly and sow bug.

The pill bug (Figure 19.1) is an arthropod belonging to the class Crustacea, which also contains lobsters, crabs, and barnacles. It is one of the few land-dwelling members of this class. You can usually find pill bugs by turning over logs and rocks in the woods or even in your backyard.

The kinetic response of pill bugs to different environmental stimuli can be investigated using an apparatus called a choice chamber. As shown in Figure 19.2, the choice chamber consists of two large petri dishes taped together with an opening in between. Each chamber can have different conditions. To determine the response of pill bugs to environmental conditions, place an equal number of pill bugs in each dish. Observe their movements and record their "choice" of chamber conditions.

Students might be interested to know that the genus name of pill bugs is Armadillium. They must have been named by a Texan! Sow bugs are actually a different genus. Either pill bugs or sow bugs can be used for this exercise. If you ordered a mixed culture from Carolina Biological Supply or field-collected your own, you probably won't even know which species is being used.

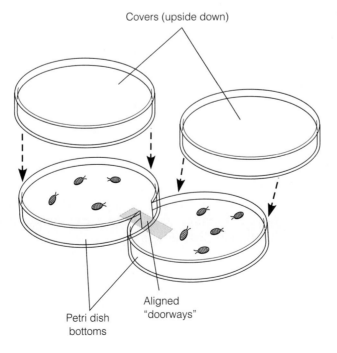

Covers (upside down)

Petri dish bottoms

Aligned "doorways"

Figure 19.2.
Choice chamber.

Hints for Successful Investigations

1. Use a soft brush when you transfer pill bugs from one container to another.

2. Once you have started an experiment, don't disturb the chamber.

3. When you get pill bugs from the stock culture, avoid picking the very small ones. They tend to ride on the backs of the larger ones instead of moving about themselves.

4. If your pill bugs all pile up in one chamber and stop at the beginning of an experiment, they are probably tired from previous experiments and are not responding to environmental stimuli. Put them in the "recovery" terrarium and replace them with a fresh batch from the stock culture.

There should be two terraria for pill bugs, one containing fresh animals and one for animals that have recently been used. Have students get their pill bugs from the "fresh" terrarium and return pill bugs to the "recovery" terrarium at the end of the lab period.

In the first activity, conditions in both chambers will be exactly the same. This exercise will make you familiar with the general behavior of the pill bugs and with the technique for recording data on their behavior.

Predict the distribution of the pill bugs after 5 minutes.

Example: Since the chambers have identical conditions, the pill bugs will be equally divided between the two.

Procedure

1. Put a piece of filter paper in a regular-sized petri dish and moisten it with a wash bottle. You will use this dish to transport pill bugs.

2. Use a soft brush to transfer 10 pill bugs from the stock culture into your dish.

3. Set your choice chambers on the bench in front of you. Environmental conditions should be the same in both chambers, so be sure they rest on a similar color background. Be careful not to let your shadow fall over either side once the pill bugs are in the chambers.

4. Use a soft brush to transfer five pill bugs from your petri dish into each chamber. Quickly put the lids over the chambers. (Put the lids on upside-down or they won't fit.) Note the position of the second hand on your watch or the wall clock.

5. At 30-second intervals, count the number of pill bugs in each chamber and note their activity level, using phrases such as "very active," "some activity," or "hardly moving" to describe what the pillbugs are doing. Record this information in Table 19.1.

6. Continue your observations for 5 minutes.

7. Return the pill bugs to your petri dish.

8. Use Figure 19.3 to graph the number of pill bugs present in Chamber 1 during the 5-minute observation. Label the axes and write a descriptive caption for the figure.

Table 19.1
Observations of Pill Bug Location and Activity Level
(five pill bugs placed in each chamber initially; chambers
have identical environmental conditions)

Minutes	Chamber 1		Chamber 2	
	Number of Pill Bugs	Activity	Number of Pill Bugs	Activity
0	5	*Very active*	5	*Very active*
0.5	3	*Most active*	7	*Most active*
1.0	2	"	8	"
1.5	3	"	7	"
2.0	2	"	8	"
2.5	3	"	7	"
3.0	5	"	5	"
3.5	7	"	3	"
4.0	8	"	2	"
4.5	5	"	5	"
5.0	7	"	3	"

Figure 19.3.

Describe and attempt to explain the general pattern of movement when pill bugs are active.

Pill bugs spend most of their time crawling along the edges of the chamber. As they crawl, they tap one antenna on the side of the container; the pill bug is probably using this to orient its movement.

How do pill bugs behave when they are inactive?

They gather together in piles, again along the edge of the chamber. When an active pill bug encounters one of these pile-ups, it tends to "stick" there.

If a pileup isn't in a preferred environment, one or more individuals become active again and the pile breaks up after a minute or two.

After 5 minutes, how were the pill bugs distributed in the two chambers? Did this distribution match your prediction?

Remember that the pill bugs tend to keep moving unless they are in a favorable environment. In this case, the environment is not ideal, so the number in each chamber will fluctuate and will probably not match the typical student's prediction that there will be five in each chamber.

9. The instructor will collect the results from all the groups in the class. Obtain the average number of pill bugs in Chambers 1 and 2 after 5 minutes and record the results in Table 19.2.

Table 19.2
Number of Pill Bugs in Each Chamber
After 5 Minutes (class average)

Chamber 1	Chamber 2

Does this number match your prediction more closely? Explain the difference.

It probably will. The larger the number of pill bugs counted, the closer the results will be to random distribution.

What were the dependent variables in this experiment?

The data that students recorded: Number of pill bugs in each chamber and activity level.

What independent variables might affect these dependent variables?

Any environmental factor that differs between the two chambers, such as light, humidity, presence of a chemical substance, color of the background, and temperature. Students may also mention that differences among pill bugs—for example, size or species— might affect pill bug activity and distribution.

EXERCISE 19.2
Agonistic Behavior in Bettas

Objectives

After completing this exercise, you should be able to

1. Define agonistic behavior and explain its adaptive significance.
2. Give examples of animal behaviors that indicate aggression and submission.
3. Explain what a fixed action pattern is and how it is triggered.
4. Describe the agonistic behavior of bettas.

Most institutions require an animal use protocol to be approved for experiments involving any kind of vertebrate. Be sure to check on regulations before you do this exercise. Bettas must be kept in separate containers: 1- or 2-L beakers can be used in place of fish tanks.

Members of the same species living in the same area are in competition for the same resources, including food, water, space, shelter, and mates. Many animals have developed **agonistic** behaviors for encounters with these rivals. ("Agonistic" is derived from a Greek word meaning struggle or contest.) Essentially, the competitors stage a contest to determine which one gets the prize. Animals may stake out territories that they then defend against intruders. Many birds do this, and often when you hear a bird singing you are listening to it advertising its territorial claim.

In most species, these contests are highly ritualized and symbolic, involving more posturing than fighting. Even in species that will engage in combat, however, the contestants first attempt to win the fight through intimidation by presenting themselves as large, fierce, and unbeatable. One participant may then perceive itself to be inferior and back down.

Give an example of an aggressive or threatening display.

> *Dogs growling and baring their teeth with their hackles raised. Cats doing the same sort of thing.*

Other examples are found among birds (for example, gulls engage in a threat display) and primates (monkeys stare, make mock charges, and slap the ground; gorillas beat their chests). Some lizards have flaps of skin that take on a bright red color, and the animal sways and bobs its head up and down.

The loser in an agonistic encounter may display submissive or appeasement behavior and thus acknowledge the other animal's superiority. This generally halts the aggressive display on the winner's part.

Give an example of submissive behavior.

> *In general, the animal reverses its aggressive display. Instead of trying to appear large, it makes itself small. It may also expose its vulnerable parts—for example, by rolling over on its back.*

What is a possible adaptive significance of agonistic behavior?

> *Controlling a territory allows an animal to reserve sufficient food resources (and perhaps a safe nesting site) for itself and its offspring. Defending the territory by agonistic behavior rather than actual combat conserves energy and avoids possible injury or death.*

In this exercise you will investigate agonistic behavior in Siamese fighting fish (Figure 19.4). The scientific name for these fish is *Betta splendens,* so they are often referred to as bettas (pronounced bay-tuh). These fish, especially the males, are territorial. The presence of an intruder triggers a display that is intended to intimidate the rival.

Figure 19.4.
Siamese fighting fish (*Betta splendens*).

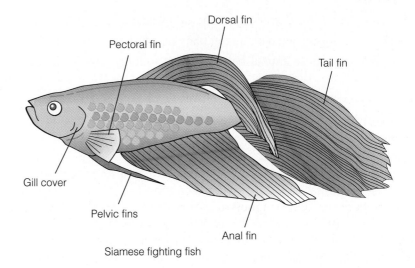

Siamese fighting fish

The agonistic behavior of the betta is inborn and follows a fixed action pattern (FAP). That is, it goes through a predictable sequence once the behavior is triggered by an innate **releasing mechanism,** or cue. Another example of a FAP is when a parent bird returns to the nest bearing food. The young birds open their mouths wide and chirp noisily. That behavior triggers the parents' feeding behavior, and the parents stuff the food down the throats of their young. A mechanical analogy for a fixed action pattern is a coffee vending machine. When you put money in (the releasing mechanism), the machine drops a cup into place and then pours coffee into it. If the cup dispenser is empty, the coffee is still poured: Once the mechanism is triggered, it goes through its entire sequence of behaviors.

What is a possible adaptive significance of having a releasing mechanism?

The behavior is only used after the trigger has occurred. In the betta example, there is no advantage to engaging in display behavior unless it can be perceived by a potential rival. Conversely, a betta that displays without an audience is wasting energy and time that could be put to better uses (such as eating).

Activity A: Solitary Behavior of the Betta

In this activity you will observe a betta at rest. Your purpose is to learn to identify the physical and behavioral characteristics of the fish when it is undisturbed so you will have a basis for comparison when the betta encounters a rival.

Procedure

1. Set the fish tank in a place where all the group members can comfortably observe the fish and where it is not near another betta. Bright-colored clothing may trigger a display, so keep your distance.

2. Observe the betta's appearance and behavior and note them below. In particular, describe the fish's color, the position of its fins, and its movement around the tank. Do not disturb the fish in any way while you are observing it.

The resting fish will have its dorsal fin folded down and its anal fin folded up. Its color may be fairly dull, especially when compared later to its aggressive state. It is not very active, remaining quietly in one place most of the time and occasionally surfacing to obtain air.

Activity B: The Betta's Reaction to Its Mirror Image

In this procedure you will show your fish its own reflection. Predict how the fish will react.

Since they have been told that the presence of another male stimulates the contest display, students will probably expect the fish to react by displaying.

Procedure

1. Hold the mirror against the side of the tank.
2. Observe how the fish responds and record your observations in the space below. Again, pay particular attention to the color, fins, and movements.

 It may take a few minutes before the fish notices the "intruder," so be patient.

Most fish will respond to their mirror images with display behavior. In the broadside display, both the dorsal and anal fins are erected and extended to their full size. The color of the fish becomes more intense. There is also a characteristic flexing of the body. In the head-on mock attack, the fish approaches its "rival" and flares its gill covers and the membrane beneath the gill covers out perpendicular to its head. Its pectoral fins beat rapidly. After each display the fish will swim away a short distance and then renew the display. Since the mirror image does exactly what the fish is doing, the fish will continue this behavior for some time.

How does the display behavior affect the appearance of the fish?

It mainly makes it appear larger.

Name another animal that uses a similar effect when confronted by an enemy.

Cats, dogs, some lizards, any animal in a threatening posture.

What are some possible releasing mechanisms for the betta's display behavior?

Example: presence of another fish (or specifically a betta), signaled by color, movement, or shape. (From the experimental conditions, we have determined that the cue must be visual.)

E X E R C I S E   1 9 . 3
Designing an Experiment

Objective

After completing this exercise, you should be able to

1. Design an original experiment to investigate some factor that affects kinesis of pill bugs or agonistic behavior of bettas.

The Instructor Notes offer some suggestions for experiments that students could do. You may want to assign each team to use either pill bugs or bettas to make the preparation more predictable. Let the prep staff know what additional materials you would like to have available for the student investigations. For example, for pill bug experiments you can have a means for varying humidity, colored paper, or various chemicals that might attract or repel pill bugs. For betta investigations, model-making materials are useful. The Prep Guide gives specific directions for some of these materials.

In Exercises 19.1 and 19.2 you learned techniques for investigating the kinetic behavior of pill bugs and agonistic behavior of bettas. In Exercises 19.3 and 19.4 your lab team will design an experiment using one of these methods, perform your experiment, and present and interpret your results. You may want to review the factors that you listed as possible causes for these behaviors to help you decide on an independent variable for your investigation.

Your instructor will tell you which animal to use for your investigation. If you are doing a pill bug experiment, you will use the choice chamber that you used in Exercise 19.1. If you are doing a betta experiment, you will use the fish that were used in Exercise 19.2. Your instructor will tell you what additional materials will be available.

Describe your experiment below.

Question or Hypothesis

Dependent Variable

Independent Variable

Explain why you think this independent variable will affect the dependent variable.

Control Treatment(s)

Replication

Brief Explanation of Experiment

Predictions

What results would support your hypothesis? What results would prove your hypothesis to be false?

Method

Include the levels of treatment you plan to use.

Design a Table to Collect Your Data

List Any Additional Materials You Will Require

EXERCISE 19.4
Performing the Experiment and Interpreting the Results

Objectives

After completing this exercise, you should be able to

1. Perform the experiment your lab team designed.
2. Present and interpret the results of your experiment.

Before you do the experiment, be sure that everyone on your lab team understands how the experiment will be performed.

Be thorough in collecting data. Don't just write down numbers; record what they mean as well. Don't rely on your memory for information that you will need when reporting on your experiment later! If you have any questions, doubts, or problems during the experiment, be sure to write them down, too.

Results

Before you begin to prepare your results for presentation, decide on the best format to use. Remember, you want to give the reader a clear, concise picture of what your experiment showed. Refer to the data presentation section of Appendix A (Tools for Scientific Inquiry) for help. If you are drawing graphs, use graph paper. Complete your tables and/or graphs before attempting to interpret your results.

Write a few sentences below *describing* the results (don't explain why you got these results or draw conclusions yet).

Discussion

Look back at the hypothesis or question you posed in this experiment. Look at the graphs or tables of your data. Do your results support your hypothesis or prove it false? Explain your answer, using your data for support.

Did your results correspond to the prediction you made? If not, explain how your results are different from your expectations and why this might have occurred.

Describe how your data are supported by information from other sources (for example, textbooks or other lab teams working on a similar problem).

If you had any problems with the procedure or questionable results, explain how they might have influenced your conclusion.

If you had an opportunity to repeat and extend this experiment to make your results more convincing, what would you do?

Summarize the conclusion you have drawn from your results.

Questions for Review

1. How do pill bugs locate appropriate environments?

    By kinesis. Pill bugs move more rapidly when in unfavorable conditions and slow down when they find a good spot.

2. If you suddenly turn over a rock and find pill bugs there, what do you expect them to be doing? If you watch the pill bugs for a few minutes, would you expect to see their behavior change?

    They'll be hunkered down, usually on the rock itself, because they've found a favorable environment. When you disturb those conditions, they'll begin moving. Note that the pill bugs are not "frightened" into movement. They move because their environment is no longer favorable.

3. Describe an experiment to test whether pill bugs are attracted to tannic acid (a chemical found in rotting wood).

    Use a choice chamber. Put filter paper moistened with water in one chamber. Put filter paper moistened with a tannic acid solution in the other chamber. Place an equal number of pill bugs in each chamber and then record their activity and distribution over a period of time.

4. How does agonistic behavior contribute to an animal's survival and reproduction?

    Agonistic behavior is a way of competing for resources and mates. If an animal is dominant, it will get more of both.

5. Most agonistic behavior involves ritualized display rather than actual combat. Give a political analogy for this behavior.

    The arms race is a perfect example. We want to have bigger and better weapons (and to advertise that we have them) in order to "deter" our rivals. The mutual display of weaponry is supposed to prevent actual combat. There are other good examples, too.

LAB TOPIC 19

Animal Behavior Instructor Notes

Most institutions have regulations governing the use of living vertebrates in laboratory experiments, even if the animals are merely observed rather than being manipulated. It will be necessary for you to file an animal use protocol to obtain permission to do Exercise 19.2. If you are not aware of the animal use regulations at your institution, check with your administration.

Organization

Students should work in teams throughout the exercises.

General Introduction (5–10 minutes)

Review the material in the Introduction to give students a context for the exercises. Emphasize the importance of studying behavior from a scientific rather than anthropomorphic point of view. You may want to review Lab Topic 1, Exercise 1.2 (Defining a Problem), which discusses testable hypotheses.

Exercise 19.1 (30 minutes)

Introduction

Define kinesis so that students understand the behavior they are studying.

Prep for Procedures

Go over the Hints for Successful Investigations on p. 19-4.

Show students how the choice chambers are constructed and make sure everyone has theirs taped together correctly.

Summary

After students have completed the pill bug exercise, collect data from all the groups so students can complete Table 19.2. Review the answers to all questions briefly and be sure students have some ideas for independent variables.

Exercise 19.2 (30 minutes)

Introduction

Explain agonistic behavior.

Prep for Procedures

Have students position their fish where it can easily be observed so it will have a few minutes to calm down.

Caution students about humane treatment of the fish. Students should not do anything to the fish except what they are instructed to do in the lab manual. They should not tap on the glass or introduce fingers or objects into the fish tank. At this point in the lab, the fish should be resting. If students disturb their fish they will have to wait longer to make observations.

Summary

After all students have completed Activities A and B, discuss the results. Students should be able to describe resting behavior and the components of display behavior. Be sure students have some ideas about possible releasing mechanisms for the agonistic display.

Exercise 19.3 (15–30 minutes)

You can assign some teams to design pill bug experiments and some teams to design betta experiments. Tell students what extra materials will be available. For example, students who are designing pill bug experiments may want a way to change humidity in the chambers, chemicals to test as pill bug attractants or repellants, or different colored papers to use as a background. Students who are designing betta experiments need to know the number and physical characteristics of the fish that are available for their use. They might also want colored paper for making models.

Have each team discuss ideas for independent variables among themselves. In general, pill bug investigations involve determining pill bug preferences for different environmental conditions. Betta investigations involve determining the releasing mechanism(s) for display behavior. Each team should then choose an independent variable, formulate a hypothesis, develop an experimental design, and fill out a proposal form.

Exercise 19.4 (60 minutes or time remaining in the lab period)

Teams work independently to test their hypotheses, but they may find it useful to share information with other teams that are working on a similar problem.

For evaluation, students can formalize and hand in their data and answer the lab manual questions for Exercise 19.4 or they can write a formal report.

For a 2-Hour Lab

Select either Exercise 19.1 (pill bugs) or Exercise 19.2 (bettas) as the demonstration exercise. After students perform the demonstration exercise, they can design and perform their own investigation as described in Exercises 19.3 and 19.4.

Possible Investigations

Pill Bugs

Pill bug preference for light or dark and moist or dry conditions can be tested. Cover one dish with a dark cloth or paper to exclude light from one chamber. To vary the humidity, put a moist filter paper in one dish to increase the humidity, and place indicating Drierite desiccant in the other dish to decrease its humidity. (Indicating Drierite changes color from blue to pink as it absorbs moisture.) Students can test a variety of chemicals—for example, insect repellants—by placing them on a filter paper in one chamber. A review question suggests tannic acid as an attractant. Preference for a background color can also be tested by setting the choice chambers on top of colored paper.

Siamese Fighting Fish

Students can use other bettas or models to test hypotheses about factors that trigger display behavior. If using other bettas, they should set the tanks close together so the fish can see each other and then make observations regarding which fish initiates the contest, which (if either) fish backs down, and the intensity of each fish's display. Give a score of 1 to the fish that wins each aspect of the encounter (for example, initiates the contest) and a 0 to the other fish. If there is not a clear winner, give each fish 0.5. After all contests, total the score for each fish to determine dominance rankings.

 Students usually want to see what happens when two fish are put in the same tank. Very likely what will happen is that one fish will injure the other. There is no purpose served by allowing the bettas to engage in actual combat. Students can collect sufficient data without subjecting the fish to possible injury.

Hypotheses can also be tested using models of variable color, size, shape, movement, or brightness. Models can be easily cut out from colored construction paper. Tape a thread to the back of the cut-out so it can be dangled beside the tank. The Prep Guide includes a list of materials for model making. Some fish do not respond to models at all, and those that do will respond with less intensity than they do to another betta. Before beginning their experiment, students should test the available fish until they find one that does display to models.

Foraging Strategy

Introduction

One essential animal behavior is foraging for food, and many structural and behavioral adaptations of animals are linked to efficient food gathering. If an animal eats other animals, it must hunt, recognize, capture, and subdue the prey, activities that clearly require complex behaviors. An animal that uses plants as food resources must be able to locate appropriate plants. This may seem to be a simple task since plants are not motile, but it isn't necessarily as easy as it sounds. Consider a bee that feeds on nectar or pollen, for example. Flowers of the right species may be located at some distance from each other, so the bee must expend energy to locate and travel between these food resources, and it must be able to recognize the flower when it finds it.

Ecologists often use models to help them understand complex interactions such as those involved in foraging behavior. The purpose of a model is to allow the scientist to simplify and define the components of a real process, and thus be able to study each component separately. It is a tool for studying the natural world in a controllable setting. You will use a type of model called a simulation to investigate aspects of foraging behavior.

Outline

Exercise 20.1: The Simulation System

Exercise 20.2: Single Prey Species

Exercise 20.3: Mixed Prey Species

Exercise 20.4: Designing an Experiment

Exercise 20.5: Performing the Experiment and Interpreting
the Results

Instructor Overview
In this lab topic, students learn a foraging simulation using pans of sand and dried beans. They perform a simulation using a single prey species, kidney beans. In a second simulation there are three prey species that have unequal caloric values (lima beans and Great Northern beans are used in addition to kidney beans) and the foragers are generalists. Questions following each exercise lead students to consider other aspects of forager-prey interactions. Each student team will then design its own experiment, which will be a variation of the basic simulation. The laboratory prompts students to organize their investigations (see Exercise 20.4) and answer a series of questions leading them to interpret their results (see Exercise 20.5). Alternatively, the instructor may require students to write lab reports.

Time Requirements
Exercise 20.1: 10 minutes.
Exercise 20.2: 25–30 minutes.
Exercise 20.3: 30–40 minutes.
Exercise 20.4: 15–30 minutes
Exercise 20.5: 40 minutes or time remaining in period

EXERCISE 20.1

The Simulation System

Objective

After completing this exercise, you should be able to

1. Describe the components of the simulation used in this lab topic and explain how they represent real processes.

In this simulation the food resources (prey) are represented by dried beans. Variety in the size and color of beans allows us to simulate having different types of food items. The food items are located in a pan of sand, which represents a patch of habitat.

We must also choose an animal to do the foraging. Ideally, the animal should be able to follow directions so that the rules of the simulation are observed. The animal should be intelligent enough to apply several variations on the basic pattern. We have therefore decided to use *you* in the role of the foraging animal.

What physical and mental abilities do humans have as foragers or predators?

> *Visual and tactile senses are used; humans are very dexterous, which makes the food handling easier. They are highly intelligent and learn easily. They may also form cooperative groups.*

What animals share some or all of these characteristics?

> *The other primates are most closely related to humans and have most of these characteristics. However, the simulation could be applied to other animals that hunt food resources visually.*

Like scientific investigations, simulations are defined by their rules. If each participant in the simulation follows the rules, we will be able to draw valid conclusions from the data.

The Rules

1. You may use only one hand to forage.
2. You may not remove *any* sand from the pan as you forage.
3. Each member of the group forages in turn. A forager has a maximum of 10 seconds to find enough prey to survive.
4. Each prey item must be placed in a dish next to the pan before the forager may capture the next prey. This simulates handling time, which is the time that the forager requires to subdue and consume its prey.
5. One team member should act as the timer/recorder/referee while another forages. After each turn, the recorder should smooth over the sand and cover any prey that have been revealed before the next person forages. The recorder also serves as the referee and should verify that the forager follows the rules. Rotate this job so that each person who forages becomes the recorder for the next group member.

6. Only the recorder may watch the person who is foraging.

7. When you have finished collecting data from a simulation, pour the sand through the sieve into the empty pan and remove any beans that remain.

E X E R C I S E   2 0 . 2
Single Prey Species

Objectives

After completing this exercise, you should be able to

1. Describe a successful foraging strategy for this simulation.

2. Give examples of natural situations to which the simulation in this exercise might apply.

To demonstrate the simulation, only one prey species, kidney beans, will be available to the foragers. You should work in teams of four or five.

Procedure

1. Disperse 30 kidney beans in the sand. All beans should be covered so they are not showing on the surface.

2. Team members will take turns foraging. You may not watch each other forage (except for the timer/recorder/referee). Each forager must find three kidney beans in a maximum of 10 seconds in order to survive. The forager should stop after finding three beans. A forager who doesn't find three beans within 10 seconds dies of starvation.

3. Record in Table 20.1 the number of prey remaining after each forager's turn and the time it actually took for each forager to find the three beans.

4. Continue taking turns foraging until each team member has died.

5. Record in Table 20.1 the total number of foraging turns taken in the simulation.

What data measure the success of the foragers?

The time it takes to find three prey; total number of foraging turns taken by the team before all foragers have died.

What happens as a food resource is depleted?

It takes longer and longer to find the necessary food; foragers begin dying.

What real situation might this simulation apply to?

One example is a seasonal crop, like seeds, whose distribution is limited. Animals can easily deplete the supply of seeds in a particular area.

Table 20.1

Results of Foraging Simulation. Each forager has 10 seconds to capture three prey. The only prey species is kidney beans.

Forager	Number of Prey Remaining	Time to Find Three
1	30	
2		

Total number of foraging turns in this simulation: _____

In nature, prey animals may have hiding places to protect them from predation. Plants, however, generally do not have this type of refuge available. Some species escape by simultaneous production of edible resources. Bamboos are a good example of this. Certain species flower and set seed only once every several years. Although they produce many seeds and seedlings are heavily predated, the predators are unable to eat all of them. (There are also animals that use this strategy.) It would be unusual for animals (except humans) to drive their prey completely to extinction.

What do animals do when food resources are depleted in nature?

If possible, they move on. The huge flocks of blackbirds that are common to some areas are a good example. They find a good patch, descend and clean it out, and then fly on. Animals will starve to death only if food is unavailable throughout their range.

How many beans were left in the sand when all the foragers died? How does this result relate to a natural situation?

There will probably be beans that have escaped detection. We have given the beans a refuge by limiting the search time. This is important in nature so that a depleted population can be renewed.

Besides observing other animals directly, how might animals avoid wasting time in an area that has already been searched by another animal?

Territoriality achieves this; an individual "reserves" the food resources of an area for himself.

From the forager's point of view, is the kidney bean an easy prey item to capture? Explain.

Kidney beans are large and therefore easy to find by touch and to pick up. Kidney beans are dark in contrast to the light sand, so they are easy to locate if any part of the bean is visible.

Observe the other prey items that are available and speculate on how the differences between them and kidney beans might make them easier or more difficult to capture.

Great Northern beans:

Smaller than kidney beans, so they are not as easy to find by touch; their light color gives them better camouflage in the sand.

Lima beans:

Larger than kidney beans, so they are easier to find by touch and easier to handle; their light color may be more difficult to spot in the sand.

Lentils:

Much smaller than kidney beans, so they are very difficult to locate by touch and more difficult to handle; their color also gives them camouflage in the sand.

How might you change the simulation to improve the forager's success in capturing prey?

Let the forager learn where prey are located by watching others forage; don't cover up the remaining beans after each turn; increase the amount of foraging time per predator; start with more prey present in the habitat; decrease the number of prey needed for the forager to survive; decrease the number of foragers per patch.

How might you change the simulation to increase prey survival?

Camouflage the prey by changing their physical characteristics as mentioned previously or by making the environment more complex (for example, by adding pebbles or inedible prey to the sand); use prey that are mobile; increase the number of prey that have to be caught for survival; let the prey reproduce.

EXERCISE 20.3
Mixed Prey Species

Objectives

After completing this exercise, you should be able to

1. Distinguish between specialist and generalist foragers and give examples of each type.

2. Give examples of natural situations to which the simulation in this exercise might apply.

In Exercise 20.2, only one type of prey was available. The forager had to find this prey or it would die. This type of forager is called a **specialist.** Many herbivorous animals specialize in eating certain types of plants. Usually specialists will eat a number of related species, but a few have extremely restricted diets. Name an example of a specialist. (Hint: Think of warm, fuzzy herbivores.)

> *The giant panda specializes in certain species of bamboo. Koalas specialize narrowly on types of eucalyptus leaves.*

Generalists, on the other hand, are able to exploit a variety of food resources. The raccoon, for example, is an opportunistic generalist. Besides its natural diet of fish, frogs, and crayfish, a raccoon eagerly consumes hot dogs, potato chips, cole slaw, and whatever else is not securely locked up by campers.

What is an advantage of being a specialist?

> *The animal can be highly adapted for locating and capturing certain types of prey. For instance, many specialist insects locate their host species by cuing on specific chemicals they contain.*

What is a disadvantage of being a specialist?

> *If your particular food resource is depleted or finds a way to evade you, you're in trouble. The population decline of the giant panda in China, for example, has been linked to a decline in the availability of the species of bamboo it eats.*

What is an advantage of being a generalist?

> *Generalists can take advantage of a variety of food resources that may be available and can adapt to changing abundance of those resources. Some birds even change their feeding habits seasonally from insects to seeds.*

What is a disadvantage of being a generalist?

> *The relationship with the prey is not as fine-tuned, so it may be more difficult to locate any particular food item. The forager must also make decisions about which item it should expend time and energy locating and capturing.*

All food is not of equal value. Some prey items are larger than others, and thus the forager obtains more food per capture. The food resources may differ in nutritional value. For example, one seed may contain carbohydrates while another seed of the same size contains fat, which has a higher caloric value per gram than carbohydrate. However, the forager's choice of food is also influenced by the abundance of the resource. Food A may contain more calories than Food B, but it may be so rare that it isn't worth the extra energy the animal must expend to find it.

Predators develop a "search image" of the prey they are seeking. We do something analogous when we are looking for something. It helps to have some feature to key in on. For example, when you're looking for a person in a crowd, it's useful to know what the person is wearing. Once a forager has a search image for kidney beans, it is easier to find that prey than switch to another.

In this simulation there are three different types of prey available: kidney beans, Great Northern beans, and lima beans. The population size is different for each species. The number of prey each forager must find in order to survive is also different. There are 40 kidney beans, 30 Great Northern beans, and 20 lima beans.

The food values of the various items are different. A forager survives if he or she can find any of the following combinations:

4 kidney beans

3 Great Northern beans

2 lima beans

3 kidney beans and 1 Great Northern bean

3 kidney beans and 1 lima bean

Are the foragers in this simulation specialists or generalists?

Generalists.

Procedure

1. Sieve the sand to remove any prey remaining from the previous simulation.

2. Disperse 40 kidney beans, 30 Great Northern beans, and 20 lima beans in the sand.

3. Take turns foraging, observing the same rules that were used in the previous simulation. Each forager must find one of the specified combinations of prey within 10 seconds in order to survive. The forager should stop after finding enough prey to survive. A forager who can't capture enough prey starves to death. After each person forages, he or she is the recorder/referee for the next forager.

4. In Table 20.2 record the number of prey remaining after each forager's turn, the number of seconds it takes each forager to capture the right combination of beans, and the total number of foraging turns before all the foragers die.

Table 20.2

Results of Foraging Simulation. Each forager has 10 seconds to capture enough prey for survival. The three prey species—kidney beans, lima beans, and Great Northern beans—have different food values.

Forager	Number of Prey Remaining			Number of Prey Captured			Time (sec)
	Kidney	Great Northern	Lima	Kidney	Great Northern	Lima	
1	40	30	20				
2							

Total number of foraging turns in this simulation: _____

What strategy did you use in deciding which prey to look for?

Example: Since kidney beans are most abundant and I am experienced in foraging for kidney beans, I looked for kidney beans first.

Would your strategy have been different if you hadn't known the initial number of each prey present? Explain.

Example: Yes. I would still have looked for kidney beans because of my past experience, but I would have paid more attention to the possibility of finding the other beans if I hadn't known there were more kidney beans than anything else.

Was one of the resources depleted first? If so, which one?

Kidney beans are likely to be depleted first. They are the most abundant and easiest to spot in the sand. It requires more kidney beans to survive than any of the other prey. Also, students may seek kidney beans first since they foraged for them in the previous exercise.

What happened after this prey species was depleted?

> *Foragers had to switch to the prey that were less abundant and more difficult to locate.*

Suppose each forager had specialized on kidney beans, lima beans, or Great Northern beans. How might the results have been different?

> *Specialists would probably have died sooner. In general, the generalist foragers live longer because they have more choices of prey. Students may test their predictions when they design their own simulations.*

E X E R C I S E   2 0 . 4
Designing an Experiment

Objective

After completing this exercise, you should be able to

1. Design a variation on the simulation used in Exercises 20.2 and 20.3 to investigate some aspect of foraging strategy.

Exercises 20.1, 20.2, and 20.3 explained and demonstrated a method of simulating foraging strategy by using beans as the prey and students as the foragers. In Exercise 20.4 your lab team will design a variation on this simulation to investigate some other aspect of foraging strategy. In Exercise 20.5 your lab team will perform the simulation and present and interpret your results. You may want to review the questions you answered for Exercises 20.2 and 20.3 to decide on an independent variable for your investigation. Possible dependent variables include the length of time required by each forager to find enough prey to survive each turn, the total number of foraging turns in a simulation, the number of prey remaining after each round of foraging, or a combination of data that are useful for evaluating your hypothesis. Use the format below to plan and describe your simulation.

For your investigation, you will use the pan of sand that you used for Exercises 20.2 and 20.3. Kidney beans, lima beans, Great Northern beans, and lentils are also supplied. Your instructor will tell you if any additional materials are available.

Since this experiment is unlike previous ones that students have designed, they may need some extra guidance to get started. The Instructor Notes offer some suggestions for this. Aspects of foraging strategy that can be investigated using this simulation are discussed in the Possible Investigations section of the Instructor Notes. The Prep Guide tells the prep staff to supply nail polish (for marking inedible prey) and pebbles (for creating a complex environment) in addition to the sand and beans. Let the prep staff know if you want to have additional items available.

Hypothesis

Dependent Variable(s)

Independent Variable

Explain why you think this independent variable will affect the dependent variable(s).

Replication

Predictions

What results will support your hypothesis? What results will prove your hypothesis to be false?

Method

Explain how you have modified the rules of the simulation.

Design a Table or Tables to Collect Your Data

EXERCISE 20.5

Performing the Experiment and Interpreting the Results

Objectives

After completing this exercise, you should be able to

1. Perform the experiment your lab team designed.
2. Present and interpret the results of your experiment.

Before you do the experiment, be sure that everyone on your lab team understands the rules of the simulation.

Be thorough in collecting data. Don't just write down numbers; record what they mean as well. Don't rely on your memory for information that you will need when reporting on your experiment later! If you have any questions, doubts, or problems during the investigation, be sure to write them down, too.

Results

Before you begin to prepare your results for presentation, decide on the best format to use. Remember, you want to give the reader a clear, concise picture of what your experiment showed. Refer to the data presentation section of Appendix A (Tools for Scientific Inquiry) for help. If you are drawing graphs, use graph paper. Complete your tables and/or graphs before attempting to interpret your results.

Write a few sentences *describing* the results. Don't explain why you got these results or draw conclusions yet.

Discussion

Look back at the hypothesis you posed in this experiment. Look at the graphs or tables of your data. Do your results support your hypothesis or prove it false? Explain your answer, using your data for support.

Did your results correspond to the prediction you made? If not, explain how your results are different from your expectations and why this might have occurred.

Describe how your data are supported by information from other sources (for example, textbooks or other lab teams working on a similar problem).

If you had any problems with the procedure or had questionable results, explain how they might have influenced your conclusion.

If you had an opportunity to repeat and extend this experiment to make your results more convincing, what would you do?

Summarize the conclusion you have drawn from your results.

Questions for Review

1. Putting beans in a pan of sand is only one way to simulate foraging. Use your imagination and think of other simulation systems that might be used.

   Forage for macaroni or other objects in grass or among styrofoam peanuts. Forage for disks of colored paper on different colored backgrounds.

2. Give some examples of how foragers or predators recognize their prey as prey.

   Foragers might recognize edible items by color, scent, or shape, or some combination of these characteristics. Predators might use scent, heat production (snakes), electrical fields (sharks), or sight, depending on the predator's most acute senses. For example, birds of prey use sight, while wolves use scent.

3. From your experience with the simulation, do you think that foragers are more efficient if they work in cooperative groups or if they work solo? Explain your answer.

   Student answers may vary depending on the simulations they performed and the results they obtained.

4. If a forager has a choice of prey items available, what factors might influence its prey selection?

   It is likely to select those prey that require the least energy to capture, just as foragers did in this simulation.

5. The ability of the prey to reproduce has a significant effect on the survival of the population. In nature, what factors might affect the ability of the members of a prey population to reproduce?

   Their biology: The number of offspring an animal or plant produces is constrained within a certain physiological limit. Food resources available to the prey species: Just as we simulated a nutritional requirement for the forager, the prey must also consume (or, in the case of plants, manufacture) enough food to maintain itself. Additional food is required for reproduction.

6. Give some examples from nature of different types of "refuges" used by prey species.

Acknowledgment

The idea of foraging for beans in pans of sand came from Hans Landel, who was a graduate student at Purdue University at the time.

LAB TOPIC 20

Foraging Strategy
Instructor Notes

Organization

Discussion must accompany the exercises so that students understand the biological bases of the simulations they perform. The data collection goes quickly. After students design their experiments (Exercise 20.4), there should be at least 40 minutes remaining for students to perform their simulations. If Exercises 20.1 and 20.2 take significantly longer than 40 minutes (total), you can have students read Exercise 20.3 to get additional ideas for investigations without actually performing the simulation.

Exercises 20.1 and 20.2 (35–40 minutes)

Introduction (5 minutes)

Give a brief introduction to foraging and the use of simulations to model complex processes. It is important for students to understand that although a simulation may seem like a game, it has a serious purpose as a tool for investigating actual phenomena.

Prep for Procedures (10 minutes)

- Go over the rules of the simulation.
- Be sure students understand how to collect data in Table 20.1.
- Divide students into teams of four or five.

Perform the Simulation in Exercise 20.2 (10–15 minutes)

Summary (10 minutes)

When students have completed Exercise 20.2, discuss the results. In particular, have students contribute ideas for variations on the simulation that can be used to test hypotheses about foraging strategy. Variations suggested by the questions at the end of Exercise 20.1 include giving the prey a refuge and varying the physical characteristics of the prey. Additional variations are suggested by the last two questions, which ask students how to improve the forager's success and how to improve the prey's survival. Choose a couple of the suggested variations and discuss how the rules might be modified to test a hypothesis. For example, to investigate the effect of prey reproduction, the rules could allow new prey to be added to the habitat after each forager's turn. The number added would depend on the number of prey remaining after the round. The independent variable for this experiment is prey reproduction rate. Exercise 20.2 uses a prey reproduction rate of 0. Dependent variables include the number of prey remaining at the end of each forager's turn, the time it takes each forager to capture enough prey to survive a turn, the prey population size, and the total number of foraging turns taken before all the foragers die. The experiment can then be extended to investigate the effect of allowing the foragers to reproduce as well as the prey.

Exercise 20.3 (30–40 minutes total)

Introduction (5 minutes)

Explain the terms specialist and generalist. In Exercise 20.2 there was only one prey species and the foragers were specialists. In Exercise 20.3 two more prey species are added and the foragers are generalists.

Prep for Procedures (5 minutes)

- Students will remain with the same teams.
- Review the list of prey combinations that foragers may use for survival.
- Be sure students understand how to collect data in Table 20.2.

Perform the Simulation in Exercise 20.3 (10–15 minutes)

Summary (10 minutes)

After all the teams have collected data for Exercise 20.3, briefly summarize their results. The purpose of the discussion is to help prepare students for designing their own experiments, so it should center more on the questions than on the numerical results.

Exercises 20.4 (15–30 minutes) and 20.5 (40 minutes or time remaining in lab period)

Introduction (10 minutes)

Several possible independent variables have been suggested in the lab (also see the Possible Investigations). Each team should choose an idea and discuss how they might modify the simulation to test a hypothesis. Students should plan to repeat the simulation several times to collect data at different levels of the independent variable, if appropriate, and to replicate the experiment. Be sure students devise tables that will be useful for collecting their data. For many variations of the simulation, it is necessary to keep track of the number of prey present in the habitat.

For a 2-Hour Lab

Discuss Exercise 20.1 to introduce the simulation system, and then have students perform Exercise 20.2 to become familiar with performing the simulation. After discussing the results, have teams design and perform their own experiments.

Possible Investigations

Change Forager Nutritional Requirements

Students can investigate the effect of metabolic demands by varying the number of prey that each forager must capture per turn for survival. The number of prey specified represents the minimum caloric intake required per unit time. A forager that has a high metabolic rate—for example, a hummingbird or a shrew—must capture more prey per unit time for survival than a forager with a low metabolic rate, such as a reptile. Variations in prey quality may also affect the number of prey that the forager must capture for survival, as discussed in Exercise 20.3.

Add a Prey Refuge/Allow Prey Reproduction

In nature, it is rare for a prey species to be eradicated, but it can happen under the basic rules of the simulation. Modify the rules so that the prey population cannot drop below a certain number—for example, 10. A related variation is to allow the prey to reproduce after each round of foraging as follows. After each forager takes a turn, calculate the number of prey remaining. (This is easily done by subtracting the total number of prey captured from the number of prey that have been dispersed in the habitat.) Then add more prey to the habitat according to the formula for reproduction that the students have devised for the simulation. For example, they might try adding one new prey for every one that remains, every two that remain, every three that remain, and so on.

Allow Prey and Forager Reproduction

In addition to prey reproduction (as just described), allow the foragers to reproduce. The formula for foragers should depend on the number of prey caught in the allotted time period. So instead of stopping after capturing enough prey for survival, the forager captures as many prey as possible. For example, a forager must catch four prey to survive. If the forager captures four additional prey, one new forager is added to the population. Students will find that as the prey population increases, so does the forager population. If the forager population becomes too large, it causes a decline in the prey population. Students may find a reproductive rate at which the two populations will cycle up and down but continue to coexist.

Allow Foragers to Learn from Others

In this simple modification of the rules, foragers are permitted to watch each other forage, testing the hypothesis that foragers will be more successful if they learn the prey distribution in the habitat.

Use Different Prey Species

Kidney beans are large and easy to spot in the sand. Students can use other prey species to test hypotheses about the effect of prey characteristics on forager success. For example, they might replace kidney beans with lentils. Using the same rules for the simulation, they might compare the average time it takes the foragers to find three beans and the total number of foraging turns (the number of foraging rounds before all foragers die of starvation) for each prey.

Vary the Prey Density

Use different initial populations of prey to determine the effect of prey density on forager success. This investigation might be an extension of an experiment using different prey type.

Compare Specialists and Generalists

Students can use the same prey mixture as in Exercise 20.3, but assign one forager to capture only kidney beans, one to capture only limas, and one to capture only Great Northern beans. The results would be compared to their results for Exercise 20.3, where the foragers were generalists.

Add Inedible Prey

Make the forager's task more difficult by adding inedible prey to the habitat. Toxic prey resemble the edible prey (that is, the same type of bean is used) but they are marked with a dot of fingernail polish (warning coloration). If the forager doesn't detect the prey's toxicity before putting it in the dish (eating it), the forager dies instantly. Compare forager success when the number of inedible prey is varied. The time spent capturing prey that turn out to be inedible and examining all prey as they are captured may make it difficult for foragers to find their quota of edible prey in the 10 seconds allowed. Students may have to extend the foraging time in order to have foragers survive long enough to produce meaningful data.

Add Complexity to the Habitat

Foragers use visual and tactile senses to locate prey. Color camouflage can afford prey some protection. It may also be adaptive for prey to live in a habitat that camouflages them against the tactile senses. Addition of items such as pebbles that feel similar to the prey can provide a more complex habitat that serves to conceal the prey to some extent. Like the simulation that uses inedible prey, students may have to adjust the foraging time to get good results for this simulation.

Introduce Territoriality or Other Changes in Competition

Some animals defend territories that effectively exclude competitors. Students can investigate how different levels of competition, represented by varying the number of foragers in the habitat, affect the success of an individual forager. Students may want to increase the minimum number of prey required for survival to simulate the energy expended defending a territory.

Tools for Scientific Inquiry

Introduction

In addition to using the particular method of approaching and solving problems that is known as scientific inquiry, scientists also use standard methods of collecting and presenting data.

Scientists use the metric system to take measurements. This appendix reviews metric measurements and techniques for using some laboratory measuring devices.

When scientists communicate the results of their experiments to others, they present numerical data in a format that is clear and concise. This appendix also explains how to present and interpret data.

Outline

Instructor Overview
This appendix introduces some of the skills that students will need to perform the labs in this manual. It includes lessons on using scientific notation and basic information about using the metric system of measurement for length, mass, volume, and temperature. Hands-on activities introduce use of triple-beam and electronic balances, pipets, and graduated cylinders. A section on data presentation includes when and how to use tables, line graphs, and bar graphs. Another exercise requires students to interpret graphs. You may want to use these activities at appropriate times during the term, or you may just have the students use them for reference.

Time Requirements
Exercise A.1: 20–25 minutes
Exercise A.2: 45–50 minutes
Exercise A.3: 40–50 minutes
Exercise A.4: 20 minutes

EXERCISE A.1
Scientific Notation

This exercise explains how to

1. Convert numbers that are expressed in scientific notation to decimals.
2. Express decimal numbers in scientific notation.
3. Multiply and divide numbers that are expressed in scientific notation.

In this lab manual, students will not need to add or subtract numbers that are in scientific notation. The procedures for carrying out those operations have therefore been omitted. If you want to teach them to your students for the sake of the completeness of the lesson, examples are given below.

If the exponents of the numbers are the same, simply add the numbers:

$3 \times 10^3 + 2 \times 10^3 = 5 \times 10^3$
  $(3,000 + 2,000 = 5,000)$

If the exponents are different they must first be changed before the numbers can be added:

$3 \times 10^3 + 2 \times 10^2 =$
  $3 \times 10^3 + 0.2 \times 10^3 = 3.2 \times 10^3$
  $(3,000 + 200 = 3,200)$

$7 \times 10^4 - 5 \times 10^4 = 2 \times 10^4$
  $(70,000 - 50,000 = 20,000)$

$7 \times 10^4 - 5 \times 10^3 =$
  $7 \times 10^4 - 0.5 \times 10^4 = 6.5 \times 10^4$
  $(70,000 - 5,000 = 65,000)$

Scientific notation is a numbering system that can easily handle the very small or very large numbers that are sometimes used in scientific investigations. The form of a number expressed in scientific notation is $a \times 10^n$, where a is a whole number or a decimal between 1 and 10 and n (the exponent) is a whole number.

For numbers greater than 1, the exponent is positive and indicates the number of decimal places to the right of the decimal point in a. Examples:

$1.2 \times 10^5 = 120,000$

$4 \times 10^{12} = 4,000,000,000,000$

$4.65 \times 10^6 = 4,650,000$

For numbers less than 1, the exponent is negative and indicates the number of decimal places to the left of the decimal point in a. Examples:

$3 \times 10^{-4} = 0.0003$

$1.7 \times 10^{-9} = 0.0000000017$

$1.0 \times 10^{-1} = 0.1$

$5.74 \times 10^{-6} = 0.00000574$

For practice, write the following numbers in decimals:

$2.3 \times 10^{-3} = 0.0023$ $6.43 \times 10^{-6} = 0.00000643$

$4.64 \times 10^2 = 464$ $1.66 \times 10^4 = 16,600$

$8.9 \times 10^{-7} = 0.00000089$ $4.3 \times 10^6 = 4,300,000$

$1.0 \times 10^5 = 100,000$ $7.2 \times 10^{-5} = 0.000072$

Write the following decimal numbers in scientific notation:

$0.0032 = 3.2 \times 10^{-3}$ $23,000 = 2.3 \times 10^4$

$1,000,000 = 1 \times 10^6$ $0.45 = 4.5 \times 10^{-1}$

$0.0000061 = 6.1 \times 10^{-6}$ $71,000,000,000 = 7.1 \times 10^{10}$

$0.01 = 1 \times 10^{-2}$ $0.000000009 = 9 \times 10^{-9}$

Multiplication

When two numbers in scientific notation are multiplied, the numbers are multiplied but the exponents are added together. Examples:

$(2 \times 10^3)(1.5 \times 10^2) = 3 \times 10^5$ (check by multiplying $2,000 \times 150$)

$(2 \times 10^{-3})(1.5 \times 10^2) = 3 \times 10^{-1}$ $(0.002 \times 150 = 0.3)$

$(2 \times 10^{-3})(1.5 \times 10^{-2}) = 3 \times 10^{-5}$ $(0.002 \times 0.015 = 0.00003)$

For practice, do the following problems:

$(3 \times 10^3)(6 \times 10^2) = 18 \times 10^5 = 1.8 \times 10^6$

$(7 \times 10^{-2})(9 \times 10^4) = 63 \times 10^2 = 6.3 \times 10^3$

$(1.2 \times 10^{-6})(3.0 \times 10^{-2}) = 3.6 \times 10^{-8}$

$(1.0 \times 10^{-8})(4.67 \times 10^9) = 4.67 \times 10^1 = 46.7$

Division

When two numbers expressed in scientific notation are divided, the exponent of the divisor is subtracted from the exponent of the dividend.

$$\frac{2 \times 10^6}{2 \times 10^4} = 1 \times 10^2 \qquad \text{Check:} \frac{2,000,000}{20,000} = \frac{200}{2} = 100$$

$$\frac{7.2 \times 10^4}{3 \times 10^2} = 2.4 \times 10^2 \qquad \text{Check:} \frac{72,000}{300} = 240$$

$$\frac{7.2 \times 10^4}{3 \times 10^{-2}} = 2.4 \times 10^6 \qquad \text{Check:} \frac{72,000}{0.03} = 2,400,000$$

$$\frac{7.2 \times 10^{-4}}{3 \times 10^{-2}} = 2.4 \times 10^{-2} \qquad \text{Check:} \frac{0.00072}{0.03} = 0.024$$

$$\frac{7.2 \times 10^{-4}}{3 \times 10^2} = 2.4 \times 10^{-6} \qquad \text{Check:} \frac{0.00072}{300} = 0.0000024$$

For practice, do the following problems:

$$\frac{1.2 \times 10^{-4}}{3 \times 10^{-3}} = 0.4 \times 10^{-1} = 4.0 \times 10^{-2}$$

$$\frac{4.8 \times 10^5}{3 \times 10^{-7}} = 1.6 \times 10^{12}$$

$$\frac{2.6 \times 10^2}{2 \times 10^7} = 1.3 \times 10^{-5}$$

$$\frac{2.1 \times 10^{-6}}{1 \times 10^2} = 2.1 \times 10^{-8}$$

E X E R C I S E   A . 2
Making Measurements

This exercise explains how to
1. Convert metric units of length, mass, and volume.
2. Use a balance.
3. Use a pipet.
4. Use a graduated cylinder.
5. Convert temperatures expressed in degrees Fahrenheit to degrees Celsius and vice versa.

Many of the labs or investigations you do in this course require you to make accurate measurements using the metric system. In this exercise you will review metric units and learn to use the measuring devices that you are most likely to encounter in the course.

Activity A: Measuring Length

The **meter** (m) is the basic unit of linear measurement in the metric system. As a comparison to American Standard units, a meter is a little longer than a yard. Longer distances are measured in **kilometers** (km). One kilometer equals 1,000 m and is approximately six-tenths of a mile. The meter is subdivided into **centimeters** (cm) and **millimeters** (mm). One centimeter is one one-hundredth of a meter (1 cm = 0.01 m). One millimeter is one one-thousandth of a meter (1 mm = 0.001 m). **Micrometers** (μm), which are also called microns, and **nanometers** (nm) are smaller subdivisions that are used for measurements of microscopic objects (see Table A.1).

Table A.1
Metric Units for Linear Measurement

Unit	Meters
Kilometer (km)	1×10^3
Meter (m)	1
Centimeter (cm)	1×10^{-2}
Millimeter (mm)	1×10^{-3}
Micrometer (also called micron; μm)	1×10^{-6}
Nanometer	1×10^{-9}

The metric side of the ruler at your station is numbered in centimeters, and each centimeter is divided into 10 mm. Compare the metric side of the ruler with the American Standard side. Approximately how many cm are there in one inch? _2.5_ How many mm in one inch? _25_ How many inches in one cm? _0.5_

Metric units can be converted by using Table A.1 to determine the conversion factor. The equation is set up so that similar units are crossed out. For example, to convert mm to cm, divide the number of m per mm (given in Table A.1) by the number of m per cm. The meter units are crossed out, leaving a factor that relates cm to mm.

$$\frac{1 \times 10^{-3} \, \text{m}}{1 \, \text{mm}} \times \frac{1 \, \text{cm}}{1 \times 10^{-2} \, \text{m}} = \frac{1 \times 10^{-1} \, \text{cm}}{1 \, \text{mm}} \qquad \text{(conversion factor)}$$

Then, to convert some number (such as 25 mm), to cm:

$$25 \, \text{mm} \; \frac{(1 \times 10^{-1} \, \text{cm})}{1 \, \text{mm}} = 25 \times 10^{-1} \, \text{cm} = 2.5 \, \text{cm}$$

(Note that the mm units cross out, leaving the answer in cm.)

This example illustrates how to convert 3.4 cm to μm:

$$\frac{1 \times 10^{-2}\,m}{1\,cm} \times \frac{1\,\mu m}{1 \times 10^{-6}\,m} = \frac{1 \times 10^{4}\,\mu m}{1\,cm} \qquad \text{(conversion factor)}$$

$$3.4\,cm\, \frac{(1 \times 10^{4}\,\mu m)}{1\,cm} = 3.4 \times 10^{4}\,\mu m = 34{,}000\,\mu m$$

Practice using metric units by doing the following conversions:

1500 m = ___1.5___ km		350 nm = ___0.35___ μm	
1.2 cm = ___12___ mm		10 km = ___10,000___ m	
100 cm = ___1___ m		175 mm = ___17.5___ cm	
600 nm = ___0.6___ μm		175 mm = ___0.175___ m	

Students who are not familiar with the metric system need to get some sense of the meaning of these units. As a homework assignment, we have students collect objects of sizes specified by the lab instructor.

Convert the measurements in Table A.2 as indicated.

Table A.2
Relative Size (in meters)

Thickness of plasma membrane	7×10^{-9} m to 10×10^{-9} m = ___7 to 10___ nm
Length of *E. coli* bacteria	1×10^{-6} m = ___1___ μm
Diameter of a typical animal cell	1×10^{-5} m to 3×10^{-5} m = ___10 to 30___ μm
Diameter of a human egg	1×10^{-4} m = ___100___ μm
Diameter of a frog egg	1×10^{-3} m = ___1___ μm
Diameter of a human eyeball	2.5×10^{-2} m = ___2.5___ cm
Length of a human esophagus	2.4×10^{-1} m = ___24___ cm
Height of an average human female	1.6 m = ___160___ cm
Length of a human small intestine	6.0 m = ___600___ cm

Activity B: Measuring Mass

The **gram** (g) is the basic unit for measuring mass in the metric system. Larger amounts are measured in **kilograms** (kg). One kilogram = 1,000 g. As a comparison to American Standard units, a kilogram is approximately 2.2 pounds. It takes approximately 28 g to make an ounce and 454 g to make a pound. Like the meter, the gram is also divided into smaller units (see Table A.3).

American Standard Equivalences
1 kg = 2.205 lb
1 oz = 28.35 g
1 lb = 453.6 g

Table A.3
Metric Units for Measuring Mass

Unit	Grams
Kilogram (kg)	1×10^3
Gram (g)	1
Milligram (mg)	1×10^{-3}
Microgram (μg)	1×10^{-6}
Nanogram (ng)	1×10^{-9}

Converting metric mass units can be done using the same conversion factors that you used for length units. For example, to convert 35 milligrams to micrograms:

$$\frac{1 \times 10^{-3}\, \cancel{g}}{1\,mg} \times \frac{1\,\mu g}{1 \times 10^{-6}\, \cancel{g}} = \frac{1 \times 10^3\,\mu g}{1\,mg}$$

$$35\,\cancel{mg}\; \frac{(1 \times 10^3\,\mu g)}{1\,\cancel{mg}} = 35 \times 10^3\,g = 3.5 \times 10^4\,mg$$

Practice using metric mass units by doing the following conversions:

60 mg = _0.060_ g   2 μg = _2000_ ng

240 g = _0.240_ kg   0.4 kg = _400_ g

1.7 mg = _1700_ μg   3 ng = _0.003_ μg

Balances that require the use of free weights are not described.

The following procedures describe the two types of instruments that are most commonly used in biology laboratories to determine weight, the triple-beam balance and the electronic balance. If your laboratory is equipped with another type of balance, your instructor will explain how to use it.

With both the triple-beam and electronic balances, a plastic or aluminum "boat" or a thin paper is used to hold the sample you are weighing. Therefore, you must set the balance to compensate for the weight of the boat or paper before weighing the sample. This is called **taring** the balance. To tare the balance, the boat or paper is placed on the balance and its weight is set at 0. Thus the weight of the container is not added to the weight of the sample.

Some students have been taught in high school to weigh the container first and then subtract its weight after the sample is weighed in it. When they try to do this after the balance has been tared, of course, they get erroneous results. Explain that taring makes preweighing the container unnecessary.

Several objects are provided at your work station, along with a weighing boat. You will practice using the balances by weighing these objects. A weighing boat would not normally be required for the type of objects used in this exercise, but you should practice using it so you will learn how to tare the balance. In other laboratories you will be weighing samples that do require a weighing boat.

The Triple-Beam Balance

A triple-beam balance is illustrated in Figure A.1.

Figure A.1.
Triple-beam balance.

Procedure

Taring the Balance

1. Be sure all the weights are set at 0.
2. Place the weighing boat on the balance pan.
3. Adjust the tare weight right or left until the arm balances at the 0 mark. (You should not have to move the weight more than a few mm.)

Weighing the Sample

1. Place the sample in the weighing boat.
2. Slide the weight on the middle beam to the right so that it fits in the first (100-g) notch.
   a. If the arm drops below the 0 mark, the item weighs less than 100 g. Slide the weight back to the left and go to step 3.
   b. If the balance arm remains above the 0 mark, the object weighs more than 100 g. Slide the weight over to the 200-g notch. If the arm then drops below the 0 mark, the object weighs between 100 and 200 g. Move the weight back to the 100-g notch. If the balance arm is still above the 0 mark, go to the 300-g notch, and so on.
3. Follow the same procedure with the back beam, which has unit markings in tens of grams.
4. The dial has unit markings in single grams and one-tenth of a gram. Turn the dial until the arm balances at the 0 mark.
5. Record the weight of the object.

The Electronic Balance

Figure A.2 shows a typical electronic balance. There are many different models of this kind of balance, so it is unlikely that the balances in your laboratory look exactly like this. Your instructor will explain any additional features on your balance model that you should be able to use.

The balances should be turned on, warmed up, and ready for use.

Double-beam and some hanging pan balances are operated very similarly to the triple-beam type of balance.

Figure A.2.
Electronic balance.

Most electronic balances have a calibration button and a button to set the desired units. We find it most convenient to preset these for the students, since they use the balances too infrequently to need to know how to do it themselves.

Most models require a warm-up period before they will weigh most accurately. Be sure the balances are plugged in when you come into lab.

The Prep Guide suggests some items that will give a range of weights.

Procedure

Taring the Balance

1. Set the weighing boat on the balance pan.
2. Press the tare button. The display should read 0.00. (The readout may be different, depending on the model of balance you're using.) The last digit may drift somewhat.

Weighing the Sample

1. Place the sample in the weighing boat.
2. Read the weight from the display.
3. Record the name of the sample and its weight below. Remember to record the units.

Practice by weighing several objects.

Object	Weight
_____	_____
_____	_____
_____	_____
_____	_____

Activity C: Measuring Volume

1 L = 1.057 qt.

The **liter** (L) is the basic unit of volume of fluids in the metric system. As a comparison to American Standard units, a liter is slightly larger than a quart. The liter is subdivided into **milliliters** (mL). One liter = 1,000 mL. Milliliters can be further subdivided into **microliters** (µL). One milliliter = 1,000 µL. (See Table A.4.)

Table A.4
Metric Units for Fluid Volume

Unit	Liters
Liter (L)	1
Milliliter (mL)	1×10^{-3}
Microliter (µL)	1×10^{-6}

Practice using metric units by doing the following conversions.

4 mL = _____0.004_____ L   10 L = _____10,000_____ mL

300 mL = __3.0×10^5__ µL   37 mL = _____0.037_____ L

0.2 L = _____200_____ mL   750 µL = _____0.750_____ mL

Pipets

To measure a small volume (10 mL or less), you will use 1-, 5-, or 10-mL pipets.

Two commonly used types of pipets are blowout and delivery pipets. With a blowout pipet, the volume is measured all the way into the tip, and accurate measurement is obtained by expelling every last drop of liquid. In a delivery pipet, the tip is unmeasured "dead space." Liquid remaining beyond the last marking should be discarded. If both types are used in your lab, get a pipet of each type and note the differences you see.

> *The lines on the blowout pipet continue into the tip. The highest number labeled on the blowout is lower than the highest number on the delivery pipet (that is, for a 5 mL pipet, the last number on a blowout is 4). There is usually a frosted band around the top of blowout pipets.*

Look at the tops of the pipets. The largest and smallest volumes you can measure are given there. For example, it may say "5 mL in 1/10." This means that the pipet measures 5 mL from the 0 line to the 5-mL line (for a delivery pipet) or the tip (for a blowout pipet). Each line in between the mL marks represents 1/10 mL.

Liquid is drawn into the pipet using a rubber bulb or some other type of filling device. In this laboratory manual we will refer to a filling device called a pipet pump, or pi-pump. Use of the pi-pump will be demonstrated in the following procedure. If your lab uses a different device, your instructor will explain how to use it.

 Never pipet by mouth. Mouth pipeting is strictly prohibited.

Pi-pumps come in different sizes. Blue pi-pumps fit 1-mL pipets. Green pi-pumps fit 5- and 10-mL pipets.

Procedure for Blowout Pipets

1. Get a 5-mL blowout pipet and a green pi-pump.
2. Attach the pi-pump to the pipet.

 Hold the pipet at the top and insert it firmly but gently into the pi-pump. Never force the pi-pump onto the pipet.

---

3. Put the pipet into the flask of red solution on your lab table. Hold the barrel of the pi-pump with four fingers, and turn the wheel clockwise with your thumb (see Figure A.3). This will suck liquid into the pipet.
4. Fill the pipet to the 0 mark.

Figure A.3.
Using a pi-pump to fill a pipet.

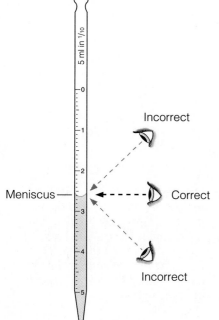

Figure A.4.
Reading the meniscus in a pipet.

5. In Figure A.4, you can see that the liquid level may be read at different angles. To read the level correctly, your eye should be directly in line with the **meniscus**, which is the crescent formed at the surface of the liquid. The *bottom* of the meniscus marks the correct volume.

6. Move the pipet to the waste container. Turn the thumbwheel counter-clockwise to release the liquid. Depress the plunger to expel the last bit.

    What volume of liquid have you measured into the container?

    5 mL.

7. Refill the pipet with red solution to the 0 mark.

8. Move the pipet to the waste container. Turn the thumbwheel slowly to release liquid down to the 3 mark. Remember to read the bottom of the meniscus.

    What volume of liquid have you measured?

    3 mL.

9. Discard the remaining liquid in the waste container.

 To avoid contamination, you should never return liquid to the container you got it from.

Procedure for Delivery Pipets

1. Get a 5-mL delivery pipet and a green pi-pump.

2. Attach the pi-pump to the pipet.

 Hold the pipet at the top and insert it firmly but gently into the pi-pump. Never force the pi-pump onto the pipet.

3. Put the pipet into the flask of red solution. Hold the barrel of the pi-pump with four fingers and turn the wheel clockwise with your thumb (see Figure A.3). This will suck liquid into the pipet.

4. Fill the pipet to the 0 mark.

5. In Figure A.4, you can see that the liquid level may be read at different angles. To read the level correctly, your eye should be directly in line with the **meniscus.** The bottom of the meniscus marks the correct volume.

6. Move the pipet to the waste container. Turn the thumbwheel slowly to release the liquid. Stop when the bottom of the meniscus reaches the 5 mL mark.

What volume of liquid have you measured into the container?

5 mL.

7. Liquid remaining in the pipet should be discarded in the waste container.

 To avoid contamination, never return liquid to the container you got it from. To expel all the remaining liquid from the pipet, push down the plunger.

8. Refill the pipet with red solution to the 0 mark.
9. Move the pipet to the waste container. Turn the thumbwheel slowly to release liquid down to the 3 mark. Remember to read the bottom of the meniscus.

    What volume of liquid have you measured?

    3 mL.

Remember to discard the remaining liquid in the waste container.

Practice measuring with pipets as directed by your instructor.

It's a good idea to give students additional practice. Just give them a series of volumes to measure. You might want to have them use a 1-mL pipet as well as the 5-mL pipet.

Graduated Cylinders

Graduated cylinders (grad cylinders) are used to measure larger amounts of fluid, although you may use a 10-mL graduated cylinder in place of a 10-mL pipet. Before you use a graduated cylinder, check to see what measurements its markings represent. For example, a typical 100-mL graduated cylinder is marked at 1-mL increments, with total volume labeled every 10 mL.

Procedure

1. Carefully pour the red solution from its container into the 100-mL grad cylinder, measuring a volume of 35 mL.

 If you are using a glass cylinder, remember to read the bottom of the meniscus. No meniscus is formed in plastic labware.

2. Add the solution from the waste container to the grad cylinder. What is the total volume now?

    At least 43 mL, depending on how much more was measured in additional practice.

3. When you are all finished with Activity C, pour the red solution from the grad cylinder back into the flask for the next team of students to use. (Since you are just practicing measurement techniques, contamination is not a concern.)

Beakers and Flasks

Most beakers and flasks have volume markings on them. However, they are not very accurate. If a procedure tells you to use 100 mL of a solution, you should measure it with a graduated cylinder. If the procedure says to use *approximately* 100 mL, the markings on a beaker or flask are close enough.

Activity D: Measuring Temperature

Scientific temperature measurements are made using the Celsius (which used to be called centigrade, "100 degree") scale, rather than the Fahrenheit scale. Conversions are made using the following formulas:

$$°C = (°F - 32)\frac{5}{9}$$

$$°F = (°C \times \frac{9}{5}) + 32$$

Table A.5 shows several comparisons between Celsius and Fahrenheit temperatures.

Table A.5
Comparison of Temperatures

	°F	°C
A very cold day	10	−12.2
Freezing point of water	32	0
A cool day	50	10
Room temperature	72	22.2
Human body temperature	98.2	36.8
Boiling point of water	212	100

Practice converting temperatures with the following problems.

145°F = _62.8°C_   28°C = _82.4°F_

16°F = _−8.9°C_   72°C = _161.6°F_

68°F = _20°C_   6°C = _42.8°F_

E X E R C I S E   A . 3
Data Presentation

This exercise explains how to
1. Distinguish between discrete and continuous variables.
2. Construct a line graph.
3. Construct a bar graph.
4. Choose the best method for presenting your data.

Students should be able to make summary presentations of their results (tables and graphs) and so should not include raw data in their lab reports. However, it is important for students to collect their data in an organized fashion (rather than on a bit of paper towel, as many are inclined to do). Also, you may need to review the students' raw data. The labs in this manual that include student-designed investigations give students a space to design a table to record their data. We have students use a separate lab notebook to record data, sketch rough graphs, and keep notes and questions on their investigations. The notebook is checked by the instructor periodically and graded for its completeness. (Of course, there are always those incorrigibles who completely miss the point and record data on a paper towel, then copy it neatly into their notebooks.)

Activity A: Tables

A student team performed the experiment that was discussed in Lab Topic 1. They tested the pulse and blood pressure of basketball players and nonathletes to compare cardiovascular fitness. They recorded the following data:

Nonathletes							Basketball Players						
	Resting Pulse			After Exercise				Resting Pulse			After Exercise		
	Trial			Trial				Trial			Trial		
Subject	1	2	3	1	2	3	Subject	1	2	3	1	2	3
1	72	68	71	145	152	139	1	67	71	70	136	133	134
2	65	63	72	142	144	158	2	73	71	70	141	144	142
3	63	68	70	140	147	144	3	72	74	73	152	146	149
4	70	72	72	133	134	145	4	75	70	72	156	151	151
5	75	76	77	149	152	153	5	78	72	76	156	150	155
6	75	75	71	154	148	147	6	74	75	75	149	146	146
7	71	68	73	142	145	150	7	68	69	69	132	140	136
8	68	70	66	135	137	135	8	70	71	70	151	148	146
9	78	75	80	160	155	153	9	73	77	76	138	152	147
10	73	75	74	142	146	140	10	72	68	64	153	155	155

If the data were presented to readers like this, they would see just lists of numbers and would have difficulty discovering any meaning in them. This is called raw data. It shows the data the team collected without any kind of summarization. Since the students had each subject perform the test three times, the data for each subject can be averaged. The other raw data sets obtained in the experiment would be treated in the same way.

Table. Average Pulse Rate for Each Subject
(Average of 3 trials for each subject; pulse taken before and after 5-min step test)

	Nonathletes			Basketball Players	
	Resting Pulse	After Exercise		Resting Pulse	After Exercise
Subject	Average	Average	Subject	Average	Average
1	70	145	1	70	134
2	67	148	2	70	142
3	67	144	3	73	149
4	71	139	4	72	151
5	76	151	5	76	155
6	74	150	6	75	146
7	71	146	7	69	136
8	68	136	8	70	146
9	78	156	9	76	147
10	74	143	10	68	155

These rough data tables are still rather unwieldy and hard to interpret. A summary table could be used to convey the overall averages for each part of the experiment. For example:

Table. Overall Averages of Pulse Rate
(10 subjects in each group; 3 trials for each subject; pulse taken before and after 5-min step test)

Pulse Rate (beats/min)		
	Before Exercise	After Exercise
Nonathletes	71.6	145.8
Basketball players	71.9	146.1

Notice that the table has a title above it that describes its contents, including the experimental conditions and the number of subjects and replications that were used to calculate the averages. In the table itself, the units of the dependent variable (pulse rate) are given and the independent variable (nonathletes and basketball players) is written on the left side of the table.

Tables should be used to present results that have relatively few data points. Tables are also useful to display several dependent variables at the same time. For example, average pulse rate before and after exercise, average blood pressure before and after exercise, and recovery time could all be put in one table.

Activity B: Graphs

Numerical results of an experiment are often presented in a graph rather than a table. A graph is literally a picture of the results, so a graph can often be more easily interpreted than a table. Generally, the independent variable is graphed on the x axis (horizontal axis) and the dependent variable is graphed on the y axis (vertical axis). In looking at a graph, then, the effect that the independent variable has on the dependent variable can be determined. "Dry mix" may help you remember which variable goes on which axis. The **d**ependent variable is sometimes called the **r**esponding variable and it goes on the **y** axis. **M**anipulated variable is a term sometimes used for the **i**ndependent variable, and it goes on the **x** axis.

Figure A.5.
Graph construction.

When you are drawing a graph, keep in mind that your objective is to show the data in the clearest, most readable form possible. In order to achieve this, you should observe the following rules:

■ Use graph paper to plot the values accurately.

■ Plot the independent variable on the x axis and the dependent variable on the y axis. For example, if you are graphing the effect of the amount of fertilizer on peanut weight, the amount of fertilizer is plotted on the x axis and peanut weight is plotted on the y axis.

■ Label each axis with the name of the variable and specify the units used to measure it. For example, the x axis might be labeled "Fertilizer applied (g/100 m^2)" and the y axis might be labeled "Weight of peanuts per plant (grams)."

■ The intervals labeled on each axis should be appropriate for the range of data so that most of the area of the graph can be used. For example, if the highest data point is 47, the highest value labeled on the axis might be 50. If you labeled intervals on up to 100, there would be a large unused area of the graph.

■ The intervals that are labeled on the graph should be evenly spaced. For example, if the values range from 0 to 50, you might label the axis at 0, 10, 20, 30, 40, and 50. It would be confusing to have labels that correspond to the actual data points (for example, 2, 17, 24, 30, 42, and 47).

■ The graph should have a title that, like the title of a table, describes the experimental conditions that produced the data.

Figure A.6 illustrates a well-executed graph.

Figure A.6.
Graph of peanut weight vs. amount of fertilizer applied.

Figure 1. Weight of peanuts produced per plant when amount of fertilizer applied is varied. (Average seed weight per plant in 100 m^2 plots, 400 plants/plot.)

The most commonly used forms of graphs are line graphs and bar graphs. The choice of graph type depends on the nature of the independent variable being graphed. **Continuous variables** are those that have an unlimited number of values between points. **Line graphs** are used to represent continuous data. For instance, time is a continuous variable over which things such as growth will vary. Although the units on the axis can be minutes, hours, days, months, or even years, values can be placed in between any two values. Amount of fertilizer can also be a continuous variable. Although the intervals labeled on the x axis are 0, 200, 400, 600, 800, and 1,000 g/100 m^2, many other values can be listed between each two intervals.

In a line graph, data are plotted as separate points on the axes, and the points are connected to each other. Notice in Figure A.7 that when there is more than one set of data on a graph, it is necessary to provide a key indicating which line corresponds to which data set.

Figure 1. Recovery rate of basketball players and nonathletes after performing a step test for 5 minutes. (Average of 10 subjects; each subject performed the test 3 times.)

Figure A.7.
Line graph representing two related sets of data.

Discrete variables, on the other hand, have a limited number of possible values, and no values can fall between them. For example, the type of fertilizer is a discrete variable: There are a certain number of types which are distinct from each other. If fertilizer type is the independent variable displayed on the *x* axis, there is no continuity between the values.

Bar graphs, as shown in Figure A.8, are used to display discrete data.

Figure 1. Average pulse rates of basketball players and nonathletes before and after performing a step test for 5 minutes. (Average of 10 subjects; each subject performed the test 3 times.)

Figure A.8.
Example of bar graph.

In this example, before- and after-exercise data are discrete: There is no possibility of intermediate values. The subjects used (basketball players and nonathletes) also are a discrete variable (a person belongs to one group or the other).

This graph could also have been constructed as shown in Figure A.9.

Figure A.9.
Alternative method of presenting data in Figure A.7.

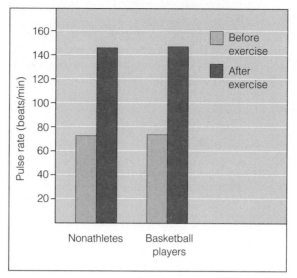

Figure 1. Average pulse rates of basketball players and nonathletes before and after performing a step test for 5 minutes. (Average of 10 subjects; each subject performed the test 3 times.)

What is the difference between the two graphs?

The first graph groups the before- and after-exercise data, and shows a side-by-side comparison of athletes and nonathletes. The second graph groups the athletes and the nonathletes and shows a side-by-side comparison of their before and after scores.

Explain why the first way would be better to convey the results of the experiment.

The investigator wants to compare athletes and nonathletes; putting those bars side by side facilitates the comparison. Comparing before-exercise pulse with after-exercise pulse for each group separately is not as useful.

Activity C: Graphing Practice

You can have students actually construct these graphs or just answer the questions.

Use the temperature and precipitation data provided in Table A.6 to complete the following questions:

1. Compare monthly temperatures in Fairbanks with temperatures in San Salvador.

   Can data for both cities be plotted on the same graph?

   Yes, that will make it easier to make a comparison.

   What will go on the *x* axis?

   Months.

Table A.6
Average Monthly High Temperature and Precipitation for Four Cities
(T = temperature in °C; P = precipitation in cm)

		Jan.	Feb.	Mar.	Apr.	May	June	July	Aug.	Sept.	Oct.	Nov.	Dec.
Fairbanks, Alaska	T	–19	–12	–5	6	15	22	22	19	12	2	–11	–17
	P	2.3	1.3	1.8	0.8	1.5	3.3	4.8	5.3	3.3	2.0	1.8	1.5
San Francisco, California	T	13	15	16	17	17	19	18	18	21	20	17	14
	P	11.9	9.7	7.9	3.8	1.8	0.3	0	0	0.8	2.5	6.4	11.2
San Salvador, El Salvador	T	32	33	34	34	33	31	32	32	31	31	31	32
	P	0.8	0.5	1.0	4.3	19.6	32.8	29.2	29.7	30.7	24.1	4.1	1.0
Indianapolis, Indiana	T	2	4	9	16	22	28	30	29	25	18	10	4
	P	7.6	6.9	10.2	9.1	9.9	10.2	9.9	8.4	8.1	7.1	8.4	7.6

Source: Pearce, E. A., and G. Smith. Adapted from *The Times Books World Weather Guide*. New York: Times Books, 1990.

How should the *x* axis be labeled?

Months (or: Time in months).

What should go on the *y* axis?

Temperatures.

What is the range of values on the *y* axis?

–19°C to 34°C.

How should the *y* axis be labeled?

Average monthly high temperature (°C).

What type of graph should be used?

Line graph. Time is a continuous variable.

2. Compare the average September temperature for Fairbanks, San Francisco, San Salvador, and Indianapolis.
   Can data for all four cities be plotted on the same graph?

   Yes, that will make it easier to make a comparison.

   What will go on the *x* axis?

   Cities.

   How should the *x* axis be labeled?

   Cities.

What should go on the *y* axis?

Temperatures.

What is the range of values on the *y* axis?

12°C to 31°C.

How should the *y* axis be labeled?

Mean temperature (°C).

What type of graph should be used?

Bar graph. Cities is a discrete variable.

3. Graph the temperature and precipitation data for San Francisco. Can both sets of data be plotted on the same graph?

    Yes. Use a double y axis (have a second y axis on the right-hand side of the graph).

    What will go on the *x* axis?

    Months.

    How should the *x* axis be labeled?

    Months (or: Time in months).

    What should go on the *y* axis?

    Temperatures on one, precipitation on the other.

    What is the range of values on the temperature axis?

    13°C to 21°C.

    How should this axis be labeled?

    Mean temperature (°C).

    What is the range of values on the precipitation axis?

    0 cm to 11.9 cm.

    How should this axis be labeled?

    Mean precipitation (cm).

    What type of graph should be used?

    Line graph. Time is a continuous variable.

EXERCISE A.4
Interpreting Information on a Graph

This exercise explains how to
1. Interpret graphs.

The purpose of this exercise is to show students how to get information from graphs. The descriptions of what each graph shows are examples; student answers will vary. It is useful to supplement this exercise with examples of graphs from other sources, such as the textbook and newspapers.

Once you understand how graphs are constructed, it is easier to get information from the graphs in your textbook as well as to interpret the results you obtain from laboratory experiments. For the graphs below, write a sentence or two describing what each graph shows, and answer the questions.

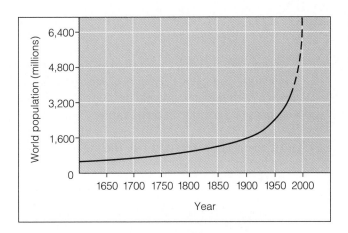

Figure A.10.
Change in world population from 1650 to 2000.

Describe what the graph shows.

The graph shows how world population has increased since 1650. The increase was slow until about 1900, and then the population grew at a much faster rate.

What was the world's population in 1900?

~1.6×10^9 (1.6 billion).

Predict the world's population in 2000.

~6.4×10^9 (6.4 billion).

Figure A.11.
Rate of an enzymatic reaction at different temperatures.

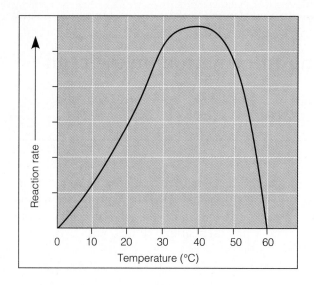

Describe what the graph shows.

> *Reaction rate increases with increasing temperature up to a certain point, and then decreases when the temperature is raised further. Both the increase and decrease are rapid (large change in rate for small change in temperature).*

At what temperature is reaction rate the highest?

> *Around 37°C to 43°C.*

Figure A.12.
Change in range of average monthly temperature as latitude changes.

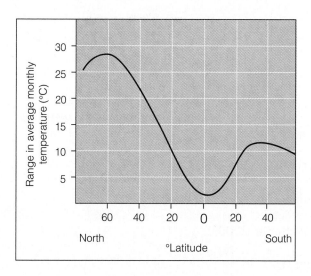

Describe what the graph shows.

> *Average temperature range is greatest at northern latitudes and least at the equator. The temperature range in southern latitudes is much smaller than in the northern latitudes.*

At what latitude does the least variation in temperature occur?

0° (equator).

Miami is at approximately 26° N latitude. From the information on the graph, what is the range in mean monthly temperature there?

~13°C.

If you know the latitude of your location, ask students to find the temperature range for it.

Minneapolis is at approximately 45° N latitude. From the information on the graph, what is the range in mean monthly temperature there?

~24°C.

Sydney, Australia is at approximately 33° S latitude. From the information on the graph, what is the range in mean monthly temperature there?

~11°C.

Figure A.13.
Absorption of light by the pigments P_r phytochrome and P_{fr} phytochrome.

Describe what the graph shows.

Both P_r and P_{fr} (which are pigments that absorb light in the visible range) show a lot of variation in their ability to absorb light in the range of 300–800 nm. Absorbance for P_r ranges from 0 to ~1.4. Absorbance for P_{fr} ranges from 0 to ~0.65.

At what wavelengths does P_r phytochrome absorb the most light?

680 nm.

At what wavelengths does P_{fr} phytochrome absorb the most light?

740 nm.

Figure A.14.
Population growth of *Paramecium aurelia* and *Paramecium caudatum* when grown separately and together.

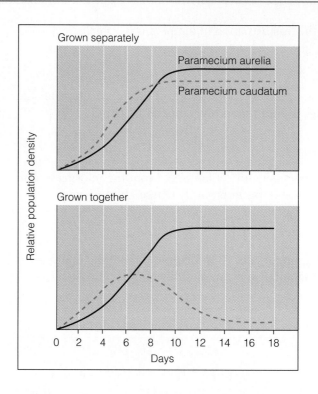

Describe what the graphs show.

The graphs show changes in population density over an 18-day period. In the top graph each species is grown by itself, and the growth curves are similar, increasing density until about Day 8. In the bottom curve the two species are grown in mixed culture. The curve for P. aurelia *is almost identical to its growth in single culture, but the population density of* P. caudatum *declines after about 6 days.*

On what day does *Paramecium aurelia* reach its maximum population density?

Day 8 or 9.

Does *Paramecium caudatum* do better when it is grown alone or when it is grown in a mixture with *Paramecium aurelia*?

P. caudatum *does much better alone.*

Questions for Review

1. Convert the following numbers to scientific notation.

   0.0032 = *3.2 × 10⁻³*   7,310,000 = *7.31 × 10⁶*

   457,200,000 = *4.572 × 10⁸*   689 = *6.89 × 10²*

   0.67 = *6.7 × 10⁻¹*   0.000483 = *4.83 × 10⁻⁴*

2. Convert the following numbers to decimals.

   7.32 × 10⁻⁴ = *0.000732*   1.34 × 10⁸ = *0.0000000134*

   3.43 × 10³ = *3430*   7.55 × 10⁻⁶ = *0.00000755*

   6.2 × 10⁵ = *620,000*   1.8 × 10⁻² = *0.018*

3. Perform the following calculations.

$$\frac{3.4 \times 10^{-2}}{1.7 \times 10^{-5}} = \textit{2.0} \times \textit{10}^{\textit{3}} \qquad \frac{1.0 \times 10^{5}}{2.0 \times 10^{3}} = \textit{0.5} \times \textit{10}^{\textit{2}} = \textit{5.0} \times \textit{10}^{\textit{1}}$$

$$\frac{8.6 \times 10^{-3}}{2.0 \times 10^{2}} = \textit{4.3} \times \textit{10}^{\textit{-5}} \qquad \frac{9.9 \times 10^{6}}{3.3 \times 10^{-4}} = \textit{3.0} \times \textit{10}^{\textit{10}}$$

   $(3.5 \times 10^{-2})(2.0 \times 10^{-5}) = \textit{7.0} \times \textit{10}^{\textit{-7}}$

   $(7.36 \times 10^{3})(1.0 \times 10^{-6}) = \textit{7.36} \times \textit{10}^{\textit{-3}}$

   $(1.6 \times 10^{4})(3.0 \times 10^{2}) = \textit{4.8} \times \textit{10}^{\textit{6}}$

   $(1.11 \times 10^{-5})(6.0 \times 10^{9}) = \textit{6.66} \times \textit{10}^{\textit{4}}$

4. Convert the following measures.

   2.8 mm = *2.8 × 10⁶*   nm   4.67 m = *4.67 × 10⁶* μm

   1.3 nm = *1.3 × 10⁻³* μm   67 cm = *6.7 × 10⁻¹* m

   12 μg = *1.2 × 10⁴*   ng   1.6 g = *1.6 × 10⁻³* kg

   300 μg = *3 × 10⁻⁴*   g   250 mg = *2.5 × 10⁵* μg

   83 mL = *8.3 × 10⁻²* L   250 mL = *2.5 × 10⁻¹* L

   175 μL = *1.75 × 10⁻¹* mL   0.5 L = *5 × 10²* mL

   75 °F = *23.9*   °C   50 °C = *122*   °F

In the following questions you will be constructing graphs without plotting data. By practicing how to construct graphs, you will learn how to graph your own data in later labs.

---

5. A team of students hypothesizes that the amount of alcohol produced in fermentation depends on the amount of sugar supplied to the yeast. They want to use 5, 10, 15, 20, 25, and 30% sugar solutions. They propose to run each experiment at 40°C with 5 mL of yeast.

   What type of graph is appropriate for presenting these data? Explain why.

   Line graph, because sugar concentration is a continuous variable.

Sketch the axes of a graph that would present these data. Mark the intervals on the *x* axis and label both axes completely. Write a title for the graph.

Figure 1. Amount of alcohol produced by fermentation with varying sugar concentrations. (Data based on one trial at each concentration. Experiments run at 40° C with 5 mL yeast in each reaction vessel.)

6. Having learned that the optimum sugar concentration is 25%, the students next decide to investigate whether different strains of yeast produce different amounts of alcohol. If you were going to graph the data from this experiment, what type of graph would be used? Explain why.

   Bar graph; yeast strain is a discrete variable.

   Sketch and label the axes for this graph and write a title.

Figure 1. Amount of alcohol produced by fermentation with different types of yeast. (Data represent average of 3 trials for each yeast type. All experiments were run at 40°C with 5 mL yeast and 5 mL 25% sugar.)

7. A team of students wants to study the effect of temperature on bacterial growth. They put the dishes in different places: an incubator (37°C), a lab room (21°C), a refrigerator (10°C), and a freezer (0°C). Bacterial growth is measured by estimating the percentage of each dish that is covered by bacteria at the end of a 3-day growth period.

   What type of graph would be used to present these data? Explain why.

   Line graph; temperature is a continuous variable.

   Sketch the axes below. Mark the intervals on the *x* axis, and label both axes completely. Write a title for the graph.

   Figure 1. Effect of temperature on bacterial growth.
   (% plate coverage by bacteria after 3-day growth period.)

8. A team of scientists is testing a new drug, XYZ, on AIDS patients. The scientists monitor patients in the study for symptoms of 12 different diseases. What would be the best way for them to present these data? Explain why.

   In a table. They are actually looking at 12 dependent variables; also, these dependent variables (disease symptoms) are themselves discrete, so graphical presentation is not practical.

9. A group of students decides to investigate the loss of chlorophyll in autumn leaves. They collect green leaves and leaves that have turned color from sugar maple, sweet gum, beech, and aspen trees. Each leaf is subjected to an analysis to determine how many mg of chlorophyll is present.

   What type of graph would be most appropriate for presenting the results of this experiment? Explain why.

   A bar graph, because leaf color and tree type are both discrete variables.

Sketch the axes for the graph below to show how you would present the data. Write a title for the graph.

Figure 1. Chlorophyll loss in autumn leaves.

Guide to Writing a Scientific Report

Purpose of the Lab Report

No matter what you are writing, you should always know who your audience is so you can write in the way that best communicates with that audience. The "reader" referred to in this guide is a person who understands science but is not as knowledgeable about your particular subject as you are. Your job is to explain your investigation and its significance and to convince the reader that your conclusions are scientifically valid. Since you are writing your lab report for a grade, your instructor is also part of your audience. Your instructor wants to evaluate how well you can communicate your background knowledge and all the components of your investigation to a wider audience, so don't assume that your reader knows as much as your instructor about your investigation. Also keep in mind that writing a good paper depends more on *how well* you designed the experiment and can present and explain the results than on *what* results you obtained. Even if your experiment did not turn out as you expected, you can still write a good lab report.

A scientific report is organized so that each phase of the investigation is described and justified, from formulating a hypothesis through reaching a conclusion about the hypothesis. The exact format used may vary slightly from one source to another. For example, in the format presented here, the Discussion and Conclusion sections are separate; in some scientific journals a combined Discussion/Conclusion section is used.

This guide describes the purpose and content of each section of a report and provides a sample paper integrated throughout as an example. The sample report is the material printed in green. Margin notes alongside the sections of the sample report point out features in each part. Read this entire appendix before you begin writing your report. Be sure to read the checklist at the end of the appendix, too. After you've written your report, review the checklist to see whether you've fulfilled all of its criteria.

 The sample report is based on a fictitious experiment that was constructed solely to illustrate the composition of a scientific report. The conclusions stated are not based on scientific data.

Sections of the Report

Each section of your report appearing after the title should have a heading.

Title

The title is a statement of the problem you are investigating. It should contain key words that indicate what information the reader will find in the paper. Your hypothesis can be used as a basis for your title.

The title should be placed on a cover page that also includes your name, your instructor's name, and the date.

As an example of turning a hypothesis into a title, consider the following hypothesis: Dark-colored hair dye is more mutagenic to *Salmonella typhimurium* bacteria than is light-colored hair dye. You could reword this for your title:

The title may seem long, but it should give enough information about the investigation for readers to decide whether they are interested in reading the contents of the report.

The Effect of Hair Dye Color on Mutagenicity of *Salmonella typhimurium* Bacteria

Introduction

The Introduction tells the reader what your investigation was about. It provides information about the biological basis for your experiment and a very brief synopsis of your experimental design.

State your hypothesis clearly at the beginning of the Introduction. Follow it with an explanation of why you and your lab team thought this hypothesis was worth investigating.

Give enough background information and references to allow the reader to appreciate the significance of your experiment, including an explanation of any unfamiliar or technical terms. You should relate your investigation to the larger issues in this area of study.

Since you are not an expert, you must cite a reference to show the source of information for any statement you make about the biological basis for your investigation. This demonstrates that you have done your homework and learned about your subject before undertaking your experiment. Citing a reference means telling the reader where you got a particular piece of information. In the text of the report, an abbreviated citation form is used—for example: Campbell, Reece, Mitchell, and Taylor (2003). A section at the end of the report called Literature Cited gives details about the source of information so the reader can locate it. For your lab reports, your sources will include the lab manual and your textbook. Your instructor may provide additional sources or suggest that you visit the library to get additional information. You may also get information from your professors or other knowledgeable people.

The Introduction should also explain briefly how you intend to go about testing your hypothesis. Finally, state your predictions concerning the outcome of your experiment.

Introduction

This experiment was performed to investigate the following hypothesis: Dark hair dye is more mutagenic in *Salmonella typhimurium* bacteria than light hair dye.

The hypothesis is stated.

One of our team members had read in the newspaper (*Greenville News,* 2002) that using dark colors of hair dye increases a person's chances of developing certain kinds of cancer. It has been shown that normal cells can be changed to cancer cells by a change or mutation in the genetic material (DNA) of the cell. A chemical or other agent that changes the genetic material is called a mutagen. Mutagens that cause normal cells to become cancerous are called carcinogens (Campbell, Reece, Mitchel, and Taylor, 2003). In our experiment, we tested the mutagenicity, or ability to cause mutations, of three hair dyes: Basic Brunette (dark color), Bombshell Blonde (light color), and Raspberry Red (a red color in between the dark and light colors).

The author explains what led her to make this hypothesis.

Including this background information helps the reader understand the biological basis for the experiment and demonstrates to the instructor that the author knows the meanings of the terms used. Notice that the citations are set off in parentheses.

We used the Ames test, which is widely used to test whether a chemical causes mutations in *Salmonella typhimurium,* a species of bacteria. According to Campbell, Reece, Mitchell, and Taylor (2003), "In general, mutagens are carcinogens." If a chemical is mutagenic in bacteria, then it is studied further to determine whether it could be mutagenic or carcinogenic in animals or humans (Atlas, 1984).

A lab manual by Sigmon (2000) explained how to perform the Ames test using a strain of *Salmonella typhimurium* that cannot make the amino acid histidine as normal *Salmonella* can do. This strain of *Salmonella* bacteria, which is called his– (his minus), can't grow unless histidine is present in agar, the substance it grows on. The part of the bacterial DNA that codes for the ability to make histidine is sensitive to mutagens. A mutagen can cause a mutation that enables the bacteria to produce histidine (Figure B.1).

In this paragraph the author demonstrates that she understands the laboratory method used to test the hypothesis. The completeness of the explanation and Figures B.1 and B.2 will help the reader understand the author's interpretation of the results when it is presented in the Discussion section.

Figure B.1.
Mutation in *Salmonella typhimurium.*

When a mutagen is present during the growth of the bacteria, then the bacteria gain the ability to produce histidine. The mutagenicity of a substance can be tested by growing the bacteria on agar that contains the suspected mutagen but doesn't contain histidine, as shown in Figure B.2.

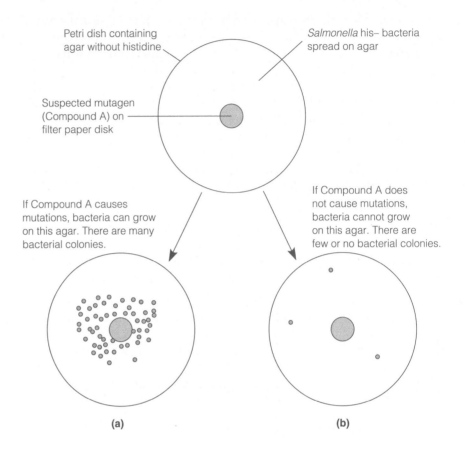

Petri dish containing agar without histidine

Salmonella his– bacteria spread on agar

Suspected mutagen (Compound A) on filter paper disk

If Compound A causes mutations, bacteria can grow on this agar. There are many bacterial colonies.

If Compound A does not cause mutations, bacteria cannot grow on this agar. There are few or no bacterial colonies.

(a)   (b)

Figure B.2.
The Ames test. *Salmonella typhimurium* is spread on agar that doesn't contain any histidine. Compound A, a suspected mutagen, is placed on a filter paper disk on the agar. If bacterial colonies appear, then Compound A is a mutagen (a). If there are few or no bacterial colonies, Compound A is not a mutagen (b).

We used the Ames test for this investigation and applied brown, red, and blonde hair dyes to the filter paper circles. The growth of the his– strain of *Salmonella typhimurium* on the dishes indicated the ability of these dyes to cause mutations.

Predictions (if . . . then statements) made.

If our hypothesis is supported, then dishes that contain the brunette hair dye will have the most growth of bacteria, showing that dark dye is most mutagenic. Dishes containing the blonde dye will have the least bacterial growth, and dishes containing the red dye will have an intermediate amount of growth.

Materials and Methods

The purpose of the materials and methods section is to give a detailed account of your experimental procedure. In order to be considered valid, the results of a scientific investigation must be able to be duplicated by other scientists working in other laboratories. It is therefore necessary to provide complete details of how your investigation was performed. In addition, scientists consult the Methods sections of published papers in order to learn techniques they can apply to their own work. So when you write this section, imagine that you are explaining what you did so that someone else can replicate your experiment exactly.

Use the following guidelines for the appropriate style in your Materials and Methods section.

- Use the past tense. Don't write as if you're giving instructions.

  INCORRECT: First you inoculate the agar plates with *Salmonella* bacteria.

  CORRECT: We inoculated the agar plates with *Salmonella* bacteria.

- Tell what you did in paragraph form. Don't write a recipe.

  INCORRECT:

  Step 1. Put 3 drops of hair dye on a filter paper.

  Step 2. Put the filter paper on the agar.

  CORRECT: We put 3 drops of hair dye on a filter paper, and then placed the filter paper on the agar.

- Be specific. Someone attempting to duplicate your experiment needs all the details.

  INCORRECT: We pipetted a sample of broth onto each plate.

  CORRECT: We pipetted a 1-mL sample of broth onto each plate.

Materials and Methods

Twelve petri dishes containing sterile agar growth medium that lacked histidine were prepared for us by the Biology Department prep staff. We used a broth that contained *Salmonella typhimurium* bacteria (strain his–) to "seed" the agar plates with bacteria. The *Salmonella* cultures were provided by the Microbiology Department. Each member of our group prepared four of the plates using the same technique. Using sterile pipets, we pipetted a 1-mL sample of the broth onto the agar in each plate. The broth was spread over the agar with a glass rod that had been sterilized by dipping the rod in alcohol and then burning off the alcohol in a flame. The rod was allowed to cool before being used to spread the broth. We learned these techniques from a lab manual (Sigmon, 2000).

All the significant details of how the experiment was done should be recounted here.

Following the standard technique for the Ames test (Sigmon, 2000), we used filter paper disks to apply the hair dye to the agar plates. We used 12 filter paper disks, each 1 cm in diameter. Using droppers, we put 1 drop of hair dye or sterile water (for a control) on each disk. Each treatment was thus replicated three times. The filter papers were air dried and then placed in the center of the petri dishes of agar. Forceps were always used to handle the filter paper.

Cite the source(s) of the techniques you used.

All the hair dyes used were Brand X. We chose Basic Brunette as the dark dye, Raspberry Red as the medium dye, and Bombshell Blonde as the light dye.

The reader should be informed about variables that must be standardized in order to have a successful experiment. If you later think of factors that should have been standardized but weren't, you should mention them in the Discussion section.

Two standardized variables in this experiment were temperature and the time that data were recorded. We put the petri dishes in an incubator set at 37°C and allowed them to incubate for 7 days. Each day at 11:30 A.M. a team member checked the plates and counted the number of bacterial colonies present. The results were recorded, tabulated, and distributed to the whole team.

Results

In the Results section you present the data in an organized, readable form. Numerical data are usually given in tables. Relationships between factors are often shown on graphs. Graphs, drawings, and anything else that is not a table is called a figure. Tables and figures should be numbered separately so that in the text you can refer to Table 1, Table 2, Figure 1, Figure 2, and so on. All tables and figures must have titles describing their contents.

Before beginning to write your Results section, you should review the data presentation section of Appendix A (Tools for Scientific Inquiry). Remember that you should not include raw data in your report. Also, results must be presented in some numerical fashion. A descriptive narrative is not acceptable.

Keep the following points in mind as you prepare your graphs:

- Use graph paper unless your graphs are computer generated.
- Label the axes completely.
- Use the entire area of the graph to display your data.
- Choose appropriate intervals and mark them evenly along the axes.
- If there is more than one set of data on the graph, be sure the reader can tell the lines apart. Include a legend (see Figure B.3).
- Each graph must have a descriptive title.

In addition to tables and figures, the Results section should include a brief paragraph that draws the reader's attention to the important pieces of data. However, you should save your explanations of *why* results are significant for the Discussion section.

Results

Author briefly points out important results that will be featured in the Discussion section.

Table B.1 shows that Basic Brunette had produced the greatest number of colonies at the end of 7 days. Figure B.3 shows that bacterial colonies on the dishes containing Raspberry Red and Bombshell Blonde only appeared on the last day, while colonies appeared on the Basic Brunette plates by Day 2 and steadily increased in number.

Table B.1
The total number of *Salmonella* his– bacterial colonies counted at the end of 7 days, using 3 replicates for each color of hair dye. The growth medium used was histidine–. Incubation was at 37°C.

Hair dye color	Number of colonies observed
Basic Brunette	36
Raspberry Red	4*
Bombshell Blonde	1
Water (control)	2

* 3 of these colonies appeared to be contaminants.

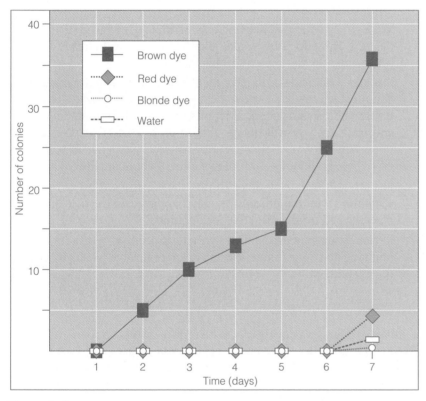

Figure B.3.
The total number of *Salmonella typhimurium* his– bacterial colonies (3 petri dishes per color) counted on each of 7 consecutive days. Incubation was at 37°C on growth medium that lacked histidine.

Discussion

In the Discussion section you should interpret the results, explain their significance, and discuss any weaknesses of the experimental methods or design. From the instructor's point of view, this is the most important section of your paper because it shows how well you understood your investigation. As you might expect, it is also the most difficult section to write.

You should complete the Introduction and Results sections before you begin writing the Discussion. Put them on the desk in front of you, along with your lab notebook open to your notes on the experiment, and begin writing your rough draft. You can outline the contents of the Discussion by taking the following steps.

1. Write down your hypothesis again. Look at the tables and/or figures you constructed for the Results section and determine whether you should accept or reject your hypothesis. (Did your experiment support your hypothesis or prove it false?)

2. Check the predictions you wrote in the Introduction section. Do your results confirm your predictions or not?

3. Write down the specific data (using the actual numbers) that led you to your conclusion about the hypothesis. If you have gotten additional results from other lab teams working on a similar problem, list that information also.

4. Write down what you know about the biology involved in your investigation. How do your results fit in with what you already know? Be sure to identify the sources of this information.

5. List any weaknesses you have identified in your experimental design. You must tell the reader how these weaknesses may have affected your results. Since your lab experiments are subject to limitations of time and facilities, you will not be able to do a "perfect" experiment. It is important for you to understand, and to acknowledge in your report, how these limitations affect the validity of your conclusions.

6. List any problems that arose during the experiment itself. Unforeseen difficulties with the procedures may have affected the data and should be described for the reader's consideration.

7. Review your experimental design and procedure. Consider how you might be able to get more specific or more reliable results by changing the experiment.

Discussion

The hypothesis is restated along with a brief statement of the investigator's conclusion.

Our experiment was designed to determine whether dark hair dye is more mutagenic to *Salmonella typhimurium* than light hair dye. Our results supported the hypothesis to the extent that Basic Brunette, a dark brown dye, was more mutagenic than Bombshell Blonde, a light hair dye. The results for Raspberry Red did not support the hypothesis.

Specific data are used in support of the conclusion. The author also explains what the data mean and compares the results to the predictions made before the experiment was performed.

As seen in Table B.1, the total number of colonies after 7 days of incubation (36) was greatest in the presence of Basic Brunette, a dark brown hair dye. This result means that the his– gene was mutated to his+ in 36 instances, supporting our hypothesis that the dark color is mutagenic. These results also confirm our prediction.

The plates containing the Bombshell Blonde hair dye and sterile water produced only one or two colonies each. Since the blonde dye did not produce any more mutations than water, which was the control, we can conclude that the blonde dye is not mutagenic. These results also support our hypothesis and prediction.

Raspberry Red hair dye produced four colonies, but three of these colonies, which were all on the same plate, were different colors from all the other colonies. Our lab instructor told us that these three colonies were contaminants, rather than *Salmonella typhimurium*, so only one of the colonies was actually mutated *Salmonella typhimurium* (Wright, personal communication). Red dye was therefore no more mutagenic than the control. This result was unexpected, since we had predicted that red dye would be intermediate in mutagenicity between brown and blonde.

The source of information obtained through conversation with a knowledgeable person should be cited as a "personal communication."

The list of ingredients on the packages of hair dye gave some insight into the results for red dye. We had assumed that the different colors of dye would have the same pigments but in different amounts. However, Basic Brunette contains two pigments called chemicals A and B, Raspberry Red contains pigments called chemicals C and D, and Bombshell Blonde contains a pigment called chemical E. If we could test just these pigments using the Ames test, we could determine whether chemicals A and B are mutagenic.

When the results are unexpected, you should try to explain why they differ from the prediction. Understanding the reasons for your results, whatever they are, is an important part of the scientific process.

For future experiments, we could also test different shades of Brand X hair color containing different pigments or different brands of hair color containing chemicals A and B.

The fact that the control plate produced two colonies suggests that some mutation takes place naturally. Starr and Taggart (1992) state that although some gene mutations are caused by mutagens, "other gene mutations are spontaneous; they are not induced by agents outside the cell." They further add that the rate of spontaneous mutations is relatively low. This supports the results from the plates containing water and Raspberry Red and Bombshell Blonde hair dyes.

The author uses this background information to show how her results fit in with what is already known on the subject.

Figure B.3 shows that five colonies had appeared in the Basic Brunette dishes by Day 2, and the number of colonies increased steadily over the entire 7 days of the experiment. The significance of this result is that the bacteria present were being mutated over time as the hair dye diffused to a wider and wider area of each dish. This means that the longer the bacteria are exposed to hair dye, the more mutations will occur. If we repeated this experiment, we would draw concentric rings around the filter paper disks and count the colonies appearing in each ring over time instead of just taking a total count of the dish.

Again, the author discusses specific data and interprets it using her knowledge of biological concepts. This leads to another suggestion for improving the experiment.

The Ames test is not a definitive test for the cancer-causing ability of any chemical. It does indicate whether a chemical is mutagenic (can cause changes in the DNA). The Basic Brunette hair dye definitely showed signs of being mutagenic to *Salmonella* DNA, especially when there is longer exposure. Raspberry Red and Bombshell Blonde hair dyes gave no evidence of being mutagenic. Our results support the study cited in the *Greenville News* (2002), which found that some hair dyes are potential carcinogens.

Author explains how the results of this investigation support published information.

Weaknesses in our experiment include the length of time allowed for incubation, the number of ingredients contained in the hair dyes, the fact that only one brand of dye was tested, and the contamination of one of the petri dishes. We could extend this experiment by testing the individual pigments in the hair dyes, using different brands of hair dye, testing the *Salmonella* his+ strain, and counting the colonies in concentric rings away from the hair dye disks.

Weaknesses of the experiment are pointed out. This shows that the author understood the method and results well enough to improve the experiment.

Conclusion

Your conclusions may be mentioned in three sections of your paper: Introduction, Discussion, and Conclusion, but they *must* be stated in the Conclusion. In the Conclusion section, try to rephrase your conclusion rather than repeat the exact wording used in a previous section. If your readers did not understand your initial version, another wording may clarify it.

The Conclusion should be brief (two or three sentences). It should repeat the significant results from your experiment, but should not contain any new information.

Conclusion

Basic Brunette caused 36 colonies of his– bacteria to grow on agar lacking histidine, so we concluded that this hair dye caused a mutation in the bacterial DNA. Raspberry Red and Bombshell Blonde dyes each produced a total of one mutated colony, so we concluded that these two dyes are not mutagenic.

Literature Cited

You must tell the reader exactly where to find the sources of information you used. In the text of the report, cite the source as (author, date) or author (date). For example:

(Author, date): A chemical that causes normal cells to become cancerous is called a carcinogen (Campbell, Reece, Mitchell, and Taylor, 2003).

Author (date): Campbell, Reece, Mitchell, and Taylor (2003) define a carcinogen as a chemical that causes normal cells to become cancerous.

At the end of the report, the Literature Cited section gives detailed information about the sources of information you used. The sources should be listed alphabetically by author. Any source that appears in your list must also be cited in the text of the report. The following examples illustrate one style. You may see slightly different citation styles used in different sources. Note that all citation styles include the same information; it is simply arranged differently.

Literature Cited

Atlas, C. *Microbiology*. New York: Macmillan Publishing Company, 1984.

Campbell, N., J. Reece, L. Mitchell, and M. Taylor. *Biology: Concepts and Connections*. San Francisco, CA: Benjamin/Cummings, 2003.

Greenville News. "New Study Links Hair Dye and Cancer." Greenville, SC: Greenville-Piedmont Publishing Company, June 4, 2002.

Sigmon, J. *Laboratory Techniques in Microbiology*. Clemson, SC: Clemson University, 2000.

Starr, C., and R. Taggart. *Biology: The Unity and Diversity of Life*, 6th ed. Belmont, CA: Wadsworth, 1992.

Wright, I. M., biology professor at Clemson University, personal communication, 2002.

Using Reference Materials Honestly

When your instructor reads your paper, he or she wants to evaluate your understanding of the biology involved in your experiment. It is essential for you to use your own words to explain your investigation rather than attempting to imitate your references. In addition, you could face a charge of plagiarism if "your" work is too similar to someone else's work.

Avoiding Plagiarism

Many students have difficulty trying to put information they've read into their own words. Typically, they change or rearrange a few words but leave the sentence essentially the same as it was written. For example, here is a sentence from a biology textbook (Campbell, Reece, Mitchell, and Taylor, 2003): "Cancer-causing agents, factors that alter DNA and make cells cancerous, are called carcinogens." A student changed this sentence to read, "Factors that cause cancer and alter DNA to make cells cancerous are called carcinogens." The student has tried to disguise this sentence from the textbook as his own work. Not only is the deception transparent to someone who has read the textbook, the awkwardness of the resulting sentence might even make the instructor wonder if the student knows what he is saying.

How can this problem be avoided?

- Take notes from your references, and then write from your notes rather than writing directly from the reference.

- Use more than one source of information for each topic so you won't get stuck on certain phrases.

- Think about what you know before you write it down. Digest and resynthesize the information you have read.

- Write down what you know quickly, without worrying about how well it's written. Later, revise your writing so a reader can understand it.

Using Quotations

Any sentence or phrase that is copied directly from a source must be placed in quotation marks.

The use of lengthy quotations indicates to the instructor that the student doesn't understand the subject well enough to explain it himself. For the lab reports you'll write for this course, it's recommended that direct quotations be no longer than one sentence.

Checklist for a Scientific Report

Title
_____Communicate the subject investigated in the paper.

Introduction
_____State the hypothesis.

_____Give well-defined reasons for making the hypothesis.

_____Explain the biological basis of the experiment.

_____Cite sources to substantiate background information.

_____Explain how the method used will produce information relevant to your hypothesis.

_____State a prediction based on your hypothesis. (If the hypothesis is supported, then the results will be . . .)

Materials and Methods
_____Use the appropriate style.

_____Give enough detail so the reader could duplicate your experiment.

_____State the control treatment, replication, and standardized variables that were used.

Results
_____Summarize the data (do not include raw data).

_____Present the data in an appropriate format (table or graph).

_____Present tables and figures neatly so they are easily read.

_____Label the axes of each graph completely.

_____Give units of measurement where appropriate.

_____Write a descriptive caption for each table and figure.

_____Include a short paragraph pointing out important results but do not interpret the data.

Discussion
_____State whether the hypothesis was supported or proven false by the results, or else state that the results were inconclusive.

_____Cite specific results that support your conclusions.

_____Give the reasoning for your conclusions.

_____Demonstrate that you understand the biological meaning of your results.

_____Compare the results with your predictions and explain any unexpected results.

_____Compare the results to other research or information available to you.

_____Discuss any weaknesses in your experimental design or problems with the execution of the experiment.

_____Discuss how you might extend or improve your experiment.

Conclusions

_____Restate your conclusion.

_____Restate important results.

Literature Cited

_____Use proper citation form in the text.

_____Use proper citation form in the Literature Cited section.

_____Refer in the text to any source listed in this section.

Acknowledgment

The sample report is based on a sample paper written by Janie Sigmon for students at Clemson University.

Credits and Trademarks

Photography Credits

Figures 5.1 and 5.9: Courtesy of Leica, Inc., Buffalo, New York 14240

Figure 8.1: Courtesy of University of Washington Department of Pathology.

Figure 9.1: Benjamin/Cummings.

Figures 10.1a, 10.1b, 10.1c: ©David M. Phillips/Visuals Unlimited.

Figure 12.5a: ©Omikron/Photo Researchers, Inc.

Figure 12.5b: ©Robert and Linda Mitchell.

Figure 12.6: ©Harold W. Pratt/Biological Photo Service.

Figure 12.7: NIH/Photo Researchers, Inc.

Figure 12.8: ©Cath Ellis, Department of Zoology, University of Hull/Science Photo Library/Photo Researchers, Inc.

Figure 12.9: Warps Scientific Establishment/Photo Researchers, Inc.

Figure 12.11: ©C. S. Murphree, Belmont U./BPS.

Figure 12.14: National Library of Medicine.

Figure 19.1: ©M.W.F. Tweedie/Photo Researchers, Inc.

Page 16-21: CALVIN AND HOBBES copyright 1988 Watterson. Reprinted with permission of UNIVERSAL PRESS SYNDICATE. All rights reserved.

Color Insert

Plate 1 *Color charts:* Karl Miyajima. *Wisconsin Fast Plants:* ©Carolina Biological Supply/Phototake NYC.

Plate 2 *Liverwort:* ©Rod Planck/Photo Researchers, Inc. *Clubmoss:* ©Runk/Schoenberger/Grant Heilman Photography. *Horsetails:* ©R. Carr/Bruce Coleman, Inc. *Gingko:* ©Runk/Schoenberger/Grant Heilman Photography. *Cycad:* ©Ken Brate/Photo Researchers, Inc.

Plate 3 *Sponges:* ©Andrew J. Martinez/Photo Researchers, Inc. *Jellyfish:* ©Paulo Curto/U.P./Bruce Coleman, Inc. *Coral:* ©David Mechlin/Phototake NYC. *Sea anemone:* ©Carl Roessler/Bruce Coleman, Inc.

Plate 4 *Nudibranchs:* ©Kjell Sandved/Bruce Coleman, Inc. *Bivalve:* ©Kjell Sandved/Photo Researchers, Inc. *Squid:* ©Bill Wood/Bruce Coleman, Inc. *Flatworm:* ©Bill Wood/Bruce Coleman, Inc. *Conus shell:* ©Michael McCoy/Photo Researchers, Inc.

Plate 5 *Shrimp:* ©James H. Carmichael/Bruce Coleman, Inc. *Scorpion:* ©Robert and Linda Mitchell. *Barnacles, brittle stars:* ©Kjell Sandved. *Wolf spider:* ©Peter J. Bryant/Biological Photo Service.

Plate 6 *Sea star, sea cucumber:* ©Kjell Sandved. *Sea urchin:* ©R.L. Sefton/Bruce Coleman, Inc. *Iguana:* ©K.H. Switak/Photo Researchers, Inc. *Sea squirt:* ©Scott Johnson/Animals Animals.

Plate 7 *Beetle:* ©Robert and Linda Mitchell. *Fly:* ©Jeff Foott/Bruce Coleman, Inc. *Bee:* ©D. Wilder. *Butterfly:* ©Lee F. Snyder/Photo Researchers, Inc.

Plate 8 *Moth, hummingbird:* ©M.P.L. Fogden/Bruce Coleman, Inc. *Bat:* ©Merlin D. Tuttle/Bat Conservation International/Photo Researchers, Inc. *Male corn plant:* ©Michael P. Gadomski/Photo Researchers, Inc. *Female corn plant:* ©Michael P. Gadomski/Bruce Coleman, Inc.

Text and Illustration Credits

The artists who contributed directly to *Laboratory Investigations for Biology* are listed on the copyright page at the front of the book. Illustrations from Campbell, Mitchell, and Reese, *Biology: Concepts and Connections* (©1994 The Benjamin/Cummings Publishing Company) that are used in *Laboratory Investigations for Biology* are the work of Nea Bisek, Mary Bryson, Barbara Cousins, Tom Dallman, Bill Glass, Illustrious, Inc., JAK Graphics, LTD, Georg Klatt, Laurie O'Keefe, Carla Simmons, Kevin Somerville, Terry Toyama, or Pamela Drury-Wattenmaker. Illustrations from Morgan and Carter, *Investigating Biology* (©1993 The Benjamin/Cummings Publishing Company) that are used in *Laboratory Investigations for Biology* are the work of Nea Bisek, Barbara Cousins, Terry Toyama, Valerie Felts, or John Norton.

The following figures are adapted from Judith Morgan and M. Eloise Carter, *Investigating Biology* (©1993 The Benjamin/Cummings Publishing Company): Figures 2.10, 2.11, 2.12, 2.13, 5.4, 5.6, 5.7, 6.8, 7.9, 9.5, 10.2, 11.4, 11.5, unnumbered 12.3, 12.12, 12.15, 13.1, unnumbered 13.3, 15.7, 17.2, unnumbered 17.4, 18.2, 18.3.

The following figures are adapted from Neil Campbell, Lawrence Mitchell, and Jane Reese, *Biology: Concepts and Connections* (Redwood City, CA: Benjamin/Cummings, 1994). Copyright ©1994 The Benjamin/Cummings Publishing Company): Figures 11.1, 15.1, 15.2, 16.1, 16.2, 16.5, 17.1, 18.1.

Figure 12.2 a & b was adapted from Neil Campbell, *Biology,* Third Edition (Redwood City, CA: Benjamin/Cummings, 1993). Copyright ©1993 The Benjamin/Cummings Publishing Company.

Preface Cartoon: Adapted from Jean Dickey, *Biology 105 Lab Manual (Clemson University),* Contemporary Publishing. Copyright ©John Norton. Reprinted by permission of the artist.

Table 2.6: E.A. Pearce and G. Smith; adapted from *The Times Books World Weather Guide,* New York: Times Books, 1990.

Figures 9.5, 9.6, 9.7: Adapted from Carolina Biological Supply, *Wisconsin Fast Plants™ Manual,* (Burlington, NC: Carolina Biological Supply, 1989). Reproduced by permission of Carolina Biological Supply, Burlington, NC 27217 and Wisconsin Alumni Research Foundation, Madison, WI.

Figures 12.1, 12.10, 12.13: Adapted from Jean Dickey, *Biology 105 Lab Manual (Clemson University),* Contemporary Publishing, pp. 158, 164, 184. Copyright ©Contemporary Publishing Company.

Trademarks

Alka-Seltzer is a registered trademark of Miles Inc.

Arm & Hammer is a registered trademark of Church & Dwight Co. Inc.

Ben-Gay is a registered trademark of Pfizer Inc.

Big Mac is a registered trademark of McDonald's Corporation.

Clorox is a registered trademark of The Clorox Company.

Dairy Ease, Lysol, and Milk of Magnesia are registered trademarks of Sterling Drug Inc.

Drano is a registered trademark of The Drackett Company.

Drierite is a registered trademark of W. A. Hammond Drierite Company.

Fast Plants is a trademark of J. M. Marketing Services, Inc.

The Food Processor is a trademark of ESHA Research Corporation.

Fruit-Fresh and Tums are registered trademarks of Beecham Inc.

Gatorade is a registered trademark of Stokely-Van Camp, Inc.

Jell-O is a registered trademark of General Foods Corporation.

Karo is a registered trademark of CPC International Inc.

Kool-Aid is a registered trademark of Perkins Products Company.

Lactaid is a registered trademark of Lactaid Inc.

Life Savers is a registered trademark of Life Savers, Inc.

Lime-A-Way is a registered trademark of Economics Laboratory, Inc.

Maalox is a registered trademark of Rorer Pharmaceutical Corporation.

Medeprin and Tylenol are trademarks of Johnson & Johnson.

Nutrasweet is a registered trademark of NutraSweet Company, Inc.

Parafilm is a registered trademark of Marathon Paper Mills Company.

Pepto-Bismol is a registered trademark of Norwich Eaton Pharmaceuticals, Inc.

PineSol is a registered trademark of American Cyanamid Company.

PROTOSLO is a registered trademark of Carolina Biological Supply Company.

Rolaids is a registered trademark of American Chicle Company.

7UP is a registered trademark of The Seven-Up Company.

Spectronic 20 is a registered trademark of Bausch & Lomb Incorporated.

Styrofoam is a registered trademark of The Dow Chemical Company.

Sweet'n Low is a registered trademark of Cumberland Packing Corp.

10-K is a registered trademark of Suntory Water Group, Inc.

Whatman is a registered trademark of Whatman Paper Limited.

Color chart for catecholase experiment

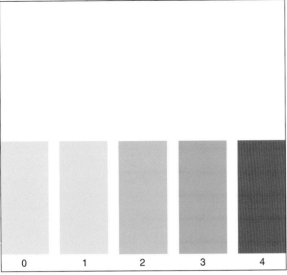

0   1   2   3   4

Compare your sample to this chart and record the number that most closely matches the intensity of your sample (Lab Topic 5).

Color chart for mitochondrial activity experiment

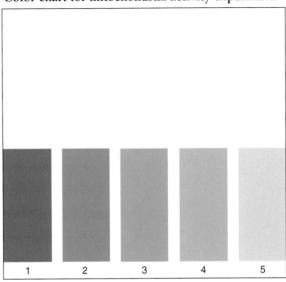

1   2   3   4   5

Compare your sample to this chart and record the number that most closely matches the intensity of your sample (Lab Topic 6).

Wisconsin Fast Plants anthocyaninless mutant seedlings next to wild-type plants (Lab Topic 9).

Wisconsin Fast Plants yellow-green mutant seedlings next to wild-type plants (Lab Topic 9).

Gingko (Lab Topic 12).

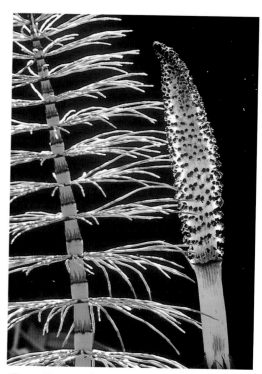

Giant Horsetail, *Equisetum teimateia,* **Clark County, WA.** (Lab Topic 12).

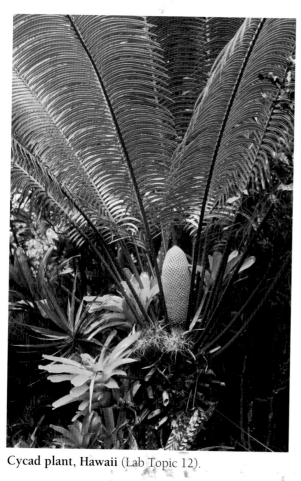

Cycad plant, Hawaii (Lab Topic 12).

Fragrant Liverwort (Lab Topic 12).

Club moss (Lab Topic 12).

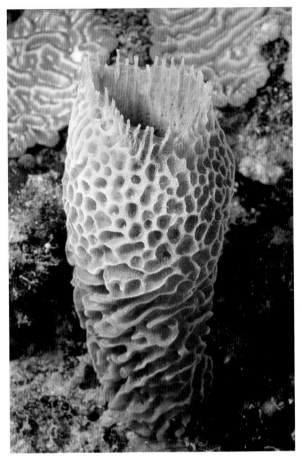

Purple sponge (Lab Topic 13).

West Coast Sea Nettles, California (Lab Topic 13).

Gorgonian coral. Papua, New Guinea (Lab Topic 13).

Translucent tentacles extend from an orange sea anemone in the Philippines (Lab Topic 13).

Hilton's Aeolid, a nudibranch (Lab Topic 13).

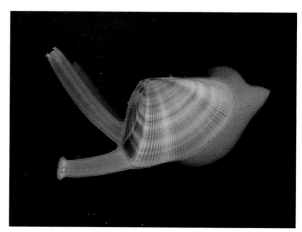

Bivalve with foot and both siphons extended (Lab Topic 13).

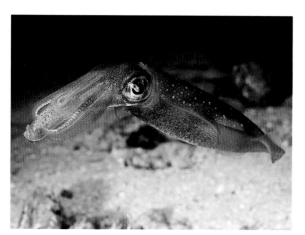

Sepioteuthis lessoniana, **a squid, Hawaii** (Lab Topic 13).

Orange boring sponge and flatworm, California (Lab Topic 13).

Geography cone snail, *Conus geographus,* **Australia** (Lab Topic 13).

Grass shrimp (Lab Topic 13).

Desert Hairy Scorpion (Lab Topic 13).

Pelagic goose barnacle, Hawaii
(Lab Topic 13).

A brittle star feeds on a blue sponge at night in the waters off the coast of Borneo in the Pacific Ocean. (Lab Topic 13).

Wolf spider (Lab Topic 13).

***Strongylocentrotus,* a sea urchin** (Lab Topic 13).

Sea star feeding on clam (Lab Topic 13).

Warty sea cucumber off Catalina Island in California (Lab Topic 13).

Snake emerging from egg (Lab Topic 13).

Tunicates (Lab Topic 13).

Hover fly on white blossom (Lab Topic 18).

Striped beetle in center of yellow poppy (Lab Topic 18).

Gulf frittilary butterfly pollinating a *Pentas* flower. *Pentas* is in the madder family, which also includes gardenias and coffee plants (Lab Topic 18).

Bee pollinating a scotch broom flower. Many members of the pea family, which includes beans and clover as well as scotch broom, have petals that form a landing platform for pollinators (Lab Topic 18).

Greater short-nosed fruit bat pollinating a banana plant (Lab Topic 18).

Yucca moth on a yucca flower. The yucca moth exacts a price for pollinator service in this highly specialized relationship. The moth lays its eggs in the ovary of the yucca flower, and its developing larvae then feed on the yucca's seeds (Lab Topic 18).

A bronzy hermit hummingbird feeds on and pollinates a red passion flower in a cloud forest (Lab Topic 18).

Wind-pollinated flower: male and female corn plants (Lab Topic 18).